DATA

D1322704

Beginning Spatial with SQL Server 2008

Alastair Aitchison

Apress®

Beginning Spatial with SQL Server 2008

ISBN-13 (pbk): 978-1-4302-1829-6

ISBN-13 (electronic): 978-1-4302-1830-2

Printed and bound in the United States of America (POD)

President and Publisher: Paul Manning
Lead Editor: Jonathan Gennick
Technical Reviewer: Evan Terry
Editorial Board: Steve Anglin, Mark Beckner, Ewan Buckingham, Gary Cornell, Jonathan Gennick,
    Jonathan Hassell, Michelle Lowman, Matthew Moodie, Duncan Parkes, Jeffrey Pepper, Frank
    Pohlmann, Douglas Pundick, Ben Renow-Clarke, Dominic Shakeshaft, Matt Wade, Tom Welsh
Coordinating Editor: Kylie Johnston
Copy Editor: Bill McManus
Compositor: Susan Glinert
Proofreader: April Eddy
Indexer: Julie Grady
Cover Designer: Anna Ishchenko

Distributed to the book trade worldwide by Springer Science+Business Media, LLC., 233 Spring Street, 6th Floor, New York, NY 10013. Phone 1-800-SPRINGER, fax 201-348-4505, e-mail orders-ny@springer-sbm.com, or visit www.springeronline.com.

For information on translations, please e-mail info@apress.com, or visit www.apress.com.

Apress and friends of ED books may be purchased in bulk for academic, corporate, or promotional use. eBook versions and licenses are also available for most titles. For more information, reference our Special Bulk Sales–eBook Licensing web page at www.apress.com/info/bulksales.

The source code for this book is available to readers at www.apress.com.

To Clare and Douglas.

Contents at a Glance

PART 1 ▦▦▦ Working with Spatial Data

PART 2 ▦▦▦ Adding Spatial Data

PART 3 ▪▪▪ Presenting Spatial Data

PART 4 ▪▪▪ Analyzing Spatial Data

PART 5 ▪▪▪ Ensuring Spatial Performance

Contents

PART 1 ■■■ Working with Spatial Data

PART 2   Adding Spatial Data

PART 3 ▪▪▪ Presenting Spatial Data

CHAPTER 10 Visualizing Query Results in Management Studio 249

PART 4 ▦▦▦ Analyzing Spatial Data

CHAPTER 11 Examining Properties of Spatial Objects 261

PART 5 ■■■ Ensuring Spatial Performance

About the Author

ALASTAIR AITCHISON has over eight years' experience as a management information consultant and has provided reporting and analysis of data for the House of Lords, the Department of Health, the Police Force, and various UK media agencies. For the past three years, Alastair has been a technical manager at Aviva, the world's fifth-largest insurance group. In this role, he has developed and promoted the use of spatial data in a range of corporate applications, such as the geographic analysis of risk patterns, plotting the success of regional sales campaigns, and understanding the impact of major weather incidents.

Alastair is a Microsoft Office Specialist Master Instructor and has delivered numerous training courses to individuals and small groups on a range of software packages.

About the Technical Reviewer

EVAN TERRY is the chief technical consultant working at the Clegg Company, specializing in data management, information architecture, and business intelligence. His past and current clients include the State of Idaho, Albertsons, American Honda Motors, and Toyota Motor Sales, USA, Inc. He is the coauthor of *Beginning Relational Data Modeling*, has published several articles in *DM Review*, and has presented at industry conferences and individual company workshops on the subjects of data quality and information management. For questions or consulting needs, Evan can be reached at evan_terry@cleggcompany.com.

Acknowledgments

Writing this book has been hard work, yet it has also been a very gratifying experience. I have no doubt that it has been made all the more enjoyable because of the immensely talented and thoroughly nice bunch of people at Apress with whom I've had the pleasure of working. I'd like to thank Kylie Johnston, who deftly managed and guided me through the book's process; Evan Terry, for editing and improving the technical content; Bill McManus, for making my writing much more elegant; and Katie Stence, who somehow managed to wrangle my artwork into a usable format. I would also particularly like to acknowledge Jonathan Gennick, my editor, who offered his unwavering support and encouragement from the very outset of this project, when the book you are now holding was just a hazy idea in my head. Thank you to all of you, and to the many other people who have contributed behind the scenes to make this book possible.

Besides the obvious contribution of those directly involved in the production of this book, I have also received a great deal of help from a number of other people. Among them, I'd like to mention my friends and colleagues at Aviva: Paul Goldsmith, Shane Tovell, and Christian Hall, who supported and promoted my passion in this area; and Andy Fisher and Anthony Payne, who taught me everything I know. I'd also like to thank Isaac Kunen and the Microsoft SQL Server Spatial team, and all the other contributors to the MSDN SQL Server Spatial forum (http://social.msdn.microsoft.com/forums/en-US/sqlspatial/threads/) who offered advice and helped generate ideas for material, and Mike Ormond and Johannes Kebeck, who graciously provided the original application on which Chapter 5 is based.

I am indebted to my family: Mari, for her grammar tips and publishing advice; and my Mum, who has selflessly supported me in everything I have ever wanted to do. Finally, I couldn't have written this book without the support of my wife, Clare, without whom I am nothing.

Introduction

The use of spatial data in information systems is hardly a new technology. Dedicated geographic information systems (GISs), such as ARC/INFO from ESRI, have been commercially available since the early 1980s. While the technological capabilities of these systems have evolved significantly over the past 25 years, their adoption has remained relatively confined within a small, specialized group of developers. One reason for this is that, because of the complex nature of spatial data, GIS systems themselves are typically complex, and require dedicated, specially trained operators. Furthermore, these systems are frequently stand-alone systems and do not integrate spatial data with central corporate data systems.

More recently, database management system providers, including Oracle, IBM (DB2), MySQL, and the PostgreSQL Global Development Group, have all released spatially enabled relational database management systems (RDBMSs). Although this has widened the adoption of spatial techniques, the spatial functionality is typically provided using an optional add-in component that requires specific product knowledge that general developers typically do not have yet. Knowledge of spatial data has therefore still largely remained limited to specialist technical fields.

With the introduction of spatial support in SQL Server 2008, Microsoft has taken a number of steps to reduce the number of barriers that, until now, have prevented mainstream developers from using spatial data:

- Spatial datatypes are included as a core component of the SQL Server 2008 database, and work "out of the box," requiring no additional components to be installed or configured.

- Spatial operations are integrated into the existing functionality of the SQL Server Database Engine, allowing developers to continue working within a familiar development environment using existing tools such as SQL Server Management Studio.

- Existing SQL Server databases can be easily enriched by adding spatial data fields to their existing structure—there is no need to migrate data onto a new platform.

- The new geometry spatial datatype conforms to accepted industry-wide standards set by the Open Geospatial Consortium (OGC).

- Spatial support is included in all versions of SQL Server 2008, including the freely available SQL Server Express Edition. As a result, even small-scale, hobbyist, and amateur programmers can start using spatial data.

In this book, I give you an introduction to working with spatial data in SQL Server 2008 that will enable you to start using these new features to add exciting and value-adding capabilities to your database applications.

Who This Book Is For

This book is aimed at developers who are being introduced to spatial data for the first time through SQL Server 2008. No previous knowledge of working with spatial data is assumed, and all topics are explained from the ground up. My intention is to explain how to use the new spatial datatypes to add additional reporting and analysis capability to your existing datasets, by demonstrating a range of practical usage examples.

How This Book Is Structured

The chapters in this book are divided into five parts. Each part introduces topics that are related to a particular aspect of spatial data, and the topics are listed in the order in which, as a newcomer to spatial data, you are likely to encounter them.

Part 1 (Chapters 1–3) introduces the fundamental concepts involved when working with any spatial data. It first covers the theoretical issues of models of the earth, coordinate references, and geodetic datums, and then describes the specific practical implementation of spatial data in SQL Server 2008. It presents a side-by-side comparison of the two new spatial datatypes, geography and geometry, and examples to demonstrate how they can be used with reference to the .NET CLR.

Part 2 spans four chapters, each of which introduces a different method to insert spatial data into a SQL Server 2008 database. Chapter 4 explains how to use the various formats that are natively supported by SQL Server (WKT, WKB, and GML), and Chapter 6 describes tools and methods that you can use to import other commonly used formats such as KML and ESRI shape-files. Part 2 includes two chapters that each cover an example of extending SQL Server functionality by using an external service. Chapter 5 explains how Virtual Earth can be used as a drawing canvas to define new spatial data, and Chapter 7 describes how to create a custom .NET assembly to access the MapPoint Find service to provide geocoding functionality in SQL Server.

Part 3 (Chapters 8–10) describes various methods of visually presenting spatial data. SQL Server 2008 has only very limited built-in spatial visualization capability, so in this part I describe how to present and visualize syndicated spatial data using the GeoRSS format, and how to build interactive front-end spatial applications using the Virtual Earth and Google Maps controls. I also describe the Spatial Results tab, which allows you to quickly examine the results of a query from within SQL Server Management Studio.

Part 4 (Chapters 11–13) introduces the range of spatial methods that can be used to query properties and relationships between spatial objects. Every method is covered in outline form, including an explanation of its purpose, a description of the context in which it can be used, and a simple code example to demonstrate its use in a real-life situation. Additionally, there are many diagrams used to illustrate the results of the most commonly used methods.

Part 5, composed of Chapter 14, covers issues related to the performance of spatial data-bases, with a focus on the important topic of spatial indices.

Prerequisites

In order to follow the code examples listed in this book, you should have a fully installed and configured instance of SQL Server 2008. All of the examples presented in this book work with any edition of SQL Server 2008, from the Enterprise Edition right down to the freely available Express Edition.

Although spatial data is supported by the SQL Server core Database Engine, this alone does not provide all the capabilities generally required for an end-to-end spatial application. For example, SQL Server has only limited capability to import common formats of existing spatial data, and has only a basic method of displaying spatial data. In order to show you how to integrate SQL Server into an end-to-end spatial application, some chapters use additional software or services. These include Shape2SQL (used in Chapter 6); Microsoft MapPoint Web Service, Microsoft Visual Basic Express Edition, and Microsoft Visual C# Express Edition (used in Chapter 7); and Virtual Earth and Google Maps (both used in Chapters 8 and 9). All of the additional software used in this book is freely available, and details of how to obtain the software are included in the relevant chapters.

Downloading the Code

This book contains numerous code examples to demonstrate the methods used in each chapter. You can download the code in a zip archive from the Source Code/Download area of the Apress web site (http://www.apress.com).

Contacting the Author

If you have any questions or comments, you can e-mail the author directly at alastair@beginningspatial.com. Alternatively, check out http://www.beginningspatial.com for additional information and resources related to this book.

PART 1

Working with Spatial Data

This part of the book introduces the fundamental principles that you need to know to use spatial data effectively in SQL Server 2008. The chapters in this part explain what spatial data is and how to define it, how SQL Server treats different kinds of spatial data using the geometry and geography datatypes, and the ways in which those spatial datatypes are implemented using the Microsoft .NET Framework CLR.

Defining Spatial Information

Spatial data analysis is a complex subject area, taking elements from a range of academic disciplines, including geophysics, mathematics, astronomy, and cartography. Although you do not need to understand these subjects in great depth to start using the new spatial features of SQL Server 2008, it is important to have a basic understanding of the theoretical concepts involved so that you use spatial data appropriately and effectively in your applications.

In this chapter you will learn how different spatial reference systems identify positions in space, and how these systems can be used to define spatial objects representing features on the earth. These concepts are fundamental to the creation of consistent, accurate spatial data, and will be used throughout the practical applications discussed in later chapters of this book.

What Is Spatial Data?

Spatial data describes the position, shape, and orientation of objects in space.

In this book, as in most common applications, we are particularly concerned with describing the position and shape of objects on the earth. This is known as *geospatial* data. Geospatial data can describe the properties of many different sorts of "objects" on the earth. These objects might be tangible, physical things, such as an office building or a mountain, or abstract features, such as the imaginary line marking the political boundary between countries.

Uses of Spatial Data

Spatial data provides information that can be used in a wide range of different areas. Some potential applications are as follows:

- Analyzing regional, national, or international sales trends

- Deciding where to place a new store based on proximity to customers and competitors

- Navigating to a destination using a Global Positioning System (GPS) device

- Allowing customers to track the delivery of a parcel

- Monitoring the routes of vehicles in a logistics network

- Optimizing distribution networks to provide the most efficient coverage of an area

- Reporting geographic-based information on a map rather than in a tabular or chart format

- Providing location-based services, such as providing a list of nearby amenities for any given address

- Assessing the impact of environmental changes, such as identifying houses at risk of flooding caused by rising sea levels

All of these examples rely on the ability of spatial data to describe the position and shape of objects on the earth in a structured, consistent way.

Representing Features on the Earth

In real life, objects on the earth often have complex, irregular shapes. It would be very hard, if not impossible, for any item of spatial data to define the exact shape of these features. Instead, spatial data represents these objects by using simple, geometrical shapes that approximate their actual shape and position. These shapes are called *geometries*.

SQL Server 2008 supports three main types of geometry that can be used to represent spatial information: Points, LineStrings, and Polygons. In this section I describe the properties of each of the three types in turn, and then I show how you can use them to represent various features on the earth.

Points

A *Point* is the most fundamental type of geometry, and is used to define a singular position in space. A Point object is zero-dimensional, meaning that it does not have length or area. Figure 1-1 illustrates a representation of a Point geometry.

Figure 1-1. *A Point geometry*

When using geospatial data to define features on the earth, a Point geometry is used to represent an exact location, which could be a street address or the location of a bank, volcano, or city, for instance. Figure 1-2 illustrates several Point geometries used to represent the locations of major cities in Australia.

Figure 1-2. *A series of Point geometries representing cities in Australia*

LineStrings

Having defined a series of two or more points in space, we then can draw straight lines connecting each point to the next point in the series, to define a *LineString*. LineStrings comprise a series of two or more distinct points and the line segments that connect them. LineStrings are one-dimensional spatial objects—they have a specified length, but do not contain any area.

LineStrings may be described as having the following additional characteristics:

- A *simple* LineString is one in which the path drawn between the points of the LineString does not cross itself.

- A *closed* LineString is one that starts and ends at the same point.

- A LineString that is both simple and closed is known as a *ring*. Even though a ring appears to represent the perimeter of a closed shape, it does not include the area enclosed within the shape—it only defines the points that lie on the line itself.

Different examples of LineString geometries are illustrated in Figure 1-3.

Figure 1-3. *Examples of LineString geometries (from left to right): a simple LineString; a simple, closed LineString (a ring); a nonsimple LineString; a nonsimple, closed LineString*

In geospatial data, LineStrings are commonly used to represent features such as **roads**, rivers, delivery routes, or contours of the earth. Figure 1-4 shows numerous LineStrings **used to** represent major rivers in France.

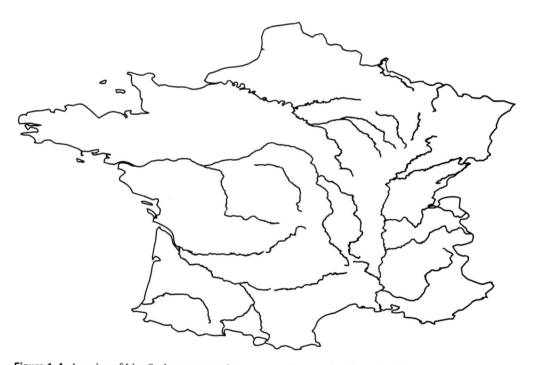

Figure 1-4. *A series of LineString geometries representing major rivers in France*

Note Some geographic information systems (GISs) make a distinction between a *LineString* and a *Line*. According to the Open Geospatial Consortium (OGC) Simple Features for SQL Specification (a standard on which the spatial features of SQL Server 2008 are largely based), a Line connects exactly two points, whereas a LineString may connect any number of points. Since all Lines can be represented as LineStrings, of these two types, SQL Server 2008 only implements the LineString geometry.

Polygons

A Polygon geometry is defined by a boundary of connected points that forms a closed LineString, called the exterior ring. In contrast to a simple, closed LineString geometry, which only defines those points lying on the ring itself, a Polygon geometry also contains all the points that lie in the interior area enclosed within the exterior ring.

Every Polygon must have exactly one external ring that defines the overall perimeter of the shape, and may also contain one or more internal rings. Internal rings define areas of space that are contained within the external ring but not included in the Polygon definition. They can therefore be thought of as "holes" that have been cut out of the main geometry.

Since Polygons are constructed from a series of one or more rings, which are simple, closed LineStrings, all Polygons themselves are deemed to be simple, closed geometries. Polygons are two-dimensional geometries—they have an associated length and area. The length of a Polygon is measured as the sum of the distances around the perimeter of all the rings of that Polygon (exterior and interior), while the area is calculated as the space contained within the exterior ring, excluding the area contained within any interior rings. Some examples of Polygon geometries are illustrated in Figure 1-5.

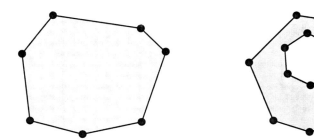

Figure 1-5. *Examples of Polygon geometries (from left to right): a Polygon; a Polygon with an interior ring*

Polygons are frequently used in spatial data to represent geographic areas such as islands or lakes, political jurisdictions, or large structures. Figure 1-6 illustrates Polygon geometries that represent the 48 contiguous states of the mainland United States.

Figure 1-6. *A series of Polygon geometries representing states of the United States*

Choosing the Right Geometry

There is no "correct" type of geometry to use to represent any given object on the earth. The choice of which geometry to use will depend on how you plan to use the data. If you are going to analyze the geographic spread of your customer base, you *could* define Polygon geometries that represent the shape of each of your customers' houses, but it would be a lot easier to consider each customer's address as a single Point. In contrast, when conducting a detailed analysis of a small-scale area for land-planning purposes, you may want to represent all buildings, roads, and even walls as Polygons that have both length and area, to ensure that the spatial data represents their actual shape as closely as possible.

Combining Geometries in a Geometry Collection

Sometimes, what could be considered a single object on the earth may be represented using a combination of several geometry objects. For instance, the Great Wall of China is not a single continuous wall, but rather it is made up of numerous separate sections of wall. As such, the overall shape of the wall may be best represented as a collection of LineStrings. Similarly, a single country spread over several islands, such as Japan, may be represented by a collection of Polygons, each one representing the shape of an individual island. When you define a single object that contains several individual geometries in this way, it is called a *Geometry Collection*. A Geometry Collection may contain any number of any type of geometries. In the specific case in which a Geometry Collection contains only multiple elements of the same type of geometry, it is referred to as a MultiPoint, MultiLineString, or MultiPolygon geometry.

DEFINING AUGUSTA NATIONAL GOLF COURSE

In order to demonstrate the different ways in which spatial data can describe the same object on the earth, let me show you a practical example. Suppose we want to store an item of spatial data describing the course at Augusta National Golf Club (in Augusta, Georgia), home of the annual US Masters Tournament.

If we were storing information for a tourist database of interesting places to visit in Georgia, it would probably suffice to describe the entire golf course using a Point geometry. This Point could describe the approximate location of the course, and would be perfectly sufficient to perform spatial calculations such as finding the distance to the closest airport, or identifying nearby places to stay.

Alternatively, we could choose to represent the course as a geometry collection containing many elements that describe the individual features of the course much more accurately: we could represent the greens and the fairways of each hole as separate Polygons; use Point objects to represent each tee; and use LineString objects to show the optimum drive off the tee. This sort of representation would be more suitable for use by a golfer who is actually playing the course, accessing spatial data via a mobile GPS system to plan their next shot.

Both of these alternative representations are equally valid—the choice simply depends on the application of the data.

Understanding Interiors, Exteriors, and Boundaries

Every geometry shape divides space into three areas relative to that geometry: the *interior*, *exterior*, and *boundary*. In the field of topological mathematics, these terms have very specific definitions, but you can think of them simply as follows:

- The interior of a geometry consists of all the points that lie in the space occupied by the geometry.

- The exterior consists of all the points that lie in the space not occupied by the geometry.

- The boundary of a geometry consists of the points that lie on the "edge" of the geometry. In SQL Server, every geometry is considered to be *topologically closed*; that is, any points that lie on the boundary of a geometry are contained within the interior of the geometry.

Every geometry specifies one or more points in their interior and exterior, although only certain types of geometry contain points in their boundaries. The classification of these different areas of space for each type of geometry follows:

Point and MultiPoint geometries: Represent singular locations, where the interior consists of the individual point(s) defined by that object. However, they do not have a defined boundary.

LineString and MultiLineString geometries: Have an interior consisting of all the points that lie on the straight line segments drawn between the defined series of points. Nonclosed LineStrings and MultiLineStrings have a boundary consisting of the points at the start and end of the LineString. However, closed LineStrings—those that start and end at the same point—do not have a boundary.

Polygon and MultiPolygon geometries: Have an interior consisting of all the points contained within the exterior ring, excluding those contained within any interior ring. The boundary of these types of geometry consists of the closed LineString that forms the exterior ring itself, together with any interior rings defined by that Polygon.

The distinction between these classifications of space becomes very important when expressing the relationship between different spatial objects, since these relationships are generally based on comparing where particular points lie with respect to the interior, exterior, or boundary of the two geometries in question. For instance, two geometries *intersect* each other if they share at least one point in common. However, they are only deemed to *touch* each other if the points that they share lie only on the boundaries of each geometry. This concept is discussed in more detail in Chapter 13.

Positioning a Geometry

After we choose an appropriate geometry (Point, LineString, or Polygon) to represent a given object, we then need to position it in the right place on the earth. We do this by relating each point in the geometry definition to the relevant real-world position it represents. For example, if we want to use a Polygon geometry to represent the US Department of Defense Pentagon building, we need to specify that the five points that define the boundary of the Polygon geometry relate to the location of the five corners of the building. So, how do we do this?

You are probably familiar with the terms *longitude* and *latitude*, and have seen them used to describe positions on the earth. If this is the case, you may be thinking that we can simply express the latitude and longitude coordinates of the relevant position on the earth for each point in the geometry. Unfortunately, it's not quite that simple.

What many people don't realize is that any particular point on the ground does not have a unique latitude or longitude associated with it. There are in fact many systems of latitude and longitude, and the coordinates of a given point on the earth will differ depending on which system is used. Furthermore, latitude and longitude coordinates are not the only way of expressing positions—there are other types of coordinates that define the location of an object without using latitude and longitude at all. In order to understand how to specify the position of your geometry on the earth, you first need to understand how different spatial reference systems work.

COMPARING RASTER TO VECTOR DATA

There are two main ways of modeling spatial information: using a vector model or using a raster model.

Vector data, discussed in this chapter, describes discrete spatial objects by defining the coordinates of geometries that approximate the shape of those features. Vector spatial information is best suited to represent discrete items of spatial data, such as the location of individual customers or warehouses, or the path of roads.

In contrast, raster data represents spatial information using a matrix of cells. These cells are arranged into a grid that is overlaid onto the surface of the earth. The value of each cell in the matrix represents a property of the underlying area covered by that grid cell. One example of raster spatial data is aerial or satellite imagery, in which case the matrix grid is the set of pixels that forms the image, and the value of any individual cell is the color of the associated pixel. However, raster data can also be used to describe any other spatial information. It is particularly suited to data that can take a continuous range of values, such as when depicting the levels of rainfall across an area of land, or the depth of an area of water.

All the spatial features in SQL Server 2008 (and therefore discussed in this book) are based on a vector model of spatial data. There is currently no built-in support for raster data in SQL Server. However, in Chapter 9, I will show you how to overlay vector shape information with raster imagery of the earth, by combining spatial data from SQL Server with the Microsoft Virtual Earth and Google Maps web services.

Describing Positions Using a Coordinate System

The purpose of a spatial reference system is to unambiguously identify and describe any point in space. This ability is essential to enable spatial data to define the positions of points that make up the various kinds of geometry used to represent features on the earth. To describe the positions of points in space, every spatial reference system is based on an underlying coordinate system. A coordinate reference is a conventional and widely accepted way of describing the position of a point from a given origin, in a given dimension. A set of n coordinates, such as $(1, 2, 3, \ldots, n)$, can therefore be used to describe the position of a point from an origin in n-dimensional space.

There are many different types of coordinate systems; when you use geospatial data in SQL Server 2008, you are most likely to use a spatial reference system based on either a geographic or projected coordinate system.

Note A set of coordinate values is called a coordinate *tuple*.

Geographic Coordinate System

In a geographic coordinate system, any position on the earth's surface can be defined using two coordinates:

> The *latitude* coordinate of a point measures the angle between the plane of the equator and a line drawn perpendicular to the surface of the earth at that point. (This is the definition of *geodetic latitude*. An alternative measure, *geocentric latitude*, is defined as the angle between the plane of the equator and a line drawn from a point on the earth's surface to the center of the earth.)

> The *longitude* coordinate measures the angle in the equatorial plane between a line drawn from the center of the earth to the point and a line drawn from the center of the earth to the prime meridian. The prime meridian is an imaginary line drawn on the earth's surface between the North Pole and the South Pole (so technically it is an *arc* rather than a line), chosen to be the line from which angles of longitude are measured.

These concepts are illustrated in Figure 1-7.

Caution Since a point of greater longitude lies further east, and a point of greater latitude lies further north, it is a common mistake for people to think of latitude and longitude as measured on the earth's surface itself, but this is not the case—latitude and longitude are angles measured from the plane of the equator and prime meridian at the center of the earth.

Coordinates of latitude and longitude are both angles, and are usually measured in degrees. In this case, longitude values measured from the prime meridian range from –180° to +180°, and latitude values measured from the equator range from –90° (at the South Pole) to +90° (at the North Pole).

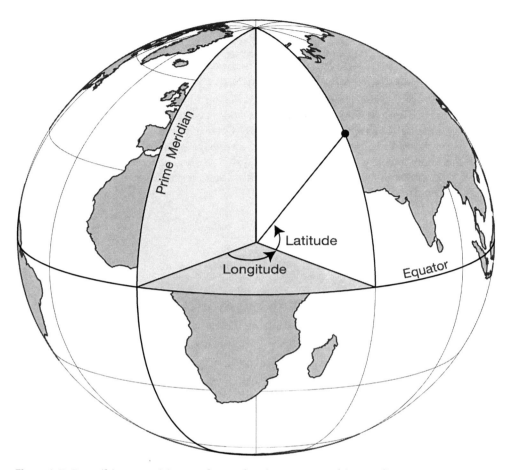

Figure 1-7. *Describing a position on the earth using a geographic coordinate system*

Longitudes to the east of the prime meridian are normally stated as positive values, or suffixed with the letter *E*. Longitudes to the west of the prime meridian are expressed as negative values, or using the suffix *W*. Likewise, latitudes north of the equator are expressed as positive values, or using the suffix *N*, whereas latitudes south of the equator are expressed as negative values, or using the suffix *S*.

There are several accepted notation methods for expressing values of latitude and longitude:

The most commonly used method is the degrees, minutes, seconds (DMS) system, also known as sexagesimal notation. In this system, each degree is divided into 60 minutes. Each minute is further subdivided into 60 seconds. A value of 51 degrees, 15 minutes, and 32 seconds is normally written as 51°15'32".

The system most commonly used by GPS receivers is to display whole degrees, and then minutes, and decimal fractions of minutes. This same coordinate value would therefore be written as 51:15.53333333.

Decimal degree notation specifies coordinates using degrees and decimal fractions of degrees, so the same coordinate value expressed using this system would be written as 51.25888889.

CONVERTING TO DECIMAL DEGREE NOTATION

When expressing geographic coordinate values of latitude and longitude for use in SQL Server 2008, you should use decimal degree notation. The advantage of this format is that each coordinate can be expressed as a single floating-point number. To convert DMS coordinates into decimal degrees, you can use the following rule:

Degrees + (Minutes / 60) + (Seconds / 3600) = Decimal Degrees

For example, the US Central Intelligence Agency's online edition of *The World Factbook* (https:// www.cia.gov/library/publications/the-world-factbook/geos/uk.html) gives the geographic coordinates for London as follows:

51 30 N, 0 10 W

When expressed in decimal degree notation, this is

51.5 (Latitude), −0.166667 (Longitude)

When converting a coordinate value from DMS to decimal degree notation, you should state the accuracy of the result with up to 15 significant figures, because this is the precision with which the converted coordinate value will be stored in SQL Server.

Projected Coordinate System

In contrast to the geographic coordinate system, which defines positions on a three-dimensional, round model of the earth, a projected coordinate system describes the position of points on the earth's surface as they lie on a flat, two-dimensional plane. A simple way of thinking about this is to consider a projected coordinate system as describing positions on a map rather than positions on a globe.

If we consider all of the points on the earth's surface to lie on a flat plane, we can define positions on that plane using familiar Cartesian coordinates of x and y, which represent the distance of a point from an origin along the x axis and y axis, respectively. In a projected coordinate system, these coordinate values are sometimes referred to as *easting* (the x coordinate) and *northing* (the y coordinate), as shown in Figure 1-8.

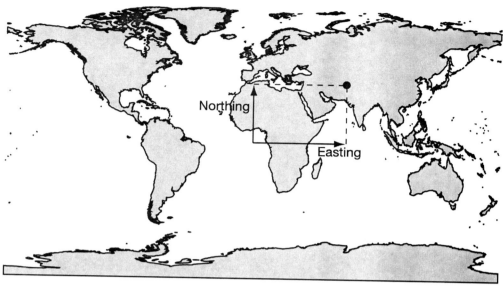

Figure 1-8. *Describing position on the earth using a projected coordinate system*

Since a projected coordinate system describes the position of an object by calculating the distance from an origin along a flat plane representing the earth's surface, northing and easting coordinate values are measured and expressed using a linear unit of measure, such as meters or feet.

Applying Coordinate Systems to the Earth

So far, we have defined two different coordinate systems that can be used to define points in theoretical space: the geographic coordinate system, which uses angular coordinates of latitude and longitude, and the projected coordinate system, which uses x and y Cartesian coordinates. However, a set of coordinates from either of these systems does not, on its own, uniquely identify a position on the earth. We need to know additional information, such as where to measure those coordinates from, in what units, and what shape to use to model the earth. For this, we need to examine the other elements of a spatial reference system—the datum, prime meridian, and unit of measurement.

Datum

A datum contains information about the size and shape of the earth. Specifically, it contains the details of a reference ellipsoid, and a reference frame. We use this information to create a geodetic model of the earth, onto which we can apply our coordinate system.

The actual shape of the earth is very complex. On the surface, we can see that there are irregular topological features such as mountains and valleys. But even if we were to remove these features and consider the mean sea level around the planet, the earth is still not a regular shape. In fact, it is so unique that geophysicists have a specific word solely used to describe the shape of the earth—the *geoid*.

When using spatial data to describe the position of geometries on the earth's surface, ideally, we would like to use coordinates that refer to positions relative to the geoid itself. However, there is no way that we can accurately model the complicated, irregular shape of the geoid, so instead we base our spatial system on an approximation of the geoid. This approximation is called a *reference ellipsoid*.

■**Note** *Geodesy* is the science of studying and measuring the shape of the earth. A *geodetic* model is therefore a model of the shape of the earth.

Reference Ellipsoid

Despite its name, a reference ellipsoid normally describes an *oblate spheroid*, which is the three-dimensional shape obtained when you rotate an ellipse about its shorter axis. When used in spatial data modeling, spheroid models of the earth are always oblate—they are wider than they are high, and resemble a squashed sphere. This is a fairly good approximation of the shape of the geoid, which bulges around the equator.

The important feature of a spheroid is that, unlike the geoid, it is a regular shape that can be exactly mathematically described by two parameters—the length of the semimajor axis (which represents the radius of the earth at the equator), and the length of the semiminor axis (the radius of the earth at the poles). This is illustrated in Figure 1-9.

■**Note** A spheroid is a sphere that has been "flattened" in one axis. An ellipsoid is a sphere that has been flattened in two axes—that is, the radius of the shape is different in the x, y, and z axes. Since ellipsoid models of the world are not significantly more accurate than spheroid models at describing the shape of the geoid, reference ellipsoids are rarely based on true ellipsoids, but rather on a simpler spheroid model.

An alternative method of stating the properties of an ellipsoid is to give the length of the semimajor axis and the flattening ratio of the ellipsoid. The flattening ratio, f, is used to describe how much an ellipsoid has been "squashed," and is calculated as

$$f = (a - b) \, / \, a$$

where a equals the length of the semimajor axis, and b equals the length of the semiminor axis.

In most ellipsoid models of the earth, the semiminor axis is only marginally smaller than the semimajor axis, which means that the value of the flattening ratio is also small—typically around 0.003. For the sake of convenience, many systems, including SQL Server 2008, use the inverse-flattening ratio of an ellipsoid instead. This is stated as $1/f$, and calculated as follows:

$$1/f = a \, / \, (a - b)$$

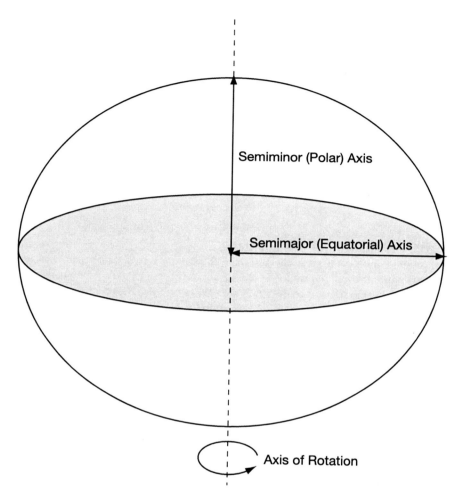

Figure 1-9. *Properties of a reference ellipsoid*

The inverse-flattening ratio of an ellipsoid model typically has a value of approximately 300. There is not a single reference ellipsoid that best represents every part of the whole geoid. Some ellipsoids, such as the WGS 84 ellipsoid used by satellite GPS systems, provide a reasonable approximation of the overall shape of the geoid. Other ellipsoids approximate the shape of the geoid very accurately over certain regions of the world, but are much less accurate in other areas. These ellipsoids are normally only applied for use in specific countries, such as the Airy 1830 ellipsoid commonly used in Britain. Figure 1-10 provides an (exaggerated) illustration of how different ellipsoid models vary in accuracy over different parts of the geoid.

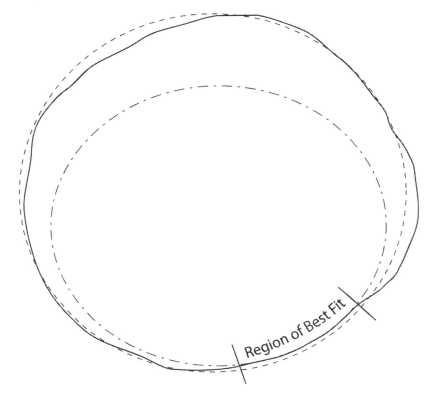

—— Geoid

- - - Ellipsoid of Best Global Accuracy

— · — Ellipsoid of Best Regional Accuracy

Region of Best Fit

Figure 1-10. *Comparison of cross-sections of different ellipsoid models of the geoid*

It is important to realize that specifying a different reference ellipsoid to approximate the geoid affects the accuracy of how well a set of coordinates that defines a geometry on that ellipsoid reflects the actual position and shape of the feature on the earth that the geometry represents. When choosing an ellipsoid to define spatial data, we must therefore be careful to use one that is suitable for the purpose of the data in question.

SQL Server 2008 recognizes a number of different reference ellipsoids that are designed to best approximate the geoid at different parts of the earth. Table 1-1 lists the properties of some commonly used reference ellipsoids that can be used.

Table 1-1. *Properties of Some Commonly Used Reference Ellipsoids*

Ellipsoid Name	Semimajor Axis (m)	Semiminor Axis (m)	Inverse Flattening	Usage
Airy (1830)	6,377,563.396	6,356,256.909	299.3249646	Great Britain
Bessel (1841)	6,377,397.155	6,356,078.963	299.1528128	Czechoslovakia, Japan, South Korea
Clarke (1880)	6,378,249.145	6,356,514.87	293.465	Africa
NAD 27	6,378,206.4	6,356,583.8	294.9786982	North America
NAD 83	6,378,137	6,356,752.3	298.2570249	North America
WGS 84	6,378,137	6,356,752.314	298.2572236	Global

Reference Frame

Remember that we are going to use our coordinate system to define positions on the reference ellipsoid as a way of approximating positions on the earth itself. Having established our ellipsoid model, we need some way to position that model so that it lines up with the right points on the earth's surface. We do this by creating a frame of reference points.

Reference points are places (normally on the earth's surface) that are assigned known coordinates in the coordinate system relative to the ellipsoid being used. By establishing a set of points of known coordinates, we can use these points to "fix" the reference ellipsoid in the right position. Once the ellipsoid is set in place based on these known points, we can apply our chosen coordinate system to obtain the coordinates of any other points on the earth, based on the ellipsoid model. Reference points are sometimes assigned to places on the earth itself; the North American Datum of 1927 (NAD 27) uses the Clarke (1866) reference ellipsoid, primarily fixed in place at Meades Ranch in Kansas. Reference points may also be assigned to the positions of satellites orbiting the earth, which is how the WGS 84 datum used by GPS systems is realized.

When packaged together, the properties of the reference ellipsoid and the frame of terrestrial reference points form a datum. The most common datum in global use is the World Geodetic System of 1984, commonly referred to as WGS 84. This is the datum used by MapPoint and Google Earth, as well as in handheld GPS systems.

Prime Meridian

As defined earlier in this chapter, the geographic coordinate of longitude is the angle in the equatorial plane between the line drawn from the center of the earth to a point and the line drawn from the center of the earth to the prime meridian. Our spatial reference therefore needs to include a definition of what line we are using for the prime meridian—the axis from which we measure our angle of longitude.

A common misconception is to think that there is a single prime meridian based on some inherent fundamental property of the earth, but this is not the case. The prime meridian of any spatial reference system is arbitrarily chosen simply to provide a line of zero longitude from which all other coordinates of longitude can be calculated. One commonly used prime meridian is the meridian passing through Greenwich, London, but there are many others. If we were to use a different prime meridian, the value of the longitude coordinate of all the points in our system would change.

Unit of Measurement

Geographic coordinates of latitude and longitude are generally measured in degrees, but may also be measured in radians, or other angular units of measure. Every spatial reference system must explicitly state the name of the unit in which geographic coordinates are measured, together with the conversion factor from the specified unit to one radian. For instance, a spatial reference system that uses coordinates measured in degrees would include the value of $\pi/180$ (approximately 0.017453293), since this equals the value of one degree when measured in radians.

When using a projected coordinate system, the individual coordinate values represent a linear distance along the earth's surface to a point. They are measured in a linear unit of measure, such as the meter, foot, mile, or yard. Any spatial reference system based on a projected coordinate system must therefore also state the linear unit of measure in which coordinate values are stated.

Projection

Remember that a projected coordinate system defines positions on the earth as they lie on a flat, two-dimensional plane, such as a map. We see two-dimensional projections of geospatial data on an almost daily basis—in street maps, in road atlases, or on our computer screens. Given their familiarity, and the apparent simplicity of working on a flat surface rather than a curved one, you would be forgiven for thinking that defining a point on the earth using a projected coordinate system is somehow simpler than doing so using a geographic coordinate system. The difficulty associated with a projected coordinate system is that, of course, the world *isn't* a flat, two-dimensional plane. In order to be able to represent it as such, we have to use a map projection.

Projection is the process of creating a two-dimensional representation of a three-dimensional model of the earth. Map projections can be constructed either by using purely geometric methods (such as the techniques used by ancient cartographers) or by using mathematical algorithms. However, whatever method is used, it is not possible to project any three-dimensional object onto a two-dimensional plane without distorting the resulting image in some way. Distortions introduced as a result of the projection process may affect the area, shape, distance, or direction represented by different elements of the map.

By altering the projection method, cartographers can reduce the effect of these distortions for certain features, but in doing so the accuracy of other features must be compromised—there is no single ideal map projection that best represents all features of the earth. Over the course of time, many projections have been developed that balance these distortions in different ways to create maps suitable for different purposes. For instance, when designing a map used by sailors navigating through the Arctic regions, a projection may be used that maximizes the accuracy of the direction and distance of objects at the poles of the earth, but sacrifices accuracy of the shape of countries along the equator.

The full details of how to construct a map projection are outside the scope of this book. However, the following sections introduce some common map projections and examine their key features.

Hammer-Aitoff Projection

The Hammer-Aitoff map projection is an equal-area map projection that displays the world on an ellipse. An *equal-area* map projection is one that maintains the relative area of objects; that is, if you were to measure the area of any particular region on the map, it would accurately represent the area of the corresponding real-world region. However, in order to do this, the shapes of features are distorted. This is illustrated in Figure 1-11.

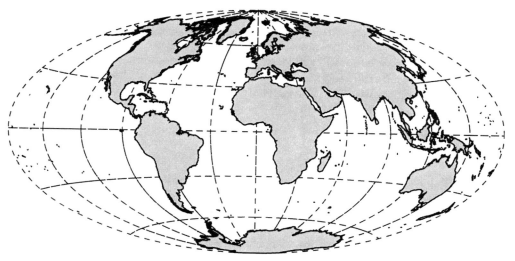

Figure 1-11. *The Hammer-Aitoff map projection*

Mercator Projection

The Mercator map projection is an example of a conformal map projection. A *conformal* map projection is any projection that preserves the local shape of objects on the resulting map.

The Mercator projection was first developed in 1569 by the Flemish cartographer Gerardus Mercator, and has been widely used ever since. It is used particularly in nautical navigation because, when using any map produced using the Mercator projection, the route taken by a ship following a constant bearing will be depicted as a straight line on the map.

The Mercator projection accurately portrays all points that lie on the equator. However, as you move further away from the equator, the distortion of features, particularly the representation of their area, becomes increasingly severe. One criticism of using this projection is that, due to the geographical distribution of countries in the world, many developed countries are depicted with far greater area than equivalent sized developing countries. For instance, examine Figure 1-12 to see how the relative sizes of North America (actual area 19 million sq km) and Africa (actual area 30 million sq km) are depicted at approximately the same size.

Despite this criticism, the Mercator projection is still commonly used by many applications, including the Google Maps web site (http://maps.google.com/).

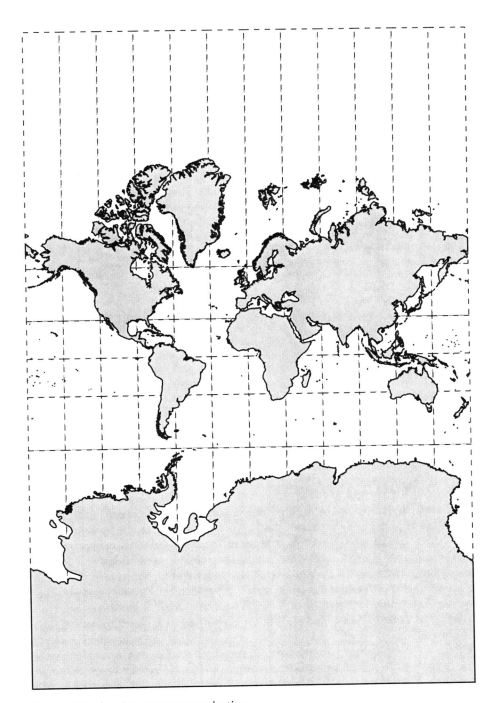

Figure 1-12. *The Mercator map projection*

Equirectangular Projection

The equirectangular projection is one of the first map projections ever to be invented, being credited to Marinus of Tyre in about 100 AD. It is also one of the simplest map projections, because the map projects equally spaced degrees of longitude on the x axis, and equally spaced degrees of latitude on the y axis.

This projection is of limited use in spatial data analysis since it represents neither the accurate shape nor area of features on the map, although it is still widely recognized and used for such purposes as portraying NASA satellite imagery of the world (http://visibleearth.nasa.gov/). Figure 1-13 illustrates a map of the world created using the equirectangular projection method.

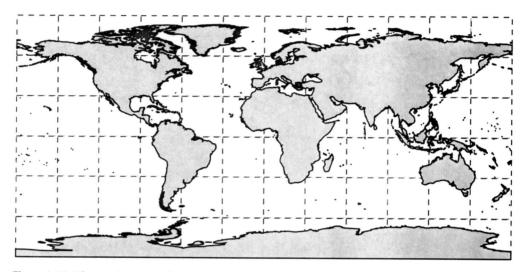

Figure 1-13. *The equirectangular map projection*

Universal Transverse Mercator Projection

The Universal Transverse Mercator (UTM) projection is not a single projection, but rather a grid composed of many projections laid side by side. The UTM grid is created by dividing the globe into 60 slices, called "zones," with each zone being 6° wide and extending nearly the entire distance between the North Pole and South Pole (the grid does not extend fully to the polar regions, but ranges from a latitude of 80°S to 84°N). Each numbered zone is further subdivided by the equator into north and south zones. Any UTM zone may be referenced using a number from 1 to 60, together with a suffix of *N* or *S* to denote whether it is north or south of the equator. Figure 1-14 illustrates the grid of UTM zones overlaid on a map of the world, highlighting UTM Zone 15N.

Within each UTM zone, features on the earth are projected using a *transverse* Mercator projection. The transverse Mercator projection is produced using the same method as the Mercator projection, but rotated by 90°. This means that, instead of portraying features that lie along the equator with no distortion, the transverse Mercator projection represents features that lie along a central north-south meridian with no distortion. Since each UTM zone is relatively narrow, any feature on the earth lies quite close to the central meridian of the UTM zone in which it is contained, and distortion within each zone is very small.

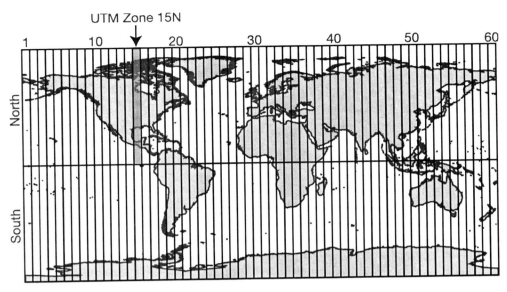

Figure 1-14. *UTM zones of the world*

The UTM projection is *universal* insofar as it defines a system that can be applied consistently across the entire globe. However, since each zone within the UTM grid is based on its own unique projection, the UTM map projection can only be used to accurately represent features that lie within a single specified zone.

Projection Parameters

In addition to the method of projection used, there are a number of additional parameters that affect the appearance of any projected map. These parameters are listed in Table 1-2.

Table 1-2. *Map Projection Parameters*

Parameter	Description
Azimuth	The angle at which the center line of the projection lies, relative to north
Central meridian	The line of longitude used as the origin from which x coordinates are measured
False easting	A value added to x coordinates so that stated coordinate values remain positive over the extent of the map
False northing	A value added to y coordinates so that stated coordinate values remain positive over the extent of the map
Latitude of center	The latitude of the point at the center of the map projection
Latitude of origin	The latitude used as the origin from which y coordinates are measured
Latitude of point	The latitude of a specific point on which the map projection is based
Longitude of center	The longitude of the point at the center of the map projection

Table 1-2. *Map Projection Parameters (Continued)*

Parameter	Description
Longitude of point	The longitude of a specific point on which the map projection is based
Scale factor	A scaling factor used to reduce the effect of distortion in a map projection
Standard parallel	A line of latitude along which features on the map have no distortion

Using Spatial Reference Systems

We have examined several components that make up any spatial reference system—a system that allows us to define positions on the earth's surface, which we can use to construct geometries representing features on the earth. Table 1-3 gives an overview of each component.

Table 1-3. *Components of a Spatial Reference System*

Component	Function
Coordinate system	Specifies a mathematical framework for determining the position of items relative to an origin.
Datum	States a model of the earth onto which we can apply the coordinate system. Consists of a reference ellipsoid (a three-dimensional mathematical shape that approximates the shape of the earth) and a reference frame (a set of points that enables us to position the reference ellipsoid to line up with the right points on the earth).
Prime meridian	Defines the axis from which coordinates of longitude are measured.
Projection[a]	Details the parameters required to create a two-dimensional image of the earth's surface (i.e., a map), so that positions can be defined using projected coordinates.
Unit of measurement	Provides the appropriate unit in which coordinate values are expressed.

[a] *Projection parameters are only defined for spatial reference systems based on projected coordinate systems.*

Through a combination of all these elements, you can use a spatial reference system to uniquely identify any point on the earth.

■**Note** In order to be able to describe positions on the earth using a projected coordinate system, a spatial reference system must first specify a three-dimensional, geodetic model of the world (as would be used by a geographic coordinate system), and then *additionally* state the parameters detailing how the two-dimensional projected map image should be created from that model. For this reason, spatial reference systems based on projected coordinate systems must contain all the same elements as those based on geographic coordinate systems, together with the additional parameters required for the projection.

Spatial Reference Identifiers

Every time we state the latitude and longitude, or x and y coordinates, that describe the position of a point in a geometry, we must also state the associated spatial reference system in which those coordinates were obtained. Without the extra information contained in the spatial reference system, a coordinate tuple is just an abstract set of numbers in a mathematical system. The spatial reference takes the abstract coordinates from a geographic or projected system and puts them in a context so that they can be used to identify a real position on the earth's surface.

However, it would be quite cumbersome to have to write out the full details of the datum, the prime meridian, and the unit of measurement (and any applicable projection) each time we wrote down a set of coordinates. Fortunately, various authorities allocate easily memorable, unique integer reference numbers that represent all of the necessary parameters of a spatial reference system. These reference numbers are called spatial reference identifiers (SRIDs).

One authority that allocates SRIDs is the European Petroleum Survey Group (EPSG), and its reference identification system is implemented in SQL Server 2008. Whenever you use any of the spatial functions in SQL Server that involve stating the coordinates of a position, you must always supply the relevant EPSG SRID as a parameter.

Tip You can view the details of all spatial reference systems administered by the EPSG registry at the following web site: `http://www.epsg-registry.org`.

Spatial References in SQL Server 2008

SQL Server 2008 stores the details of all supported geodetic spatial reference systems in a special system table called `sys.spatial_reference_systems`. Every row in this table corresponds to a unique spatial reference system that you can use to define spatial data in SQL Server 2008.

In order to see the list of supported geodetic spatial reference systems, execute the following code in a SQL Server Management Studio query window:

```
SELECT
  *
FROM
  sys.spatial_reference_systems
```

Table 1-4 lists and describes each column of the `sys.spatial_reference_systems` table shown in the results.

Table 1-4. *Columns of the sys.spatial_reference_systems Table*

Column Name	Description
spatial_reference_id	The integer identifier used within SQL Server 2008 to refer to this system
authority_name	The name of the authority that defines this reference
authorized_spatial_reference_id	The identifier allocated by the authority to refer to this system
well_known_text	The parameters of the spatial reference system, expressed in well-known text format
unit_of_measure	A text description of the unit used to express linear measurements in this system, such as distance and length
unit_conversion_factor	A scale factor for converting from meters into the unit of measurement

The sys.spatial_reference_systems table only includes those spatial reference systems based on geographic coordinates supported by SQL Server. In addition to the spatial reference systems listed in this table, you can also define data using *any* projected spatial reference system, as you will see in the next chapter.

Note Currently, the only authority used to define spatial references in SQL Server 2008 is the EPSG, and all internal SRIDs are based on the EPSG numbering system. As a result, the value of the internal spatial_reference_id for any system is the same as the authorized_spatial_reference_id.

Expressing Spatial References in the Well-Known Text Format

Within the sys.spatial_reference_systems table, SQL Server stores the relevant details of each spatial reference using the Well-Known Text (WKT) format, which is an industry-standard format for expressing spatial information defined by the OGC. The WKT description of the spatial reference is stored as a text string in the well_known_text column.

To illustrate how spatial references are represented in WKT format, let's examine the properties of the EPSG:4326 spatial reference, by running the following query:

```
SELECT
  well_known_text
FROM
  sys.spatial_reference_systems
WHERE
  authority_name = 'EPSG'
  AND
  authorized_spatial_reference_id = 4326
```

The following is the result (with line breaks and indents added to make the result easier to read):

```
GEOGCS[
  "WGS 84",
  DATUM[
    "World Geodetic System 1984",
    ELLIPSOID[
      "WGS 84",
      6378137,
      298.257223563
    ]
  ],
  PRIMEM["Greenwich", 0],
  UNIT["Degree", 0.0174532925199433]
]
```

Let's examine this result, to identify each of the component elements of a spatial reference system:

Coordinate system: The first line of a WKT spatial reference is a keyword to tell us what sort of coordinate system is used. In this case, GEOGCS tells us that EPSG:4326 uses a geographic coordinate reference system. If a spatial reference system is based on projected coordinates, then the WKT representation would instead begin with PROJCS. Immediately following the declaration of the type of coordinate system is the name of this spatial reference. In this case, we are describing the "WGS 84" spatial reference.

Datum: The values following the DATUM keyword provide the parameters of the datum. The first parameter gives us the name of the datum used. In this case, it is the "World Geodetic System 1984" datum. Then follow the parameters of the reference ellipsoid. In this spatial reference, we are using the "WGS 84" ellipsoid, with a semimajor axis of 6,378,137 m and an inverse-flattening ratio of 298.257223563.

Prime meridian: The PRIMEM value tells us that this system defines Greenwich as the prime meridian, where longitude is 0.

Unit of measurement: The spatial reference specifies that the units of angular measurement are expressed as "Degree". The value of 0.0174532925199433 is a conversion factor required to convert to the appropriate units. This represents the value of $\pi/180$, required to convert angular measurements from radians into degrees.

Contrasting a Geographic and a Projected Spatial Reference

Let's compare the result in the preceding section to the WKT representation of a spatial reference system based on a projected coordinate system. The following example shows the WKT representation of the UTM Zone 10N reference, a projected spatial reference system used in North America. The SRID for this system is EPSG:26910.

```
PROJCS[
  "NAD_1983_UTM_Zone_10N",
  GEOGCS[
    "GCS_North_American_1983",
    DATUM[
      "D_North_American_1983",
      SPHEROID[
        "GRS_1980",
        6378137,
        298.257222101
      ]
    ],
    PRIMEM["Greenwich",0],
    UNIT["Degree", 0.0174532925199433]
  ],
  PROJECTION["Transverse_Mercator"],
  PARAMETER["False_Easting", 500000.0],
  PARAMETER["False_Northing", 0.0],
  PARAMETER["Central_Meridian", -123.0],
  PARAMETER["Scale_Factor", 0.9996],
  PARAMETER["Latitude_of_Origin", 0.0],
  UNIT["Meter", 1.0]
]
```

Notice that the spatial reference for a projected coordinate system contains a complete set of parameters for a geographic coordinate system, embedded within brackets following the GEOGCS keyword. The reason is that a projected system must first define the three-dimensional, geodetic model of the earth, and then specify several additional parameters that are required to project that model onto a plane.

Comparing Spatial Reference Systems

By now, you should have a good appreciation of the fact that a given point on the earth may be represented using many different sets of coordinates, each one corresponding to a particular spatial reference system. (Conversely, the same set of coordinate values can refer to different places on the earth depending on the spatial reference system from which the coordinates were obtained.) Whenever we define an item of spatial data in SQL Server to represent an object on the earth, three bits of information are required:

- The type of geometry used to represent the object (e.g., Point, LineString, Polygon)

- The coordinates of each of the points that define that geometry, expressed in decimal degree notation (e.g., 37.215, –57.5)

- The unique identifier of the spatial reference system from which those coordinates were obtained (e.g., 4326)

To demonstrate this in practical terms, let's compare how a particular feature on the earth—Loch Ness, in Scotland—can be expressed in different spatial reference systems. For this example, we will use a geographic coordinate system based on the WGS 84 datum, and a projected coordinate system based on the National Grid of Great Britain.

WGS 84

WGS 84 is the most commonly used geodetic spatial reference system and is used by GPS systems. It is based on the WGS 84 ellipsoid, which gives a reasonable approximation over the whole surface of the geoid. It is referenced by the EPSG reference 4326. In the WGS 84 system, Loch Ness can be represented by a Point object positioned at the following coordinates:

- Geometry: Point

- Latitude/longitude coordinates: (57.3, –4.5)

- SRID: 4326

National Grid of Great Britain

Many countries have defined their own grid systems for referencing coordinates of positions that lie exclusively within that country. Great Britain, Ireland, New Zealand, Malaysia, Singapore, the Netherlands, and Sweden all have defined national grid systems that are commonly used to express coordinates for local positioning within those countries.

The National Grid of Great Britain uses a projected coordinate system based on the transverse Mercator projection. The transverse Mercator projection is similar to the Mercator projection, but instead of accurately portraying features lying on the equator of the earth, it has been rotated so that the map accurately portrays features lying along a given meridian—a line running north-south between the poles of the earth. This makes the transverse Mercator projection more suitable for mapping tall, thin countries such as Great Britain. The datum is the Ordnance Survey of Great Britain 1936 (OSGB 36), using the Airy 1830 ellipsoid, which is the ellipsoid that provides the best fit for the geoid over this region. The "true" origin on which the projection is based has a latitude of 49° north and longitude 2° west. However, coordinates in the grid system are actually stated from a "false" origin situated 400 km west and 100 km north of the true origin. Using the false origin as the point from which coordinates are measured ensures that coordinate values for any point in Great Britain are always positive. The SRID of this system is EPSG:27700.

The grid system is constructed by overlaying a series of 100 km by 100 km squares, starting from the false origin, that cover the land surface of Great Britain. Each of these squares is given a two-letter identifier. To refer to any point in the system, you state the identifier of the square that the point lies in, together with the easting and northing coordinates of the point measured in meters from the bottom-left corner of the square. Alternatively, easting and northing coordinates can be expressed in absolute meters east and north from the false origin. As such, Loch Ness could be represented in this system as follows:

- Geometry: LineString

- Easting/northing coordinates: (238172, 808732), (261620, 839938)

- SRID: 27700

The comparison between the coordinate values obtained from the National Grid of Great Britain and WGS 84 is illustrated in Figure 1-15.

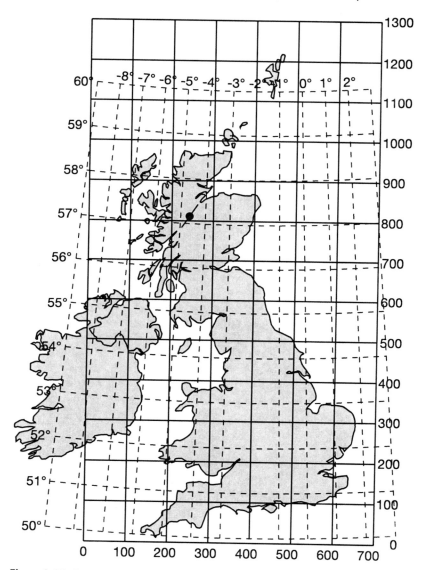

- - - - WGS84 (EPSG: 4326)

——— National Grid of Great Britain (EPSG: 27700)

Figure 1-15. *Comparing representations of Loch Ness in two different spatial reference systems*

Note these are only two examples—this same feature could be described by many other sets of coordinates in different spatial reference systems.

Summary

After reading this chapter, you should understand how spatial data can be used to describe the properties of features on the earth:

- Spatial data creates representations of features on the earth by defining regular shapes that approximate the shape and position of those features. The three basic types of shape that can be used in SQL Server 2008 are Points, LineStrings, and Polygons. When used in geospatial data, these shapes are called geometries.

- Each of these geometries can be defined by specifying the coordinate values of a series of points that make up the overall geometry.

- To define the coordinates of the position of a point on the earth, we use a spatial reference system.

- A spatial reference system consists of a coordinate system (which describes a position using either projected or geographic coordinates), a datum (which describes a model representing the shape of the earth), the prime meridian (which defines the origin from which units are measured), and the unit of measurement. When using projected coordinates to describe a point, the spatial reference system also defines the properties of the projection used.

- A geographic coordinate system defines the position of objects using angular coordinates called latitude and longitude, which are measured from the equator and the prime meridian, respectively.

- A projected coordinate system defines the position of objects using Cartesian coordinates, which measure the x and y distance of a point from an origin. These are also referred to as easting and northing coordinates.

- Whenever you state a set of coordinates representing a point, it is essential that you also give details of the associated spatial reference system. The spatial reference system defines the additional information that allows us to apply the coordinate reference to identify a point on the earth.

- For convenience, spatial reference systems may be specified by a single integer identifier— known as a spatial reference identifier (SRID).

- Details of all the geodetic spatial reference systems supported by SQL Server 2008 are contained within a system table called `sys.spatial_reference_systems`.

CHAPTER 2

■ ■ ■

Implementing Spatial Data in SQL Server 2008

In the last chapter, I introduced you to the theory behind spatial reference systems, and explained how different types of systems describe features on the earth. In this chapter, you'll learn how to apply these systems to store spatial information using the new spatial datatypes in SQL Server 2008.

Understanding Datatypes

Every variable, parameter, and column in a SQL Server table is defined as being of a particular datatype. This tells SQL Server what sort of data values will be stored in this field, and how they can be used. Some commonly used datatypes are listed and described in Table 2-1.

Table 2-1. *Some Common SQL Server Datatypes*

Datatype	Usage
char	Fixed-length character string
datetime	Date and time value, accurate to 3.33ms
float	Floating-point numeric data
int	Integer number between -2^{31} ($-2,147,483,648$) and $2^{31} - 1$ ($2,147,483,647$)
money	Monetary or currency data
nvarchar	Variable-length, Unicode character string

SQL Server 2008 introduces two new datatypes specifically intended to hold spatial data: geography and geometry (see Table 2-2).

Table 2-2. *Spatial Datatypes Introduced in SQL Server 2008*

Datatype	Usage
geography	Geodetic vector spatial data
geometry	Planar vector spatial data

Although both datatypes can be used to store spatial data, they are distinct from each other and are used in different ways. Whenever you define an item of spatial data in SQL Server 2008, you must also choose whether to store that information using the geometry datatype or the geography datatype.

Note The word geometry has two different meanings in this book. To avoid confusion, I use geometry (with no special text formatting) to refer to a Point, LineString, or Polygon that is used to represent a feature on the earth, and I use `geometry` to refer to the geometry datatype. This convention will be used throughout the rest of the book.

Comparing Spatial Datatypes

There are several similarities between the two spatial datatypes:

- They can both represent spatial information using a range of geometries—Points, LineStrings, and Polygons.

- Internally, both datatypes store spatial data as a stream of binary data in the same format.

- When working with items of data from either type, you must use object-orientated methods based on the .NET Framework (discussed in more detail in the next chapter).

- They both implement many of the same standard spatial methods to analyze and perform calculations on data of that type.

However, there are also a number of important differences between the two spatial datatypes, as outlined in Table 2-3. You must choose the appropriate datatype to reflect how you plan to use spatial data in your database.

Table 2-3. *Comparison of the geometry and geography Datatypes*

Property	geometry Datatype	geography Datatype
Shape of the earth	Flat	Round
Coordinate system	Projected (or natural planar)	Geographic
Coordinate values	Cartesian (x and y)	Latitude and longitude
Unit of measurement	Same as coordinate values	Defined in `sys.spatial_reference_systems`
Spatial reference identifier	Not enforced	Enforced
Default SRID	0	4326 (WGS 84)
Size limitations	None	No object may occupy more than one hemisphere
Ring orientation	Not significant	Significant

The significance of these differences will be clearer after you read about the features of the two datatypes in more detail.

The geography Datatype

The most important feature of the geography datatype is that it stores *geodetic* spatial data, which takes account of the curved shape of the earth. When you perform operations on spatial data using the geography datatype, SQL Server uses angular computations to work out the result. These computations are calculated based on the ellipsoid model of the earth defined by the spatial reference system of the data in question. For example, if you were to define a line that connects two points on the earth's surface in the geography datatype, the line would curve to follow the surface of the reference ellipsoid. Every line drawn between two points in the geography datatype (whether that line is a segment of a LineString geometry, or an edge of a Polygon ring) is actually a *great elliptic arc*—that is, the line on the surface of the earth formed by the plane containing both points and the center of the reference ellipsoid. This is illustrated in Figure 2-1.

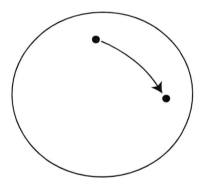

Figure 2-1. *Calculations on the geography datatype account for curvature of the earth.*

Caution Do not be misled by the name of the geography datatype. Like the geometry datatype, it too stores geometry shapes representing features on the earth.

Coordinate System

The geography datatype is based on a three-dimensional, round model of the world, so you must use a geographic coordinate system to specify the positions of each of the points that define a geometry in this datatype. Remember that, when using a geographic coordinate system, the coordinates of these points are expressed using angles of latitude and longitude.

Unit of Measurement

Since the geography datatype defines points using angular measurements of latitude and longitude, the coordinate values are usually measured in degrees. These angular coordinates are useful for expressing the location of points, but are not that helpful for expressing the distance between points or the area enclosed within a set of points. For example, using the spatial reference system EPSG:4326, we can state the location of Paris, France as a point at 48.87°N, 2.33°E. Using the same system, the location of Berlin, Germany could be described as 52.52°N, 13.4°E. However, if you wanted to know the distance between Paris and Berlin, it would not be very helpful for me to say that they are 11.65° apart, stating the answer in degrees. You would probably find it much more useful to know that the distance between them is 880 km, or 546 miles.

To account for this, when you perform calculations on any items of spatial data using the geography datatype, the results are returned in the linear unit of measurement specified in the unit_of_measure column of the sys.spatial_reference_systems table for the relevant spatial reference system. For example, to check the unit of measurement used by the EPSG:4326 spatial reference system, you can run the following T-SQL query:

```
SELECT
  unit_of_measure
FROM
  sys.spatial_reference_systems
WHERE
  authority_name = 'EPSG'
  AND
  authorized_spatial_reference_id = 4326
```

The result of this query is as follows:

```
metre
```

This tells us that the results of any linear calculations performed against data stored in the geography datatype and defined using the EPSG:4326 spatial reference system will be stated in meters. To calculate the distance between Paris and Berlin based on the coordinates given earlier, you can execute the following T-SQL code:

```
DECLARE @Paris geography = geography::Point(48.87, 2.33, 4326)
DECLARE @Berlin geography = geography::Point(52.52, 13.4, 4326)
SELECT @Paris.STDistance(@Berlin)
```

The result will be expressed in meters:

```
879989.866996421
```

■**Note** Don't worry if you can't follow the syntax of the preceding query—I'll cover this in the next chapter. You just need to know that this query creates two geography points, representing Paris and Berlin, and then calculates the shortest distance between them.

Because most spatial reference systems are based on metric units, distances calculated using the geography datatype are usually expressed in meters, and areas in square meters.

Spatial Reference ID

Every time you store an item of data using the geography datatype, you must supply the appropriate SRID corresponding to the spatial reference system from which the coordinates were obtained. SQL Server 2008 uses the information contained in the spatial reference system to apply the relevant model of curvature of the earth in its calculations, and also to express the results of any linear methods in the appropriate units of measurement. The supplied SRID must therefore correlate with one of the supported spatial references in the sys.spatial_reference_systems table.

If you were to supply a different SRID when storing an item of geography data, you would get different results from any methods using that data, since the calculations would be based on a different set of geodetic parameters.

Size Limitations

Due to technical limitations, SQL Server imposes a restriction on the maximum size of a single object that can be stored using the geography datatype. The effect of this restriction is that every geometry using the geography datatype, whether created as a new item of data or the result of any calculation, must fit inside a single hemisphere. In this context, the term *hemisphere* does not refer to a predetermined area of the globe, such as the Northern Hemisphere or Southern Hemisphere, but rather refers to one-half of the earth's surface, centered about any point on the globe. If you try to create an object that exceeds this size, or perform a calculation whose result would exceed this size, you will receive the following error:

```
Microsoft.SqlServer.Types.GLArgumentException: 24205: The specified input does not
represent a valid geography instance because it exceeds a single hemisphere.
Each geography instance must fit inside a single hemisphere.
A common reason for this error is that a polygon has the wrong ring orientation.
```

To work around this limitation, you can break down large geography objects into several smaller objects that each fit within the relevant size limit. For example, rather than having a single Polygon object representing the entire ocean surface of the earth, you can define multiple Polygons that each represent an individual sea or ocean. When combined together, these smaller objects represent the overall ocean surface.

Note The size limit imposed on the geography datatype applies not only to geometries that contain an area greater than a single hemisphere, but also to any geometry that contains points that do not lie in the same hemisphere. Thus, for example, you cannot create a MultiPoint geometry using the geography datatype that contains two points representing the North Pole and South Pole.

Ring Orientation

Look again at the error message shown in the previous section. It states that a common reason for invalid geography instances is that "…a polygon has the wrong ring orientation." What does this mean? Remember from Chapter 1 that a ring is a closed LineString, and that Polygon geometries are made up of one or more rings that define the boundary of the area contained within the Polygon. *Ring orientation* refers to the "direction," or order, in which the points that make up the ring of a Polygon are stated.

The geography datatype defines features on a geodetic model of the earth, which is a continuous, round surface. Unlike the image created from a map projection, it has no edges—you can continue going in one direction all the way around the world and get back to where you started. This becomes significant when you define the points of a Polygon ring since, when using a round model, there is ambiguity as to which side of the ring contains the area included within the Polygon. Consider Figure 2-2, which illustrates a Polygon whose exterior ring is a series of points drawn around the equator. Does the area contained within the Polygon represent the Northern Hemisphere or the Southern Hemisphere?

Figure 2-2. *Ambiguous Polygon ring definition using the geography datatype*

To resolve this ambiguity, when you define the points of a Polygon using the geography datatype, SQL Server 2008 treats the area on the "left" of the path drawn between the points of a ring as being contained within the interior of the Polygon, and excludes any points that lie on the "right" side of the ring. In the example given in Figure 2-2, if you were to imagine walking along the path of the ring in the direction indicated, the area to your left would be north, so the Polygon illustrated represents the Northern Hemisphere. Another way of thinking about this is to imagine looking directly down at a point on the earth from space. If it is enclosed by a ring of points in a counterclockwise direction, then that point is contained within the Polygon (since it must lie on the left of the path of that ring). If, instead, it appears to be encircled by a ring of points in a clockwise direction, then that point is not included in the Polygon definition.

Caution If you define the points of a small Polygon ring in the wrong direction, the resulting object would be "inside out"—encompassing most of the surface of the earth, and only excluding the small area contained within the linear ring. This would break the size limitation that no geography object can cover more than one-half of the earth's surface, and would cause an error. When you are containing an area within a Polygon, be sure to define the points in a counterclockwise direction, so that the area to be included is on the left of the path connecting the points.

What about ring orientation for interior rings, which define areas of space cut out of a geometry? The classification of "interior" and "exterior" cannot easily be applied to rings defined on the continuous, round surface of the geography datatype. In fact, a Polygon in the geography datatype may contain any number of rings, each of which divides space on the globe into those points included within the Polygon, and those points that are excluded. Every one of these rings could be considered to be an exterior ring or an interior ring. The key rule to remember is that the area on the left of the path drawn between points of a ring is contained within the Polygon, and the area on the right side is excluded. Therefore, to define an area of space that should be excluded from a polygon, you should enclose it in a ring of points specified in *clockwise* order— so that the area is contained on the right-hand side of the path of the ring. To illustrate this, Figure 2-3 demonstrates the appropriate ring orientation of a Polygon in the geography datatype containing two rings. The arrows illustrate the orientation of the points in each ring, and the area enclosed by the Polygon is shaded in gray.

Figure 2-3. *Ring orientation of a Polygon containing two rings*

The geometry Datatype

In contrast to the geography datatype, the geometry datatype treats spatial data as lying on a flat plane. As such, the results of all spatial calculations, such as the distance between points, are worked out using simple geometrical methods. This flat-plane approach is illustrated in Figure 2-4.

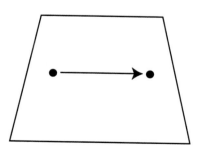

Figure 2-4. *Calculations on the planar geometry type operate in a flat plane.*

Coordinate System

Since the geometry datatype works with spatial data on a flat, two-dimensional plane, the position of any point on that plane can be defined using a single pair of Cartesian (x, y) coordinates. The geometry datatype can be used to store coordinates from any one of the following types of coordinate system:

Projected coordinates: The geometry datatype is ideally suited to storing projected coordinates, where each x and y coordinate pair represents the easting and northing coordinate values obtained from a projected spatial reference system. In this case, the process of projection has already mapped the angular geographic coordinates onto a flat plane, onto which the methods of the geometry datatype can be applied.

Geographic coordinates: "Unprojected" geographic coordinates of latitude and longitude can be assigned directly to the y and x coordinates, respectively, of the geometry datatype. Although this may seem like an unprojected geographic coordinate system, it is actually still an example of a projected system, because it is the method used to create an equirectangular projection.

Naturally planar coordinates: These coordinates could represent any geometric spatial data that can be expressed in x and y values, but are not associated with a particular model of the earth. Examples of such data might be collected from a local survey or topological plans of a small area where curvature is irrelevant, or from geometrical data obtained from computer-aided design (CAD) packages.

Unit of Measurement

When using the geometry datatype, the Cartesian coordinates of a point represent the distance of that point from an origin along a defined axis, expressed in a particular unit of measurement. Since the geometry datatype uses simple planar calculations based on these coordinate values, the results of any computations using the geometry datatype will be expressed in the same units of measurement as the coordinate values themselves. For instance, the northing and easting coordinates of many projected coordinate systems are expressed in meters. This is the case for the Universal Transverse Mercator (UTM) system and many national grid reference systems. If you use the geometry datatype to store spatial data based on coordinates taken from any of these systems, lengths and distances will also be measured in meters. If you were to calculate an area using the geometry datatype, the result would be the square of whatever unit was used to define the coordinate values—in this case, square meters. If, however, you were to store coordinates from a projected spatial reference system measured in feet, then the results of any linear calculations would also be expressed in feet, and areas in square feet.

> ▪**Caution** Earlier, I told you that the geometry datatype could be used to store "unprojected" geographic coordinates of latitude and longitude, directly mapped to the y and x coordinates. However, remember that latitude and longitude are angular coordinates, usually measured in degrees. If you use the geometry datatype to store information in this way, then the distances between points will also be measured in degrees, and the area enclosed within a Polygon will be measured in degrees squared. This is almost certainly not what you want, so exercise caution when using the geometry datatype in this way.

Spatial Reference ID

Since geometry data does not consider any curvature of the earth and does not rely on the unit of measurement stated in the SRID, supplying a different SRID does not make any difference to results obtained using the geometry datatype. This can be a tricky concept to grasp. In the last chapter I told you that that any pair of coordinates—projected or geographic—must be stated with their associated SRID so that they can refer to a point on the earth. If we are using the geometry datatype to store coordinates from a projected coordinate system, how come it doesn't make a difference what SRID is provided?

The answer is that the SRID *is* required in a projected coordinate system to initially determine the coordinates that uniquely identify a position on the earth. However, once we have derived those values, all further operations on that data can be performed using basic geometrical methods. Any decisions concerning how to deal with the curvature of the earth have already been made in the process of defining the coordinates that describe where any point lies on the projected image.

For example, when using the geometry datatype, the distance between a point at coordinates (0,0) and a point located at (30,40) will always be 50 units, whatever spatial reference system was used to obtain those coordinates, and whatever units they are expressed in. The actual features on the earth *represented* by the points at (0,0) and (30,40) will be different depending on the system in question, but this does not affect the way that the geometry data is used in calculations.

In order to perform operations using spatial objects of the geometry type, it makes no difference what spatial reference system the coordinates of each point were obtained from, *as long as they were all obtained using the same system.*

Ensuring Consistent Metadata

Even though it does not make a difference to the results, when using the geometry datatype to store Cartesian data based on a projected coordinate system, you should still specify the relevant SRID to identify which spatial reference was used to derive those coordinates. The spatial reference system includes the important additional information that makes those coordinates relate
to a particular position on the earth. Explicitly stating the SRID with every set of coordinates ensures not only that you retain this important metadata, but also that you do not accidentally try to perform a calculation on items of spatial data defined using different spatial references, which would lead to an invalid result.

■ **Note** The sys.spatial_reference_systems table only contains details of geodetic spatial references, since these are required to perform calculations using the geography datatype. To find the appropriate SRID for a projected coordinate system, you can look it up on the following web site: http://www.epsg-registry.org/.

Storing Nongeodetic Spatial Data

Since the geometry datatype stores planar coordinates and uses standard Euclidean methods to perform calculations for which no SRID is necessary, it can also be used to store any spatial data that can be described using x and y coordinates, which do not necessarily relate to any particular model of the shape of the earth. This is useful for contained, small-scale applications, such as storing the position of items in a warehouse. In this case, positions can be defined using x and y coordinates relative to a local origin—they do not need to be expressed using a projected coordinate system applied to the whole surface of the earth.

The geometry datatype can also be used to store any other naturally planar geometrical data that can be represented using a coordinate system. For instance, if you had a database that stored the details of components used in a manufacturing process, you could use a field of the geometry datatype to record the shape of each component.

When using the geometry type to record spatial data in this way, you should define any geometry features using SRID 0. This tells SQL Server that the coordinates are not derived from a particular spatial reference system, and so they are treated as coordinate values with no specific unit of measurement.

Ring Orientation

Consider Figure 2-5, which illustrates a Polygon ring representing the Northern Hemisphere defined on a planar surface, as used by the geometry datatype.

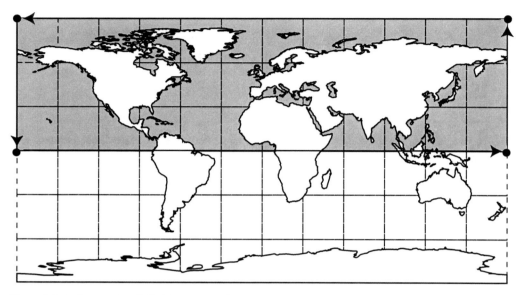

Figure 2-5. *Polygon ring definition using the geometry datatype*

Unlike the equivalent Polygon using the geography datatype (illustrated in Figure 2-2), the area contained by the Polygon in Figure 2-5 is unambiguous. Even if the ring orientation was reversed, so that the points formed a clockwise ring, the area enclosed by the ring would still represent the Northern Hemisphere. In the geometry datatype, ring orientation—the direction in which the points of a Polygon ring are specified—is unimportant. In practical terms, the ring defined by the point coordinates

(50,30), (52,30), (52,31), (50,31), (50,30)

is treated identically to how it would be treated if it were specified by the following points:

(50,30), (50,31), (52,31), (52,30), (50,30)

When defining interior rings containing areas of space cut out of the Polygon, these too may be specified in clockwise or counterclockwise fashion, so long as they are completely contained within the exterior ring, and do not cross one another or contain one another.

Choosing the Right Spatial Datatype

Having looked at the key differences between the two datatypes, you are probably wondering which one you should use, and in what situations. These are important questions, and although there are not necessarily any definitive answers, the following list gives you some general guidelines:

- If you have latitude and longitude data (such as from a GPS device, from Google Earth, or from elsewhere on the Web), use the geography datatype, normally using the default 4326 SRID.

- If you are using projected coordinate data (e.g., collected from a flat map), use the geometry datatype with an SRID to represent the map projection and datum used.

- If you are using x, y data that is not particularly defined relative to the earth, use the geometry datatype with SRID 0.

In addition to these general rules, there are a number of additional factors, described in this section, that you should consider to help inform your decision. Choosing the right datatype is the first step to ensuring that you have an effective design for your spatial database, so be sure to review these elements carefully before making your decision.

Consistency

In order to perform operations using multiple items of spatial data in SQL Server 2008, all of the data must be defined using the same spatial reference system, and stored using the same datatype. It is not possible to combine geometry and geography data in the same query, nor is it possible to perform operations on items of the same datatype defined using different SRIDs. If you attempt to do so, SQL Server will return a NULL result.

If you already have existing spatial data that you would like to integrate into your system, you should therefore use a datatype suitable for the format in which that data has been collected. For instance, if you have projected data collected from the National Grid of Great Britain system, you should store the data in a geometry field, using the SRID 27700. If you are using latitude and longitude coordinate data collected from a GPS system, then you should choose a geography type, with SRID 4326.

Note Remember that the SRID provides information about the system in which coordinate values have been defined—it does not dictate the system itself. You therefore cannot simply supply a different SRID value to existing coordinates to express them in a different spatial reference. In order to convert coordinates from one spatial reference system into another, you must reproject the data. SQL Server 2008 does not natively support any reprojection methods, but numerous third-party tools are available to do this, such as OGR2OGR, part of the OGR Simple Feature Library (http://www.gdal.org/ogr/index.html).

Accuracy

The geometry datatype uses a flat-earth model based on a projection of the earth's surface onto a plane. As explained in the last chapter, the process of projecting any three-dimensional surface in this way always leads to some distortion in how features are represented, which may affect their area, shape, distance, or direction. This means that, using the geometry datatype, the results of certain spatial operations will also be distorted, depending on the projection used and the particular part of the earth's surface on which the calculation is based.

Generally speaking, the greater the surface area being projected, the more distortion that occurs. Although over small areas the effects of these distortions are fairly minimal, for large-scale or global applications, they can significantly impact the accuracy of any results obtained using the geometry datatype when compared to results obtained using the geography datatype (which is not distorted by projection). In many applications that cover only a small spatial area, such as those contained within a particular state of the United States, the results of calculations performed using the geometry datatype on the relevant state plane projection will be sufficiently accurate. However, over larger distances, the computations based on a planar projection become less accurate, and the geography datatype becomes a more suitable choice.

The End(s) of the World

Extreme examples of the distortion occurring as a result of projection arise because, unlike the earth itself, a projected map has *edges*. When you are storing projected spatial data using the geometry datatype, you must give special consideration to situations in which you need to define data that crosses these edges. This typically occurs whenever a geometry, or result of a calculation, crosses the 180th meridian or one of the poles of the earth.

To demonstrate how these distortions affect calculations using the geometry datatype, consider how you might calculate the shortest straight-line route between Vancouver and Tokyo. Using the flat geometry datatype, based on a projection centered on the Greenwich prime meridian, the result would look like that shown in Figure 2-6.

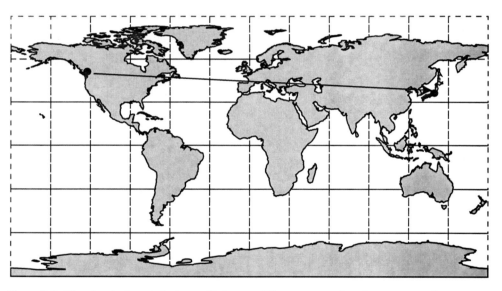

Figure 2-6. *The shortest route between Tokyo and Vancouver using the geometry datatype*

In contrast, the geography datatype uses a continuous, round model of the earth, which is unaffected by the edges introduced as a result of projection. The answer obtained for the shortest route between Tokyo and Vancouver using the geography datatype would instead look something like that shown in Figure 2-7.

Figure 2-7. *The shortest route between Tokyo and Vancouver using the geography datatype*

By comparing the routes illustrated in Figure 2-6 and Figure 2-7, you can see that the result of the geometry datatype, because it cannot cross the edge of the map, depicts a much longer route, crossing almost all of America, Europe, and Asia. In contrast, the result obtained using the geography datatype shows the shortest route to be across the Pacific Ocean, which represents the accurate answer based on the real, round earth.

A further demonstration of these issues occurs when trying to define geometry instances that extend across the edges of the map in a given projection. Figure 2-8 highlights a Polygon geometry representing Russia, as depicted in an equirectangular map projection centered on the Greenwich prime meridian.

Notice that although most of the polygon is contained in the Eastern Hemisphere, the most north-easterly part of Russia (the region of Chukotka) actually crosses the edge of the map, to appear in the Western Hemisphere. Using the geometry datatype based on this projection, it would not be possible to represent Russia using a single Polygon geometry. Instead, you would need to use a MultiPolygon geometry containing two elements to represent each shape created where the edge of the map had caused the original shape to be divided into two.

Both of the problems demonstrated in this section could be mitigated to some extent by choosing an appropriate projected spatial reference system for the geometry datatype, in which the particular geometry in question does not cross the edges of the map. An example of such a projection is illustrated in Figure 2-9. However, while this might avoid the issue for one particular case, it does not solve it completely—even if a new map is created based on a different projection, there will always be features that are located along *its* edges instead.

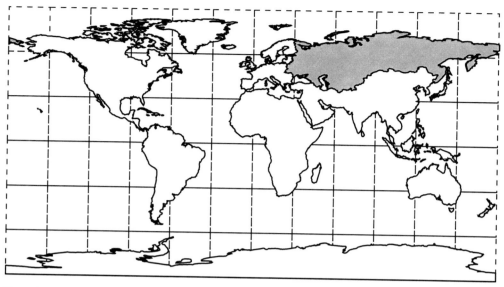

Figure 2-8. *Polygon geometry representing Russia crossing edges of a projection in the geometry datatype*

If you expect to deal with situations where geometries will have to cross the edges of a projected map, then the geography datatype would be a better choice in which to store your data.

Presentation

Since the geography datatype operates on a three-dimensional model of the earth, if you want to present the results of any geography data on a map, you first need to project them. As previously discussed, this introduces distortion. In the example given in Figure 2-7, although the geography datatype accurately works out the shortest straight-line route connecting two points, if we were to display this result on a projected map, this "straight" line may appear distorted and curved. The exact effect of this distortion will differ depending on the particular properties of the map projection used.

Figure 2-9 depicts the same route between Vancouver and Tokyo as shown in Figure 2-7, but projected onto a map using the Mercator method, centered on the meridian at 150° longitude. Notice how the route calculated using the geography datatype, which appeared to be a direct path when plotted on a globe, appears to curve upward on a flat map due to the effects of distortion caused by projection.

■**Note** You will often see great elliptic arcs—the shortest path between two points in the geography datatype—depicted as curved lines (such as shown in Figure 2-9) on maps used by airlines to illustrate the approximate route taken by airplanes between destinations.

Conversely, since the `geometry` datatype is based on data that has already been projected onto a plane, no further distortion needs to be introduced to express the results on a map—"straight" lines in the `geometry` datatype remain straight when drawn on a map (provided that the map is projected using the same projection as the spatial reference system from which the points were obtained).

If you will be using data to represent straight lines between points, and you want those lines to remain straight when displayed on a map, then you should choose the `geometry` datatype, and select an appropriate spatial reference corresponding to the map on which those results will be displayed.

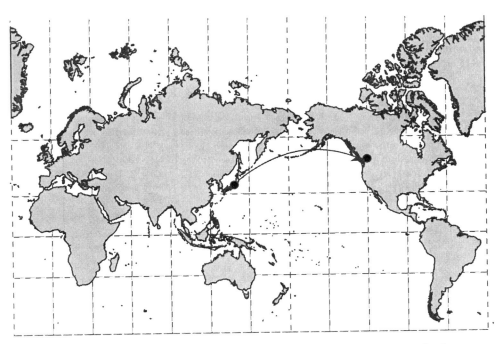

Figure 2-9. *The shortest route between Tokyo and Vancouver using the geography datatype, as projected using the Mercator projection*

Performance

Performing spherical computations uses more computing resources than does performing Cartesian computations. As a result, performing spatial calculations on the ellipsoidal model used by the `geography` datatype may take longer to compute than the equivalent operations using the `geometry` datatype. This will only affect methods where the `geography` datatype has to calculate metrics based on the geodetic model of the earth, such as distances, lengths, or areas of geometrical objects. When you use any methods that return properties of objects that do not need to take account of the model of the earth, such as counting the number of points in an object or describing the type of geometry used to represent a feature, there is no difference in performance between the `geography` and `geometry` types.

Standards Compliance

According to its web site, http://www.opengeospatial.org/, the Open Geospatial Consortium
(OGC) is "a non-profit, international, voluntary consensus standards organization that is leading
the development of standards for geospatial and location based services." The OGC adminis-
ters a number of industry-wide standards for dealing with spatial data. By conforming to these
standards, different systems can ensure core levels of common functionality, which means
that spatial information can be more easily shared between different vendors and systems.
In October 2007, Microsoft joined the OGC as a principal member, and the spatial datatypes
implemented in SQL Server 2008 are largely based on the standards defined by the OGC:

> The geometry datatype conforms to the OGC Simple Features for SQL Specification v1.1.0
> (http://www.opengeospatial.org/standards/sfs) and implements all the required
> methods to meet that standard.

> The geography datatype implements many of the same methods as the geometry datatype,
> although it does not completely conform to the required OGC standards.

> As such, if it is important to you to use spatial methods in SQL Server 2008 that adhere to
> accepted OGC standards, or if you want to ensure compatibility with another system based on
> those standards, you should use the geometry datatype.

How Spatial Data Is Stored

geometry and geography are both variable-length datatypes. This means that, in contrast to a
fixed-length datatype such as int or datetime, the actual amount of storage required for an
item of spatial data depends on the complexity of the object that the data describes.

SQL Server 2008 stores the information contained in the geography and geometry datatypes
as a stream of binary data. Each stream begins with a header section that defines basic information such as the type of geometry being described, the spatial reference system used, and
the overall number of points in the object. This header is immediately followed by the coordinate
values of each x and y (or longitude and latitude) coordinate in the geometry, represented in
8-byte binary format. The more points that an object has in its definition, the longer this binary
stream will be, and therefore the more storage space that it will require.

Note SQL Server 2008 stores each coordinate as a double-precision binary floating-point number, following
the specifications of the IEEE Standard for Binary Floating-Point Arithmetic (IEEE 754-2008). This format is
able to store floating-point numbers with 15 significant digits of precision—which is generally sufficient to
describe any position with submillimeter accuracy.

The following list gives examples of the storage space required to store some common
items of spatial data:

- A Point geometry defined with only two coordinates always occupies 22 bytes of storage space.

- A LineString between two points, containing the minimum of four coordinates (latitude and longitude, or x and y values of the start point and the end point), occupies 38 bytes.

- A Polygon geometry occupies a variable amount of space depending on the number of points that make up that Polygon. Any interior rings defined by a Polygon also increase the space required to store that Polygon.

There is no specific maximum size for the data storage space used by a geometry or geography object. However, SQL Server 2008 has an overall restriction on any kind of large object, which is limited to a size of $2^{31} - 1$ bytes. This is the same limit as is applied to datatypes such as varbinary(max) and varchar(max), and equates to approximately 2GB for each individual item of data. You would need to store a very complex geometry object to exceed this limit! If necessary, remember that complex geometries can be broken down into a number of individual objects that each fit within the allowed size.

Tip You can use the T-SQL DATALENGTH function to find out the number of bytes used to store the value of any item of geometry or geography data.

Converting Between Datatypes

Remember that the geometry datatype can be used to store geometries defined using a projected coordinate system, a geographic coordinate system, or a naturally planar coordinate system. The geography datatype, in contrast, can only be used to store geometries defined using a geographic coordinate system. Since both spatial datatypes can be used to store geographic coordinate data, any item of data stored using the geography datatype can also be expressed using the geometry datatype instead. However, you cannot simply use the CAST or CONVERT function to convert data between the two types. If you try to do so, by running the query

```
DECLARE @geog geography
SET @geog = geography::STGeomFromText('POINT(23 32)',4326)
SELECT CAST(@geog AS geometry)
```

you will receive the following error:

```
Msg 529, Level 16, State 2, Line 5
Explicit conversion from data type sys.geography to sys.geometry is not allowed.
```

Notice that the error states that explicit conversion is not *allowed*—this is a deliberate restriction imposed by SQL Server to ensure that you understand the implications of working with each datatype, and that you do not casually swap data between them. There are occasions, however, when it is helpful to be able to convert data between datatypes, since there are functions available for the geometry datatype that are not available when working with data stored using the geography datatype. In order to convert from geography to geometry, we can instead take advantage of the fact that the value of an item of data in either datatype can be represented as a binary stream. In the following example, the value of the geometry variable, @geom, is set based on the binary stream representation of the geography variable @geog:

```
DECLARE @geog geography
SET @geog = geography::STGeomFromText('POINT(23 32)',4326)
DECLARE @geom geometry
SET @geom = geometry::STGeomFromWKB(@geog.STAsBinary(), @geog.STSrid)
```

While every item of geography data can be expressed using the geometry datatype, not every item of geometry data can be expressed as an item of geography data. In order to convert a value from the geometry datatype to the geography datatype, the x and y coordinate values of the existing geometry instance *must* represent longitude and latitude coordinates taken from a supported geodetic spatial reference system. The data must also conform to all the other requirements of the geography datatype, such as ring orientation and size limitations. If these conditions are all met, you can create a geography instance, @geog, from the binary representation of a geometry instance, @geom, as follows:

```
DECLARE @geom geometry
SET @geom = geometry::STGeomFromText('POINT(23 32)',4326)
DECLARE @geog geography
SET @geog = geography::STGeomFromWKB(@geom.STAsBinary(), @geom.STSrid)
```

If, however, you are storing northing and easting coordinates from a projected system, or other nongeodetic data, that data can only be stored using the geometry datatype, and cannot be converted to the geography datatype.

░**Caution** There are relatively few cases in which an item of spatial data can be converted between the geometry and the geography datatypes, and even when it is technically possible, it does not always make logical sense to do so. If you find the need to convert between datatypes, it might be an indication that the design of your spatial data is incorrect.

Spatially Enabling Your Tables

Having chosen the appropriate spatial datatype, you must add a column of that datatype to the SQL Server table in which you plan to store your spatial data. There are two cases to consider: either you are creating a new table, or you are adding a column to an existing table.

Creating a New Table

There are no special attributes or features required to enable spatial data to be stored in a SQL Server database—all that is required is a table that contains at least one geography or geometry column. Since both the geography and geometry column types are already registered, system-defined datatypes, you can use a normal T-SQL CREATE TABLE statement to create a table containing a field of datatype geography or geometry, as follows:

```
CREATE TABLE [dbo].[Cities] (
    [CityName] [varchar](255) NOT NULL,
    [CityLocation] [geometry] NOT NULL
)
GO
```

This example creates a table containing two columns—CityName, which can hold a 255-character variable-length string, and CityLocation, which can be used to hold the spatial data relating to that city, using the geometry datatype.

Adding to an Existing Table

One of the benefits of using the spatial features of SQL Server 2008 is that new geometry or geography columns can be added to existing tables, enabling spatial information to be seamlessly integrated alongside existing items of data.

Suppose that we have an existing table, Customer, that contains the following fields of customer information:

```
CustomerID int,
FirstName varchar(50),
Surname varchar (50),
Address varchar (255),
Postcode varchar (10),
Country varchar(32)
```

Now suppose that we want to add an additional spatial field to this table to record the location of each customer's address. No problem—geography and geometry fields can be added to existing tables just like any other by using an ALTER TABLE statement, as follows:

```
ALTER TABLE [dbo].[Customer]
ADD CustomerLocation geography
GO
```

By extending the table in this way, we have enabled the possibility of using spatial methods in conjunction with our existing customer data, to find answers to questions such as how many customers we have within a certain area, and how far a particular customer lives from their closest store.

Enforcing a Common SRID

The coordinates defining a point only make sense in the context of a given spatial reference system. Although SQL Server 2008 allows you to store data objects from different spatial references in the same column, when performing functions, all of the spatial data items must be defined using the same spatial reference (i.e., have the same SRID). Trying to perform a calculation using coordinates from different systems can be compared to trying to add amounts of money denominated in different currencies:

25 Dollars + 12 Pounds = 37 ... what?!

To prevent you from accidentally performing a nonsensical calculation such as this, SQL Server will not allow you to perform any operations using data defined in different spatial references. The result of any method that attempts to do so will be NULL.

If you know that you will only be storing data based on one particular spatial reference system in a column, you can enforce the same SRID on all items in that column by adding a *constraint*. You can do this by using the ADD CONSTRAINT statement, as follows:

```
ALTER TABLE [dbo].[Customer]
ADD CONSTRAINT [enforce_srid_geographycolumn]
CHECK (CustomerLocation.STSrid = 4326)
GO
```

This example creates a constraint called enforce_srid_geographycolumn, which ensures that every spatial object inserted into the CustomerLocation field of the Customer table is defined using the SRID 4326. If you attempt to insert a geography object based on a different SRID, you will receive the following error:

```
Msg 547, Level 16, State 0, Line 1
The INSERT statement conflicted with the CHECK constraint
"enforce_srid_geographycolumn".
The conflict occurred in database "Spatial",
table "dbo.Customers", column 'CustomerLocation'.
The statement has been terminated.
```

As a result, you will be able to perform calculations using any two items of data from this column, knowing that they will be defined based on the same spatial reference system.

Summary

In this chapter, you learned how SQL Server 2008 implements spatial data using the geometry and geography datatypes:

- The geography datatype uses geodetic spatial data, accounting for the curvature of the earth.

- The geometry datatype uses planar spatial data, in which all points lie on a flat plane.

- There are a number of factors that you should consider when choosing the appropriate datatype for a given application, including precision, presentation, standards compliance, and consistency with any existing sources of data.

- Internally, SQL Server stores spatial data represented as a stream of binary values.

- You can add spatial columns to an existing table using ALTER TABLE, or create a new spatially enabled table using the CREATE TABLE T-SQL syntax.

- You can add a constraint to a column of spatial data to ensure that only data of a certain SRID can be inserted into that column.

CHAPTER 3

███

Working with Spatial Data in the .NET Framework

The geography and geometry datatypes both utilize the functionality provided by the .NET Framework common language runtime (CLR). The .NET Framework CLR was first introduced to SQL Server in SQL Server 2005 to provide a range of additional functionality and to extend the ways in which SQL Server could access and manipulate data. SQL Server 2008 takes integration with the .NET CLR one step further, by relying on the .NET CLR to implement core functions, including any operations using the spatial datatypes of geometry and geography.

By using .NET, SQL Server is able to access and work with spatial data more efficiently than would be possible using traditional Transact-SQL alone. However, .NET also introduces a range of new concepts and language syntax with which you may not be familiar. In this chapter I will introduce the principles of working within an object-oriented environment such as .NET, the way in which .NET syntax differs from conventional T-SQL syntax, and how these factors specifically relate to using the new geography and geometry datatypes in SQL Server 2008.

What Is the .NET Framework?

The .NET moniker has been applied to a range of separate, though related, Microsoft technologies: Visual Basic .NET (VB .NET), an object-oriented programming language based on Visual Basic in which .NET code can be written; ASP.NET, a web development framework used to create dynamic web sites and web applications; and the .NET Framework, which is the subject of this chapter.

The .NET Framework is a software component that forms a core part of the Windows Vista and Windows Server 2008 operating systems. It can also be installed as an optional add-in to previous Microsoft operating systems, including Windows XP and Windows Server 2003. SQL Server 2008 requires the .NET Framework to operate, and version 3.5 of the .NET Framework is installed as part of the SQL Server 2008 installation process.

■**Note** The .NET Framework is not distributed with SQL Server 2008 Express Edition. Before installing Express Edition, you must first download and install the .NET Framework from `http://www.microsoft.com/net`.

The .NET Framework itself contains several elements. The two main components are as follows:

Base Class Library (BCL): A library that provides all .NET applications with a shared set of methods to perform common programming tasks such as reading and writing to files, accessing external resources over a network, serializing and encoding data, providing presentation and user interface components, and querying structured data such as relational databases or XML sources. These methods are all highly optimized, and readily available for use by any .NET application. As a result, any applications that run on the .NET Framework can reuse an existing, consistent, and efficient approach to achieving these tasks, without needing to define their own proprietary methods.

Common language runtime (CLR): The execution environment in which .NET code is run. Code written for execution using the .NET CLR (called *managed code*) is platform-independent, so prior to execution, the CLR must first compile it into the native machine language of the system on which it is operating. This is known as just-in-time (JIT) compilation. In addition to simply executing the compiled code, the CLR environment also takes care of issues such as monitoring memory usage, performing garbage collection, managing threads, and controlling application security.

When installed on an operating system, these two components of the .NET Framework provide the method of execution and the essential resources required to enable any .NET applications to be run on that system.

■**Note** .NET applications may be developed using a range of programming languages that conform to the Common Language Specification (CLS). Two commonly used languages for .NET development are C# and VB .NET. Prior to execution, CLS-compliant code written in any supported language is first compiled into a common, platform-independent format called Common Intermediate Language (CIL), which can be passed to the CLR for JIT compilation and execution specific to the system on which the .NET Framework is running.

How .NET Is Hosted

.NET applications are normally executed in the .NET Framework CLR process hosted by the operating system. However, when you use .NET managed code in SQL Server 2008, SQL Server actually hosts the runtime environment within its own platform layer—SQLOS—and the CLR shares the Database Engine's process space. When implemented in this way, the CLR is referred to as the SQLCLR.

By hosting the SQLCLR itself, SQL Server 2008 ensures close integration between the SQL Server query processor and the runtime engine that executes any .NET managed code within SQL Server. This ensures high-performance transitions between the two platforms. It also ensures that SQL Server remains in control of allocating resources to the CLR, so that it governs the memory and processing time given to any .NET processes. The SQLCLR is completely contained within the integrity and security model provided by SQL Server.

The SQLCLR is a core component of SQL Server 2008, and the SQLCLR process is always loaded as part of SQL Server. The SQL Server Database Engine uses the functionality provided by the CLR seamlessly to provide a range of functions, which includes performing operations using data in the geography and geometry datatypes.

MICROSOFT.SQLSERVER.TYPES.DLL

The .NET code required by the geometry and geography datatypes in SQL Server 2008 is contained in a dedicated assembly called Microsoft.SqlServer.Types.dll. You can find this assembly within the 100\SDK\Assemblies subdirectory of the directory in which SQL Server is installed.

SQL Server automatically imports this assembly on startup. However, you can also manually import this assembly into other .NET applications, which enables you to use the same spatial datatypes and methods as SQL Server itself, but from within a .NET application executed by the normal .NET CLR rather than the SQLCLR. Because Microsoft.SqlServer.Types.dll contains all the required code to implement spatial functionality, once you've imported this assembly, you can even create your own stand-alone .NET spatial applications that don't depend on SQL Server at all. In fact, Microsoft.SqlServer.Types.dll contains some methods that you can access via .NET that aren't available directly in SQL Server itself (such as the SqlGeometryBuilder and SqlGeographyBuilder classes, which you can use to programmatically create new geometry and geography instances, respectively).

The following three code listings illustrate and compare how you can use the methods contained within Microsoft.SqlServer.Types.dll in SQL Server, in a Visual Basic.NET console application, and in a C# console application. In each case, the code listed creates a new geometry Point at coordinates (10,20), using SRID 0, and then returns the WKT representation of that geometry.

SQL Server

```
DECLARE @MyGeometry geometry
SET @MyGeometry = geometry::Point(10, 20, 0)
SELECT @MyGeometry.ToString()
```

Visual Basic .NET

```
Imports Microsoft.SqlServer.Types
Module MyModule
  Sub Main()
    Dim MyGeometry As New SqlGeometry()
    MyGeometry = SqlGeometry.Point(10, 20, 0)
    Console.Write(MyGeometry.ToString())
  End Sub
End Module
```

C#

```
using Microsoft.SqlServer.Types;
class MyClass
{
  static void Main(string[] args)
  {
    SqlGeometry MyGeometry = SqlGeometry.Point(10, 20, 0);
    System.Console.Write(MyGeometry.ToString());
  }
}
```

The result of all three methods is exactly the same:

```
POINT (10 20)
```

This aim of this book is to examine the spatial functionality provided *within* SQL Server 2008, so all the examples I'll give will demonstrate the spatial methods directly available from SQL Server itself. However, you should remember that, in many cases, you can achieve the same objective by using the spatial methods provided by Microsoft.SqlServer.Types.dll in a .NET application outside of SQL Server.

Why Use .NET for Spatial Functionality?

Like most relational database management systems, SQL Server is primarily based on Structured Query Language (SQL). SQL is a widely used, set-based programming language, and is the recognized standard method used for retrieving and processing data from a database system. You probably already are familiar with the basic structure of a SQL query, such as shown in the following listing:

```
SELECT
  ColumnName1,
  Function(ColumnName2) AS Alias,
FROM
  TableName
WHERE
  Condition1 = True
  AND
  Condition2 = 'Value'
```

Although SQL is a standard approved by both the International Standards Organization and the American National Standards Institute, many database vendors implement their own proprietary versions of SQL, which implement additional functionality on top of the core elements defined in the SQL standard. The particular implementation of SQL used in SQL Server 2008 is called Transact-SQL, or T-SQL.

T-SQL was originally developed by Microsoft and Sybase, who extended the basic set-based SQL standard by adding a number of elements to support procedural logic, such as loops, variables, and conditional branching. These additions make T-SQL a very powerful, *quasi*-procedural language that has been optimized to perform bulk operations against large amounts of data normally stored in relational databases. However, as you've already seen in Chapter 2, there are several differences between the spatial data stored by the geography and geometry types compared to conventional numeric or character data. A single item of spatial information describes a number of different properties of the feature it represents, has a complex structure, and may contain a significant amount of data. This makes dealing with spatial data more complicated than using other types of data. Even with the enhancements it offers over the SQL standard, T-SQL is simply not designed to handle the complex types of information necessary to describe spatial data.

In order to be able to access the information in the geometry and geography datatypes effectively, SQL Server 2008 therefore leverages the power of the .NET CLR instead. The .NET CLR uses an object-oriented approach to storing and manipulating data, which is ideal for describing the multifaceted nature of spatial data. Additionally, working with spatial data involves relatively complex mathematical computations and procedural logic that would be hard to achieve using predominantly set-based Transact-SQL, but are relatively simple using a modern CLS-compliant language, such as C#. Whenever SQL Server implements a datatype using the functionality of the .NET CLR, such as the geography and geometry datatypes, these are called CLR datatypes.

Note Because SQL Server 2008 incorporates both T-SQL and .NET, there are many scenarios where you will face a choice of which platform to use to implement a particular data operation. Even though the methods employed will be different, it is generally possible to achieve the same desired result using either platform. Each has its own advantages and disadvantages—set-based operations tend to perform better using T-SQL, while procedural operations, recursive code, and constructs such as arrays tend to work better in .NET. Although spatial functionality could be achieved using T-SQL alone, SQL Server 2008 uses .NET as a more appropriate and effective choice for handling this type of data.

CLR USER-DEFINED TYPES

Support for datatypes based on the .NET CLR was first incorporated into SQL Server 2005. At that time, developers were able to use the .NET Framework to create their own user-defined types (UDTs), and access data in these types using object-oriented methods. However, before SQL Server 2008, there were not any system-defined datatypes that relied on the .NET CLR in this way.

SQL Server 2008 actually comes with three preregistered .NET CLR datatypes: geometry, geography, and hierarchyid. These datatypes all work in much the same way as UDTs, except that they are system-defined types that are already registered and ready for use in SQL Server.

While UDTs provided a useful way for users to be able to extend the system-defined scalar datatypes with their own custom types, they were of limited use for storing complex data in SQL Server 2005, since each item of data stored in a UDT could only contain 8KB of data. SQL Server 2008 extends the size limit of any .NET CLR datatype to 2GB. This increased size limit applies to UDTs as well as to the geography and geometry system-defined types.

Applying Principles of Object Orientation

There are a number of important principles that apply to any object-oriented programming language, including those targeted at development for the .NET Framework. Following these principles affects the ways in which you use data stored using CLR datatypes such as geography and geometry compared to other types of data in SQL Server 2008. If you have used other object-oriented programming languages in the past, such as Java or C++, you may already be familiar with the concepts of data abstraction, encapsulation, inheritance, and polymorphism. However, if you are not, or if you need a refresher, then read on!

Data Abstraction

Most common SQL Server datatypes (such as char, int, datetime, and float) are *scalar*—that is, each item of data only holds a single value at a time. That value completely describes a single piece of information—a customer's address, a product reference number, or the date on which a transaction took place, for example. In contrast, we know that in order to fully describe a single item of spatial data, a number of distinct pieces of information are required—the type of geometry used to represent that feature, how many points that geometry contains, the coordinates of each of those points, and the spatial reference from which those coordinates were obtained.

When storing an item of spatial data representing a feature on the earth, the geography and geometry datatypes store these abstract pieces of data together in a compound item of data called an *object*. Each object contains several data members, with each representing the value of one particular property of that item of spatial data. When combined together in a single package, these individual data members form an object representing the entire feature. The process of describing the properties of real-world features as a collection of individual items of data is the principle of *data abstraction*.

There are different ways of abstracting the same real-world item into component elements. For instance, even though they can be used to describe the same feature on the earth, a geography object and a geometry object may contain different data members. This concept is illustrated in Figure 3-1.

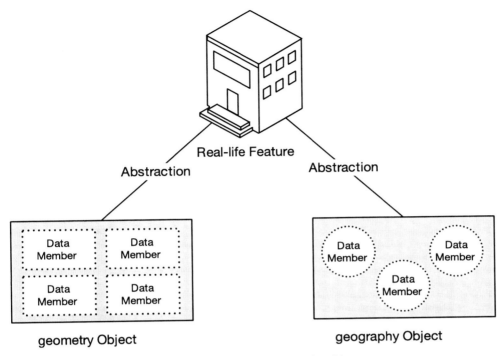

Figure 3-1. *Data abstraction in the geometry and geography object types*

Encapsulation (Data Hiding)

In addition to containing a number of different data members representing different properties of a feature, each object also contains a number of *methods* to specify how the data contained in those data members can be accessed.

Normally in SQL Server, we consider the items of data stored in a database as being separate from the functions that can be performed on that data. For instance, the COUNT function, which is a function defined by the T-SQL language, can be applied to count the number of items in any column of data, whatever datatype the values in that column represent.

In contrast, when using an object-oriented language like .NET in SQL Server, the datatype of a column not only determines the individual elements of the data stored in that column, but also specifies what methods can be used to access that data.

.NET methods are like T-SQL functions—they perform operations on items of data. What makes methods different from functions is that the *only* way of directly accessing the data contained within an object is by using the methods specified by that object itself. This is the principle of *encapsulation*, or data hiding—the individual data members contained within an object of data using the geometry or geography datatype are kept hidden from the rest of the system, and the only way of accessing them is by using one of the specific methods provided by that type of object. The concept of encapsulation is illustrated in Figure 3-2.

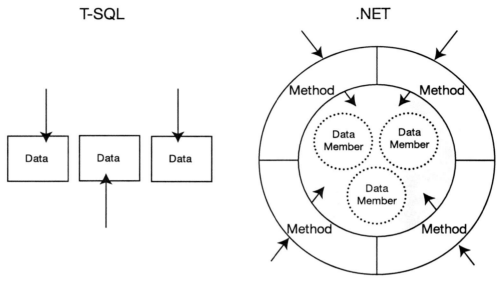

Figure 3-2. *Using methods to access encapsulated data contained within an object*

The main advantage of encapsulation is that, by preventing access to the data contained in an object by any other means than those methods defined by the object itself, it is easier to ensure the integrity of data contained in the object. Data members contained within an object cannot be subject to any externally defined processes over which the object has no control.

The geography and geometry datatypes provide methods that can be used to perform a wide range of actions using the data contained within objects of that type, or combine and compare that data with different objects. These methods are described in detail in Chapters 11–13 of this book.

Inheritance

Although every object created from the geography or geometry datatype is based on the fundamental properties of that datatype, not all objects of a given datatype are the same. We know that each item of geometry or geography data actually defines a particular kind of geometry object—a Polygon, a LineString, a Point, or a multielement collection of those geometries. Each of these specific classes of object is derived from either the geometry or geography generic, abstract datatype, and inherits the properties and behavior of its parent datatype. This is the principle of *inheritance*.

In addition to properties inherited from its parent datatype, each class of object also defines a number of properties that are specific to that particular class. Any child object created from this object will inherit these properties as well, making each successive class of derived object more specific than its parent.

The inheritance tree of objects in the geography datatype, demonstrating which object types are derived from other types, is shown in Figure 3-3.

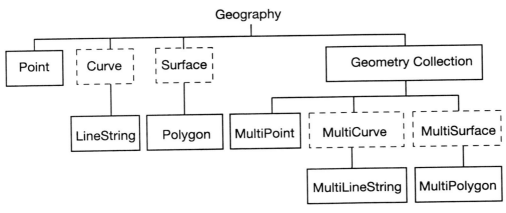

Figure 3-3. *The inheritance hierarchy of objects in the geography datatype. Instantiable types—those types from which an instance of data can be created in SQL Server 2008—are shown with a solid border.*

Note that in addition to Points, LineStrings, Polygons, and multielement types with which you are now familiar, the inheritance tree contains classes of objects based on additional geometry types—Curve, Surface, MultiCurve, and MultiSurface. The objects in the first category, Points, LineStrings, and Polygons, are *instantiable* types—that is, these are the specific subtypes of spatial data that you can create in SQL Server from the abstract geometry and geography datatypes. The second category of objects, Curves and Surfaces, cannot be created. However, you *can* create instances of the LineString and Polygon types that are derived from these types, and therefore inherit their properties. For instance, the Curve is the generic one-dimensional geometric object created from a sequence of points. The LineString object inherits the generic properties of the Curve, but specifies the additional property that a LineString must be composed of *straight-line* segments joining the Points.

Why does SQL Server include Curves and Surfaces if they can't be created—why not just have LineStrings and Polygons descended directly from the parent geography or geometry datatype? Remember that the spatial features in SQL Server are largely based on the standards set out in the Open Geospatial Consortium's Simple Features for SQL Specification. To conform to this standard, the object hierarchy model must include the Curve and Surface subclasses. Additionally, by including the object classes from the outset, SQL Server may be easily extended to make these types of objects instantiable in the future.

■**Note** When you create an object of a given datatype, it is referred to as an *instance* of that type. The process of creating a new object of geometry or geography data is therefore known as *instantiation*.

The inheritance hierarchy of objects for the geometry datatype mirrors that of the geography datatype shown in Figure 3-3. However, since each class of object only inherits the relevant methods of the type from which it is created, a Point object created from the geography datatype will have a set of methods different from that of a Point object created from the geometry datatype. Although many of the methods provided by the two spatial datatypes are similar, they are not identical. You should therefore not assume that simply because you can use a certain method on an item of data of the geometry datatype, you will be able to use an equivalent method on an item of data of the geography datatype.

Polymorphism

The principle of inheritance tells us that the geography and geometry datatypes each define a number of methods of accessing data, and that each Point, LineString, and Polygon object inherits the methods available from the particular datatype from which it is derived. However, each type of object does not necessarily implement those methods in the same way. The principle of *polymorphism* (from the Greek, meaning "many forms") means that you can call the same function against different types of objects and get different behavior in each case.

For instance, when you call the STLength() method against a LineString object, you get the length of the line. When you call the STLength() method against a Polygon object, you get the total length of all defined rings. Applying polymorphism means that, under the covers, different objects may implement the same method in different ways, but you do not need to worry about exactly how this occurs—when you invoke the method, you get the appropriate response from that object for the situation in question.

Instantiating Spatial Objects

When you set the value of a scalar variable, or insert an item of data into a scalar column type, the approach taken is quite simple—you can use the SET statement or the INSERT statement, and pass it the new value that the variable should hold. For instance, consider the following examples:

```
DECLARE @myInt int
SET @myInt = 5

DECLARE @myTable table (myString varchar(32))
INSERT INTO @myTable (myString) VALUES ('This is a string')
```

However, since an item of data in the geography or geometry type is an object with several individual members, you cannot simply assign the value of an item of spatial data so easily. Instead, whenever you create a new item of spatial data in SQL Server, you must invoke a *static method.*

Using Static Methods

Whereas most methods are inherited by, and applied to, individual instances of a datatype, static methods act upon the geography or geometry datatype as a whole. You cannot create a new object by applying a method on that particular object itself, because it doesn't exist yet. Instead, to instantiate a new object, you use a static method belonging to the appropriate datatype. Both datatypes define a number of static methods that can be used to create new instances of objects of that type.

In order to explicitly state which datatype the method belongs to, you specify a static method by preceding the name of the method with the datatype name in every case, separated using two colons (::). For example, STGeomFromText() is a static method, belonging to the geography datatype, which can be used to create a new item of geography data. To use this method, you would invoke it using the following syntax:

```
geography::STGeomFromText()
```

Note that the geometry datatype also implements a method called STGeomFromText(). If you wanted to use this method instead, you would invoke it in the following way:

```
geometry::STGeomFromText()
```

The geometry and the geography datatypes each implement 19 static methods, which can be used to create different kinds of objects of that type, based on different kinds of input. These are discussed in detail in the next chapter.

Applying a Static Method of the Appropriate Datatype

It is important to remember that every static method can only instantiate an object of the same datatype as the datatype of the method itself. Therefore, when you use a static method to create an object that is to be inserted into a geography or geometry column, or set as the value of a parameter, you must use a static method belonging to the relevant datatype.

Consider the following example, which creates a table variable containing a geometry column, and then inserts data into that column from the result of the STGeomFromText() static method of the geometry datatype:

```
DECLARE @myTable table (
  FeatureName varchar(32),
  FeatureGeometry geometry
)

INSERT INTO @myTable VALUES (
  'Statue of Liberty',
  geometry::STGeomFromText('POINT(-74.045 40.69)', 4326)
)
```

At this stage, don't be too concerned about the parameters contained in the brackets—they will be covered in the next chapter. The important point to note is that, since we created a table variable with a geometry column, to insert data into that column we must use a static method belonging to the geometry datatype. In this case, we use the STGeomFromText() method. If you are keen to know, in this example the parameters state that the method should create a Point at coordinates –74.045° longitude and 40.69° latitude, specified using the spatial reference system denoted by the SRID EPSG:4326. This is the approximate location of the Statue of Liberty.

Now suppose instead that you were to provide the same parameters to the STGeomFromText() method of the geography datatype, as shown in the following query:

```
DECLARE @myTable table (
  FeatureName varchar(32),
  FeatureGeometry geometry
)

INSERT INTO @myTable VALUES (
  'Statue of Liberty',
  geography::STGeomFromText('POINT(74.045 40.69)', 4326)
)
```

This would result in the error

```
Msg 206, Level 16, State 2, Line 2
Operand type clash: sys.geography is incompatible with tempdb.sys.geometry
```

because the resulting object created by the geography::STGeomFromText() static method is an instance of a geography object, which cannot be inserted into a column of the geometry datatype.

Invoking Spatial Instance Methods

Having seen how to create new items of geography and geometry data using static methods, let's now look at how we can use .NET CLR instance methods to manipulate that data. Unlike static methods, which are defined by and applied to either the geometry or geography datatype as a whole, an *instance method* applies to an individual instance of geography or geometry data. As with static methods, there are a number of differences between the .NET approach and the equivalent T-SQL syntax.

T-SQL Function Syntax

The generic syntax for applying a function to a column of data in T-SQL is as follows:

```
Function(Column, [Parameter])
```

To demonstrate this syntax in action, the following example shows how you can use the T-SQL LEFT function to return the leftmost 15 characters from a string value held in a varchar(32) variable:

```
DECLARE @myString varchar(32)
SET @myString = 'This is an example of the left function'
SELECT LEFT(@myString, 15)
```

The result is as follows:

```
This is an exam
```

.NET CLR Instance Method Syntax

In contrast to the T-SQL syntax, when using .NET, the method name comes *after* the instance to which it is being applied. The name of the instance and the method name are separated by a dot (.), and the name of a method is always immediately followed by a pair of opening and closing round brackets. If a method takes a parameter, it is supplied within the brackets. This generic syntax is demonstrated as follows:

```
Instance.Method( [Parameter] )
```

For instance, the ToString() method in .NET can be used to return a text representation of an object. It takes no parameters. It is therefore called as follows:

```
SELECT
  Column.ToString()
FROM
  TableName
```

Compare this to the T-SQL CONVERT function to convert a value to a variable-length character string, which provides similar functionality:

```
SELECT
  CONVERT(varchar(32), Column)
FROM
  TableName
```

Caution Unlike T-SQL functions, the .NET CLR methods of the geometry and geography datatypes in SQL Server 2008 are case sensitive, so be careful to use Instance.ToString(), not Instance.toString().

Note that, in contrast to the static methods previously described, you do not need to explicitly specify whether the method belongs to the geometry or geography datatype. This is because the method acts on a particular instance of an object, which in itself has already been defined as an instance of an object derived from one of the two types, and has inherited the relevant methods.

Chaining Multiple Method Calls

In many cases, it is desirable to apply one method to an instance, and then take the result of that method as the input for a further method. Rather than apply the second method to the instance itself, you therefore apply it to the result of the first method. The correct syntax in this case is to list each method after the value to which it should be applied, again using a dot before each method, and opening/closing round brackets after each method name. Methods are applied in left-to-right order, starting with the method that is directly applied to the instance itself, and then working outward. For instance, consider the following example:

```
Instance1.STUnion(Instance2).STArea()
```

In this example, the `STUnion()` method is first to be applied to `Instance1`. The `STUnion()` method combines two geometries—the geometry that is represented by the instance on which it is applied, and a second geometry that is supplied as a parameter to the method. In this case, the result of this method therefore represents the union of `Instance1` and `Instance2`. The `STArea()` method is then applied to the resulting geometry, which returns the area of the combined shape.

Accessor and Mutator Methods

In most cases, the .NET CLR methods provided by the `geography` and `geometry` datatypes are read-only—they can be used to access the data members within an object to return the results, but do not change any values of that object. Methods of this type are known as *accessor* methods. In contrast, methods that change the value of data members contained within an object are called *mutator* methods.

Accessing Properties

Most methods provided by the `geometry` and `geography` datatypes perform some operations on the members of an object before returning the result. For example, `STArea()` calculates the area contained within a geometry; `STLength()` calculates the length of a geometry; and `STOverlaps()` determines whether two geometries overlap. However, in some cases, we may simply want a method that retrieves (or sets) the value of an individual data member within an object. We cannot access the value directly—data encapsulation prevents us from doing so—but we don't really need the method to provide any additional functionality other than retrieving a single value of a data member within the object. In these cases, an object may provide access to its data members as a *property* of that object. A property can be thought of as a method that does nothing more than provide an exposed interface to a particular data member. Properties still retain the benefits of encapsulation by not allowing direct access to the item, but provide a transparent way by which the value can be read, or written.

Syntax

To access the properties of a `geometry` or `geography` object, you use a syntax that is similar to, although simpler than, the syntax that you use for methods. You first state the name of the object, followed by a dot, followed by the name of the property. However, properties do not take any arguments, so you do not include the round brackets after the property name. The generic syntax is therefore:

```
Instance.Property
```

For example, the `STX` property of a geometry Point object returns the value of the x coordinate of that Point:

```
SELECT
  GeometryColumn.STX
FROM
  TableName
```

As with methods, the properties of a particular item of spatial data are inherited from the parent datatype of which that object is an instance. The STX property used in this example can ... e geometry datatype. Why is this property only defined by ... nd not the geography datatype? Remember that Points in the ... by using coordinates of x and y, but rather by using angular ... de—therefore, it wouldn't make sense for the geography ... epresenting an x coordinate. Instead, the geography datatype ... ty, called Long, which provides access to the longitude coor-

·ite Properties

... y and geography datatypes are read-only. For instance, you ... ordinate of a geometry Point by setting the STX property of ... change the properties of points defined within a geometry, you must instead use a static method to redeclare the value of the entire object.

However, the STSRid property of both the geometry and geography datatypes is unusual in that it enables you to both retrieve and set the spatial reference identifier (SRID) of an item of spatial data. As such, the SRID of an item of geography or geometry data can be set to a new value by using a SET T-SQL statement as follows:

```
SET Instance.STSRid = 4326
```

Combining T-SQL and .NET CLR Methods

You may have noticed in the examples already used in this chapter that while the .NET CLR complements and extends the functionality of Transact-SQL already used by SQL Server, it does not replace it. .NET provides the procedural methods by which geography and geometry data is accessed, but queries using spatial data in SQL Server are still based around familiar T-SQL language statements such as SELECT, INSERT, UPDATE, and DELETE. When working with spatial data in SQL Server 2008, you will therefore need to write queries containing both T-SQL and .NET syntax.

You learned earlier in this chapter that, according to the principle of encapsulation, the only way to directly access the data contained within an object of geometry or geography data is to use the relevant CLR method. However, once that method has retrieved the relevant results, they are returned to T-SQL, where they can be used in conventional T-SQL functions.

For instance, to calculate the area of a Polygon object stored using the geometry datatype, we can use the STArea() method. The result returned by this method is a floating-point number, representing the total area of the Polygon geometry. We can use this result in the ORDER BY clause of a standard T-SQL SELECT statement to sort a result set by area. To demonstrate this, consider the following query, which selects the three largest US states by area:

```
SELECT TOP 3
  State_Name,
  State_Geometry.STArea() AS Area
FROM
  US_States
ORDER BY
  State_Geometry.STArea() DESC
```

The results are as follows:

State_Name	Area
Alaska	1702788.113
Texas	692436.1872
California	423236.0281

Using Open Geospatial Consortium Methods

Many of the .NET CLR methods implemented by the spatial datatypes in SQL Server 2008 are based on standard methods defined by the Open Geospatial Consortium. Whenever the geography or geometry datatype implements one of these standard methods, the method name is prefixed by the letters "ST". Examples of such OGC-compliant methods are STBuffer() and STUnion(). When using any of these methods, the results obtained will be consistent with the standards of the OGC Simple Features for SQL Specification, v1.1.0, available at http://www.opengeospatial.org/standards/sfs.

In addition to the OGC standard spatial methods, both spatial datatypes provide a number of extended methods that provide additional functionality on top of the OGC standards. These methods are named according to their function, without any additional prefix. Examples of extended methods are the Reduce() and MakeValid() methods.

Note The letters "ST" used to prefix OGC-defined methods actually stand for "Spatio-Temporal"—that is, they are methods designed to perform calculations of space *and* time. However, SQL Server 2008 does not implement the time-based aspects of any methods (yet), so instead you may find it helpful to remember "ST" as denoting a STandard method.

Handling Exceptions in the CLR

SQL Server 2008 implements error handling through the use of a TRY... CATCH construct. This construct is an example of one of the procedural additions that T-SQL provides over the SQL standard. When using a TRY... CATCH construct, one or more T-SQL statements are wrapped in a TRY block. If an error occurs when these statements are executed, or from within a stored procedure called within the TRY block, an error is *thrown*, and execution is transferred instead to another group of T-SQL statements contained within a CATCH block. The CATCH block catches the error, and can then determine how to deal with it.

In order to choose an appropriate course of action, statements contained within the CATCH block have access to a number of functions that give information related to the error that caused the CATCH block to be executed. These functions are error_number(), error_severity(), error_state(), error_message(), error_procedure(), and error_line().

To demonstrate the use of TRY... CATCH using spatial methods, let's try to use the static method STPolyFromText() to create a Polygon of the geometry datatype whose exterior ring is not closed:

```
BEGIN TRY
  DECLARE @polygon geometry
  SET @polygon = geometry::STPolyFromText('POLYGON((0 0,10 0,10 10,0 10,2 2))',2285)
END TRY
BEGIN CATCH
  SELECT 'The method failed because of error number ' +
  CAST(ERROR_NUMBER() AS varchar(32))
END CATCH
```

In this case, the parameters provided to the STPolyFromText() method specify that a Polygon should be created, the exterior ring of which starts at coordinate (0,0), but ends at coordinate (2,2). The result is as follows:

```
The method failed because of error number 6522
```

Now let's try to generate another error. This time, we'll try to use the STLineFromText() static method to create a LineString object using the geography datatype that only contains one point:

```
BEGIN TRY
  DECLARE @line geography
  SET @line = geography::STLineFromText('LINESTRING(2 0)',17453)
END TRY
BEGIN CATCH
  SELECT 'The method failed because of error number ' +
  CAST(ERROR_NUMBER() AS varchar(32))
END CATCH
```

This time, the result is . . .

```
The method failed because of error number 6522
```

. . . huh? Why does the value of ERROR_NUMBER() remain the same even though we have caused a different error, based on the geography datatype rather than the geometry datatype? To answer this question, we can look up the meaning of error 6522 in the sys.sysmessages table:

```
SELECT
  description
FROM
  sys.sysmessages
WHERE
  error = 6522
  AND msglangid = 1033
```

Note that the sys.sysmessages table contains multilingual descriptions for all possible generated errors in SQL Server, so we add the msglangid = 1033 condition to obtain the English description of the error. The result is as follows:

```
A .NET Framework error occurred during execution of user-defined routine
or aggregate "%.*ls": %ls.
```

What does this mean? Remember that SQL Server hands over all processing of spatial methods to the SQLCLR. If an exception occurs when working with the geometry and geography datatypes in SQL Server, this initially raises an error in the .NET environment in which the SQLCLR operates, rather than directly in the T-SQL environment. The SQLCLR itself then raises a further error in the T-SQL environment—which it does under error code 6522. Whenever you generate an error using the geometry or geography datatype, you actually generate two error responses—the .NET error code generated by the SQLCLR, and a T-SQL error code of 6522. Unfortunately, ERROR_NUMBER() only holds the value of the T-SQL error, so whenever you encounter an error using the geometry or geography datatype, this will be 6522 in all cases, which is just the generic T-SQL error raised every time there is an error encountered by a method using the .NET Framework.

What we would really like to be able to do is retrieve the value of the original .NET error code generated so that we can use that in our CATCH block to determine a course of action based on the specific error. Fortunately, there is an alternative way to do this.

When the SQLCLR encounters an exception, it generates an error message, consisting of a verbose description of the exception together with the contents of the stack (the list of individual tasks that the CLR was performing). The details of the T-SQL error 6522 are then appended to this error message, which is accessible by using the ERROR_MESSAGE() function. When called in a CATCH block, ERROR_MESSAGE() returns an nvarchar(2048) string containing the complete text of the error message that caused the CATCH block to be run, including both SQLCLR and T-SQL error messages.

To demonstrate this, let's use the first example again to create a Polygon with an unclosed external ring, except this time we'll select the value of ERROR_MESSAGE():

```
BEGIN TRY
  DECLARE @polygon geometry
  SET @polygon = geometry::STPolyFromText('POLYGON((0 0,10 0,10 10,0 10,2 2))',2285)
END TRY
BEGIN CATCH
  SELECT ERROR_MESSAGE()
END CATCH
```

The results are as follows:

```
A .NET Framework error occurred during execution of user-defined routine or
aggregate "geometry":
   System.FormatException: 24119: The Polygon input is not valid because the start
and end points of the exterior ring are not the same. Each ring of a polygon must
have the same start and end points.
   System.FormatException:
      at Microsoft.SqlServer.Types.GeometryDataBuilder.EndFigure()
      at Microsoft.SqlServer.Types.OpenGisWktReader.ParseLineStringText
(FigureAttributes attributes)
      at Microsoft.SqlServer.Types.OpenGisWktReader.ParsePolygonText()
      at Microsoft.SqlServer.Types.OpenGisWktReader.ParsePolygonTaggedText()
      at Microsoft.SqlServer.Types.OpenGisWktReader.ReadPolygon()
      at Microsoft.SqlServer.Types.SqlGeometry.STPolyFromText(SqlChars
polygonTaggedText, Int32 srid)
```

Notice that this message contains three distinct parts:

- The description of `Msg 6522` that we have already seen—the T-SQL error that tells us that there has been a problem in a method using the SQLCLR.

- `System.FormatException: 24119` is the reference number relating to a specific .NET error. The error also has a helpful description of the particular cause of the error—in this case, because the start and end points of the exterior ring of the Polygon are not the same.

- A description of the processes on the stack at the time the error was generated—in this case, the error occurred when attempting to use the `GeometryDataBuilder.EndFigure()` method.

If we repeat this same process with our second example, we can see that the .NET error returned when trying to create our invalid LineString is

```
System.FormatException: 24117: The LineString input is not valid because it does not
have enough distinct points. A LineString must have at least two distinct points.
```

By distilling the contents of `ERROR_MESSAGE()`, we can therefore take specific action based on the exception raised within the .NET Framework environment.

Summary

In this chapter, you saw how SQL Server 2008 uses the .NET Framework when using the geometry and geography datatypes:

- The .NET Framework CLR is an integral component of SQL Server 2008, and is used to provide the functionality of the geometry and geography CLR datatypes.

- There are a number of differences between the predominantly set-based T-SQL language and the object-oriented .NET environment. These differences affect the ways in which you use geometry and geography compared to other SQL Server datatypes.

- CLR datatypes such as geometry and geography are based on an object-oriented programming model, which applies the principles of data abstraction, encapsulation, inheritance, and polymorphism.

- Static methods act upon a datatype, and can be used to create new items of geometry and geography data.

- Instance methods act upon an individual item of geometry or geography data, and can be used to perform spatial operations using the data contained within that instance.

- Errors that occur using the geometry and geography datatypes are raised within the SQLCLR environment. Details of those errors can be retrieved using the T-SQL ERROR_MESSAGE function.

You have seen how using the .NET CLR enables SQL Server 2008 to access complex spatial data using the geography and geometry datatypes. In later chapters of this book, we will examine other ways in which we can leverage the power of .NET to extend the capabilities of SQL Server 2008, by using the functions provided by the Base Class Library to perform transformations of XML-based spatial data, and by accessing web-based geocoding services to obtain the coordinate positions of locations.

PART 2

Adding Spatial Data

This part of the book explains various ways by which you can add spatial data to your SQL Server 2008 database. Chapter 4 introduces each of the underlying static methods that are used to create any items of data from known coordinates. Chapter 5 demonstrates a technique that harnesses external resources to help derive those coordinates—using Microsoft's Virtual Earth Map Control. Chapter 6 discusses other common data formats in which spatial data can be stored, and provides examples of how data stored in these formats can be imported into SQL Server. Finally, Chapter 7 shows how you can extend SQL Server to provide geocoding functionality—automatically obtaining coordinates of an address by accessing the Microsoft MapPoint Web Service via .NET.

CHAPTER 4

▪ ▪ ▪

Creating Spatial Data Objects

In the last chapter, I introduced the concept of static methods and explained how they can be used to instantiate objects of the geography and geometry datatypes. In this chapter we will examine the different static methods available for each datatype, and compare how they can be used to create new items of spatial data.

Note Most of the code samples in this chapter declare local variables, such as @Point, to store the resulting instances created by static methods. However, you can insert the results of a static method directly into a geometry or geography column in a table by invoking the method as part of an INSERT statement.

Choosing an Appropriate Static Method

The geography and geometry datatypes both provide a number of different static methods for creating spatial data objects. The appropriate method to use in a particular situation depends on the following factors:

- *The type of geometry object you are trying to create*: Some methods can only be used to create particular types of geometries. For instance, the STLineFromWKB() and STLineFromText() methods can only create instances of LineString geometries.

- *How you will describe the properties of this geometry*: All of the methods require you to supply the spatial representation of a geometry using one of three different standard formats: Well-Known Text (WKT), Well-Known Binary (WKB), or Geometry Markup Language (GML). You must use an appropriate method depending on which representation format you choose to describe the geometry. For example, to create any geometry instances from a GML representation, you must use the GeomFromGml() method.

- *Whether you want to create an item of* geometry *or* geography *data*: The geography and geometry datatypes both provide their own implementations of each static method. Since a static method can only create instances of objects of the same datatype as the datatype of the method itself, you must choose the method belonging to the appropriate datatype for the type of object you want to create. For instance, to create a geometry Point object from WKT, you should use the geometry::STPointFromText() method. To create a geometry Point object from WKT, you should use the equivalent method of the geography datatype instead, geography::STPointFromText().

Table 4-1 shows the methods that can be used to create different geometry instances. All of these methods are implemented by both the geography and geometry datatypes.

Table 4-1. *Supported Methods to Instantiate Spatial Objects from Different Representation Formats*

Type of Object	Well-Known Text	Well-Known Binary	Geography Markup Language
Point	STPointFromText()	STPointFromWKB()	GeomFromGml()
LineString	STLineFromText()	STLineFromWKB()	GeomFromGml()
Polygon	STPolyFromText()	STPolyFromWKB()	GeomFromGml()
MultiPoint	STMPointFromText()	STMPointFromWKB()	GeomFromGml()
MultiLineString	STMLineFromText()	STMLineFromWKB()	GeomFromGml()
MultiPolygon	STMPolyFromText()	STMPolyFromWKB()	GeomFromGml()
Geometry Collection	STGeomCollFromText()	STGeomCollFromWKB()	GeomFromGml()
Any supported type	STGeomFromText() Parse()	STGeomFromWKB()	GeomFromGml()

Whichever language is used to represent the spatial object, and whichever type of object is being created, all of the static methods discussed in this chapter share the same basic syntax. This generic syntax is as follows:

```
datatype::method(geometryrepresentation, srid)
```

The four elements of this syntax are as follows:

- datatype specifies whether the static method belongs to the geography or the geometry datatype, and therefore determines the datatype of the resulting instance created by the method.

- method is the name of the method to create the geometry. This must be one of the methods listed in Table 4-1.

- geometryrepresentation is a valid representation of the geometry to be created. This representation must be expressed in the appropriate format expected by the chosen method. The representation is a character string (WKT and GML formats) or binary stream (WKB) that contains all of the information required to define the geometry in question.

- srid is an integer value representing the identifier of the spatial reference system that was used to define the coordinates in the geometryrepresentation parameter passed to the method.

Each format in which geometries can be represented, and each method that can be used to create instances of objects from those representations, has its own advantages and disadvantages. However, the decision of which method to use to create an item of geography or geometry data is only significant at the point of creation, and becomes irrelevant once the item is created. Having used a particular method to create an item of spatial data, that item of data is exactly the same as if it had been created using any of the other possible methods.

LATE BINDING

The reference to a geometry or geography object specified by any of the static methods is only resolved at runtime, when the query that invokes the method is executed. This is known as *late binding*. One effect of late binding is that, until the query is executed, the SQLCLR does not know what type of object will be created by a static method. As a result, SQL Server cannot parse a representation to check whether it would produce valid data. For instance, the following query compiles correctly in SQL Server Management Studio without generating any parsing error:

```
SELECT geometry::STGeomFromText('This is not a real geometry')
```

However, on execution, this query will fail because the representation passed to the method is not valid WKT, as required by the STGeomFromText() method. Be careful when you specify the representations passed to any static methods, since errors may not show up until you attempt to execute those queries.

Creating Geometries from Well-Known Text

Well-Known Text is one of the standard formats defined by the Open Geospatial Consortium (OGC) for the exchange of spatial information. It is a simple, text-based format that is easy to examine and understand. You have already seen an example of WKT in Chapter 1; it is the format SQL Server 2008 uses to store the parameters of supported geodetic spatial reference systems in the well_known_text column of the sys.spatial_reference_systems table. In that context, WKT was used to describe the properties of a spatial reference system. However, WKT can also be used to express individual geometry objects *within* a spatial reference system. For example, the following code line demonstrates how WKT can be used to define a LineString between the points at coordinates (12,20) and (30,44):

```
LINESTRING(12 20,30 44)
```

Some of the advantages of the WKT format are

- It is a simple, structured format that is easy to store and share between systems.

- Since it is text based, visibly identifying the information conveyed in a WKT representation is easy.

However, WKT has the following disadvantages:

- As with any text-based representation, it is not possible to precisely state the value of certain floating-point coordinate values obtained from binary methods. The inevitable rounding errors introduced when attempting to do so will lead to a loss of precision.

- Since SQL Server must parse the text in a WKT representation to create the relevant spatial object, creating objects from WKT can be slower than when using other methods.

Owing to its easy readability and relative conciseness, the WKT format is commonly used to demonstrate and share spatial data with other users, and is the format used in most of the examples of this book. It is also the format used in SQL Server 2008 Books Online, the Microsoft SQL Server online documentation, at http://msdn.microsoft.com/en-us/library/ms130214.aspx.

SQL Server 2008 implements specific static methods for creating each of the basic types of geometry shapes from WKT—Points, LineStrings, and Polygons—as well as methods to create homogenous multielement and heterogeneous Geometry Collection object types. There are also generic methods that can create any of the supported kinds of geometry object from WKT. Let's examine each method in more detail.

Creating a Point from WKT

The WKT syntax to represent a Point from Cartesian coordinates, such as from a projected coordinate reference system, is as follows:

```
POINT(x y)
```

The equivalent WKT syntax for a Point specified in geographic coordinates is

```
POINT(longitude latitude)
```

In each case, the representation begins with the POINT keyword, followed by the relevant coordinate values of that point contained within round brackets. Notice that the coordinate values are separated by a *space*, not a comma.

■**Caution** In everyday language, it is common to refer to coordinates of latitude and longitude (in that order). However, when defining a geographic coordinate pair in WKT, the longitude coordinate comes first, then the latitude coordinate. Be sure that you specify your coordinates in the right order!

In order to create a geography or geometry Point object from this WKT representation, you can use the STPointFromText() method of the appropriate datatype. For example, to create a Point object representing Edinburgh using the geography datatype, with coordinates of –3.19 degrees longitude and 55.95 degrees latitude, expressed using the spatial reference system identified by SRID 4326, you can run the following query in SQL Server Management Studio:

```
DECLARE @Edinburgh geography
SET @Edinburgh = geography::STPointFromText('POINT(-3.19 55.95)', 4326)
```

You will receive the following message:

```
Command(s) completed successfully.
```

The variable @Edinburgh now holds the Point instance created by this method.

Let's try another example. This time, we'll try to create a point representing Glasgow at coordinates (258647,665289) using the spatial reference system 27700. To do this, try running the following query:

```
DECLARE @Glasgow geography
SET @Glasgow = geography::STPointFromText('POINT(258647 665289)', 27700)
```

Whoops—this time we get the following error message:

```
Msg 6522, Level 16, State 1, Line 2
A .NET Framework error occurred during execution of user-defined routine
or aggregate "geography":
System.ArgumentException: 24204: The spatial reference identifier (SRID) is not
valid. The specified SRID must match one of the supported SRIDs displayed in the
sys.spatial_reference_systems catalog view.
```

Remember that the geography datatype can only be used to store geographic coordinates from a recognized geodetic spatial reference system. However, in this case we used the STPointFromText() method of the geography datatype but supplied coordinates based on SRID 27700, which is a spatial reference system based on projected coordinates.

To create an instance of a spatial object based on these coordinates, you can still use the same WKT syntax, but instead you must use the STPointFromText() method belonging to the geometry datatype, as follows:

```
DECLARE @Glasgow geometry
SET @Glasgow = geometry::STPointFromText('POINT(258647 665289)', 27700)
```

Because we are now using the geometry datatype, which stores planar data such as that obtained from a projected spatial reference system, the Point instance can successfully be assigned to the @Glasgow variable, indicated by the following message:

```
Command(s) completed successfully.
```

Caution Remember that WKT is only a text representation of a geometry or geography object—it is not the value of the object itself. You cannot insert a WKT string directly into a geometry or geography column like this: INSERT INTO GeographyColumn VALUES ('POINT(100 40)'). Instead, you must pass the WKT string as a parameter to a static method such as STPointFromText().

Z AND M COORDINATES

We know that to define the position of a point on the earth's surface, we only need two coordinates—x and y (in a projected coordinate reference system), or latitude and longitude (in a geographic coordinate reference system). However, in addition to these two coordinates, a Point object in WKT may optionally be defined with additional z and m coordinates:

- The z coordinate represents the height, or elevation, of a point. Just as positions on the earth's surface are measured with reference to a horizontal datum, the height of points above or below the surface are measured using a vertical datum. The z coordinate of a point can measure the height above sea level, the height above the underlying terrain, or the height above the reference ellipsoid, depending on which vertical datum is used.

- The m coordinate stores the "measure" value of a point. This coordinate can be used to represent any additional properties of a point that can be expressed as a double-precision number. For instance, if you are recording the spatial properties of time-based data, you could use the m value of a point to represent the time at which the measurement was taken. Or, if the point is a position lying on a route, the m coordinate could be used to store the distance of that point along the route.

To represent a point in WKT containing z and m coordinates, you use the following syntax:

```
POINT (x y z m)
```

Or, if using geographic coordinates, use the following syntax:

```
POINT (longitude latitude z m)
```

The static methods provided by SQL Server 2008 based on WKT syntax, such as STPointFromText(), support the creation and storage of z and m values as part of a POINT definition. However, these coordinates are *not* used when performing calculations. For instance, when calculating the distance between the points located at (0 0 0) and (3 4 12), SQL Server calculates the result as 5 units (the square root of the sum of the difference in the x and y dimensions), and not 13 (the square root of the sum of the difference in the x, y, and z dimensions).

Creating a LineString from WKT

The WKT syntax for a LineString containing *n* points is as follows:

```
LINESTRING(x1 y1, x2 y2, … , xn yn)
```

Like the WKT syntax for a Point, the representation begins with a keyword specifying the type of geometry to be created—LINESTRING—followed by the coordinates of each point in the geometry contained in round brackets. The coordinate values defined for a given point are separated by a space, and the sets of coordinates representing each point in the LineString are separated by a comma.

We can use the WKT representation of a LineString in combination with the STLineFromText() method to create LineString objects. Let's start by creating a simple LineString geometry connecting two points, representing the two ends of Sydney Harbour Bridge. For this example, we'll use the geography datatype, and define the coordinates of the start and end points using SRID 4326.

```
DECLARE @SydneyHarbourBridge geography
SET @SydneyHarbourBridge =
  geography::STLineFromText(
    'LINESTRING(
      151.209 -33.855,
      151.212 -33.850
    )',
    4326
  )
```

Now let's try a more complicated LineString, connecting five points. For this example, I'll also change the code to demonstrate how you can insert the resulting geometry into the geometry column, GeometryColumn, of a table Geometries, as follows:

```
INSERT into Geometries (
  NameColumn,
  GeometryColumn
)
VALUES (
  'Linestring connecting five points',
  geometry::STLineFromText(
    'LINESTRING(
      53.4 -2.99,
      53.5 -3.15,
      53.47 -4.66,
      53.40 -5.11,
      53.34 -6.25
      )',
    0
  )
)
```

This example demonstrates how you can make longer and more complicated LineStrings by simply adding the coordinates of further points into the WKT representation, separated by a comma in each case.

Creating a Polygon from WKT

The WKT syntax for a Polygon, containing z rings with n points in each ring, is as follows:

```
POLYGON(
  (ax1 ay1, ax2 ay2, … , axn ayn, ax1 ay1),
  (bx1 by1, bx2 by2, … , bxn byn, bx1 by1),
  …
  (zx1 zy1, zx2 zy2, … , zxn zyn, zx1 zy1)
)
```

As in all the previous WKT examples, the representation is contained within round brackets following the initial keyword—in this case POLYGON. Within these round brackets, the coordinates of each ring of the Polygon are contained within their own additional set of brackets. Since the rings of a Polygon are closed LineStrings, the WKT syntax for each ring is the same as the syntax used for a LineString definition: the individual coordinate values within a coordinate tuple are separated by a space, and the sets of coordinates are separated by a comma.

The first set of coordinates, (ax1 ay1, ax2 ay2, … , axn ayn, ax1 ay1), describes the points that define the exterior ring of the Polygon. Following the exterior ring definition, the Polygon may optionally define any number of internal rings: (bx1 by1, bx2 by2, … , bxn byn, bx1 by1), (zx1 zy1, zx2 zy2, … , zxn zyn, zx1 zy1), and so forth. Every internal ring definition follows the same syntax as for the exterior ring, contained within round brackets, and separated from the previous ring definition by a comma. It is important to remember that each ring must be closed; so, within each ring, the first coordinate tuple and the last coordinate tuple must be equal.

▨**Note** When using the geometry datatype, the external ring defines the overall perimeter of the area contained within a Polygon, while each internal ring defines an area of space cut out of the Polygon ("holes"). When using the geography datatype, the distinction between "external" and "internal" rings is not significant— every ring defines an area of space included in the Polygon, and an area excluded. Therefore, when using the geography datatype, the order in which rings are listed in the WKT representation is not important.

Even if you are describing a Polygon that only contains one ring, the points of that ring must still be contained within their own set of round brackets. In this case, the point coordinates appear within double brackets after the POLYGON keyword definition, as follows:

```
POLYGON((x1 y1, x2 y2, … , xn yn, x1 y1))
```

▨**Caution** A Polygon is composed of closed rings, so the first set of coordinates and the last set of coordinates of each ring must be the same.

The SQL Server method to create a geometry or geography Polygon object from WKT is called STPolyFromText(). Let me show you how to use this method to create a Polygon geometry representing the US Department of Defense Pentagon building, using the geography datatype and SRID 4326:

```
DECLARE @Pentagon geography
SET @Pentagon = geography::STPolyFromText(
  'POLYGON(
    (
      -77.0532238483429 38.870863029297695,
      -77.05468297004701 38.87304314667469,
      -77.05788016319276 38.872800914712734,
```

```
        -77.05849170684814 38.870219840133124,
        -77.05556273460388 38.8690670969385,
        -77.0532238483429 38.870863029297695
    ),
    (
        -77.05582022666931 38.8702866652523,
        -77.0569360256195 38.870737733163644,
        -77.056732177773439 38.87170668418343,
        -77.0554769039154 38.871848684516294,
        -77.05491900444031 38.87097997215688,
        -77.05582022666931 38.8702866652523
    )
  )',
  4326
)
```

This Polygon definition contains two rings. Although each ring defines a five-sided pentagonal shape, each ring contains six points in its definition, since the point at the start/end of the geometry is stated twice. Also notice that, since this example uses the `geography` datatype, the points of the first ring, which define the outer edge of the building, are defined in counterclockwise order, whereas the points in the second ring, which enclose the central courtyard omitted from the geometry, are defined in clockwise order. (For more information on the significance of ring ordering when using the `geography` datatype, refer to Chapter 2.)

Creating a MultiPoint from WKT

A MultiPoint is a collection of several Point geometries in a single object. To represent a MultiPoint object in WKT, you first declare the `MULTIPOINT` element name, followed by a comma-separated list of the coordinate tuples of each geometry included in the instance, in the same manner as they would be represented in a single-element Point object (that is, the individual coordinate values of each Point in the collection are stated one after another in `x y z m` order, or `longitude latitude z m` order, separated by spaces).

For instance, the following is a WKT representation of a MultiPoint object, containing three Points:

```
MULTIPOINT(21 2, 12 2, 30 40)
```

Since each point in WKT may contain between two and four coordinate values (depending on whether the optional z and m coordinates are defined), you must place the comma delimiter correctly to separate each coordinate tuple. Compare the preceding example with the following code, which uses the same coordinate values but instead creates a MultiPoint instance containing two Point geometries, each one specifying x, y, and z coordinates:

```
MULTIPOINT(21 2 12, 2 30 40)
```

The WKT representation of a MultiPoint geometry can be supplied to the `STMPointFromText()` method to instantiate a new MultiPoint object. For example, the following code demonstrates the use of the `STMPointFromText()` method to create a MultiPoint geometry containing three Points, representing the great pyramids of Khafre, Khufu, and Menkaure at Giza, defined using SRID 32636:

```
DECLARE @Pyramids geometry
SET @Pyramids =
  geometry::STMPointFromText(
    'MULTIPOINT(319640 3317580, 319980 3317940, 319400 3317200)',
    32636
)
```

Creating a MultiLineString from WKT

The WKT representation of a MultiLineString geometry is formed by a comma-separated list
of individual LineString geometries. However, since the WKT syntax of a LineString geometry
already contains comma delimiters between each Point, each LineString contained within the
MultiLineString collection must be surrounded by round brackets to distinguish the points
contained within that LineString from other LineString elements in the MultiLineString.

The following example demonstrates the syntax for a MultiLineString instance containing
two LineString elements, the first containing three Points, and the second containing two Points:

```
MULTILINESTRING((10 20, 3 4, 43 42),(44 10, 20 40))
```

To create an instance of a geometry MultiLineString from this representation, based on
SRID 20539, you can use the STMLineFromText() method, as follows:

```
DECLARE @MultiLineString geometry
SET @MultiLineString =
geometry::STMLineFromText(
  'MULTILINESTRING((10 20, 3 4, 43 42),(44 10, 20 40))',
  20539
)
```

Creating a MultiPolygon from WKT

The WKT representation of a MultiPolygon instance begins with a declaration of the MULTIPOLYGON
keyword. As in the MultiLineString, each Polygon element within a MultiPolygon is contained
within an additional set of round brackets, and separated by a comma. The following is a WKT
representation of a MultiPolygon element containing two Polygons, each consisting of an exterior
ring only:

```
MULTIPOLYGON(((10 20, 30 40, 44 50, 10 20)),((5 0, 20 40, 30 34, 5 0)))
```

The method to create a MultiPolygon instance from WKT is STMPolyFromText(). This can
be used with the preceding representation using SRID 0, as follows:

```
DECLARE @MultiPolygon geometry
SET @MultiPolygon =
geometry::STMPolyFromText(
  'MULTIPOLYGON(((10 20, 30 40, 44 50, 10 20)),((5 0, 20 40, 30 34, 5 0)))',
  0
)
```

Caution Do not get confused between the WKT representation of a MultiPolygon containing two Polygons and a Polygon containing two rings. A MultiPolygon, such as `MULTIPOLYGON(((10 20, 30 40, 44 50, 10 20)),((35 36, 37 37, 38 34, 35 36)))`, defines the points contained within two distinct areas of space, whereas a Polygon containing two rings, such as `POLYGON((10 20, 30 40, 44 50, 10 20), (35 36, 37 37, 38 34, 35 36))`, defines a single area of space from which the points contained within a second defined area have been excluded.

Creating a Geometry Collection from WKT

A Geometry Collection is a multielement object, but unlike the homogenous multielement types MultiPoint, MultiLineString, and MultiPolygon, a Geometry Collection may contain multiple different types of geometry in a single object.

To form the WKT representation of a Geometry Collection, you use the `GEOMETRYCOLLECTION` keyword, followed by the WKT representation of each individual object to be contained in the collection, separated by commas. The associated method to create an instance of a Geometry Collection from this representation is `STGeomCollFromText()`. To demonstrate how to use this method, the following code creates a new Geometry Collection containing a Polygon and a Point object, using the `geometry` datatype and SRID 0:

```
DECLARE @GeometryCollection geometry;
SET @GeometryCollection = geometry::STGeomCollFromText(
 'GEOMETRYCOLLECTION(
  POLYGON((5 5, 10 5, 10 10, 5 5)),
  POINT(10 10))',
  0
)
```

Note The multielement types MultiPoint, MultiLineString, and MultiPolygon are all specific geometry types derived from the generic Geometry Collection class.

Creating Any Kind of Geometry from WKT

Each of the methods discussed so far can only be used to create a particular type of geometry object from WKT. For example, `STPointFromText()` instantiates a Point object, and `STPolyFromText()` instantiates a Polygon object. In addition to these geometry-specific methods, SQL Server 2008 provides two generalized methods that can create any kind of object from well-formed WKT. These methods are `STGeomFromText()` and `Parse()`. Both methods can be used with either the geometry or geography datatype.

`STGeomFromText()` is the OGC standards-compliant method that can create any kind of spatial object from WKT. So, instead of using the geometry-specific methods `STPointFromText()` and `STLineFromText()` in the following example,

```
DECLARE @myTable TABLE (
  GeographyColumn geography
)
INSERT INTO @myTable (Geographycolumn) VALUES
  (geography::STPointFromText('POINT(-122.34 47.65)', 4326))
INSERT INTO @myTable (Geographycolumn) VALUES
  (geography::STLineFromText('LINESTRING(32.51 -23.34, 33.98 -12.10)', 4326))
```

you could alternatively use the STGeomFromText() method in both cases, as follows:

```
DECLARE @myTable TABLE (
  GeographyColumn geography
)
INSERT INTO @myTable (Geographycolumn) VALUES
  (geography::STGeomFromText('POINT(-122.34 47.65)', 4326))
INSERT INTO @myTable (Geographycolumn) VALUES
  (geography::STGeomFromText('LINESTRING(32.51 -23.34, 33.98 -12.10)', 4326))
```

Since the same STGeomFromText() method can be used to create any sort of valid object from WKT, you may be wondering why you shouldn't simply use STGeomFromText() in every situation, instead of using the more specific STPointFromText(), STLineFromText(), and so on.

The answer to this question is that, whereas the object-specific methods parse the WKT representation supplied to ensure that it represents valid data for the type of object created by the method, STGeomFromText() cannot be as strict in validating the input supplied. There is therefore more danger that you could create a badly formed or incorrect type of object using the STGeomFromText() method. If you know in advance that you will only be creating a certain sort of geometry, it is better to use the type-specific method designed for that particular geometry object. However, in some situations STGeomFromText() is more useful, such as when you reuse the same function to create many different types of geometries, or when you do not know in advance what sort of geometry is being created.

■**Note** Although the generalized method is called ST*Geom*FromText(), it can be used for both the geometry and geography types.

In addition to providing the OGC-compliant STGeomFromText() method, the geometry and geography datatypes also provide the Parse() method. Parse() operates in exactly the same way as STGeomFromText(), except that it assumes a default SRID for the coordinates based on the datatype of the method. For the geometry datatype, Parse() assumes a SRID of 0. For the geography datatype, Parse() uses SRID 4326—the EPSG code of the spatial reference system used by GPS satellite positioning devices and many other common applications. If you are creating objects defined using the default spatial reference system for the datatype you are using, you may find it more convenient to use the Parse() method instead, omitting the SRID parameter as follows:

```
DECLARE @LineString geography
SET @LineString = geography::Parse('LINESTRING(120 50, 128 52)')
```

This code defines a LineString geometry between the points at 120°E 50°N and 128°E 52°N, defined using the EPSG:4326 spatial reference system.

Representing an Existing Geometry As WKT

Remember that WKT is a *representation* of a spatial object—it is a text string that can be supplied as a parameter to any of the previously described methods in order to create an instance of a spatial object in SQL Server. However, it is not the actual value stored in a geography or geometry column or variable.

When you use any of the static methods based on WKT, such as the STGeomFromText() method, SQL Server 2008 parses the WKT representation provided to create a binary object representing that same information. It is this binary object that is returned by the method and stored as an item of geography or geometry data.

For instance, if you run the following query,

```
DECLARE @Point geography
SET @Point = geography::STGeomFromText('POINT(-122.34 47.65)', 4326)
SELECT @Point
```

the result you get is not the string 'POINT(-122.34 47.65)', but the following binary value instead:

```
0xE6100000010C3333333333D34740F6285C8FC2955EC0
```

In order to express an existing item of spatial data in WKT format, SQL Server 2008 provides three methods: STAsText(), STAsTextZM(), and ToString(). Table 4-2 gives an overview of each of these methods, which will then be explored in more depth in the following sections.

Table 4-2. *Methods to Retrieve the WKT Representation of Instances of Spatial Data*

Method	Description
STAsText()	This is the OGC standard method for expressing any kind of geometry instance in the WKT format. Each point within the geometry in the resulting output is represented by only two coordinates (x/y or longitude/latitude).
AsTextZM()	This method retrieves the WKT representation of a geometry, with each point of the geometry containing up to four coordinate values (x/y or longitude/latitude, z [if defined], and m [if defined]).
ToString()	This method is inherited and implemented by every type of object in .NET. When used on an instance of geography or geometry data, this method, like AsTextZM(), returns the WKT representation containing z and m coordinate values if defined.

■**Note** The WKT representation returned by the STAsText(), ToString(), and AsTextZM() methods contains the coordinates of each point of the geometry, but does not include the SRID of the spatial reference system from which they were obtained. You can retrieve this information using the STSrid() property instead. This is discussed in more detail in Chapter 11.

The STAsText() Method

STAsText() is the OGC-compliant method for returning the WKT string representation of an item of data. Note that, in contrast to the static methods used to create data *from* WKT, STAsText() acts upon a particular existing instance of an object. To retrieve the WKT representation of every row of data in the geometry column of a table using STAsText(), the syntax is as follows:

```
SELECT
GeometryColumn.STAsText()
```

STAsText() only returns the x and y (or longitude and latitude) coordinates defining the points of an instance. Although the method can be used against an object whose points also contain z or m coordinate values, they will not be returned in the output. Consider the following example:

```
DECLARE @Point geometry
SET @Point = geometry::STPointFromText('POINT(30 20 10 5)', 0)
SELECT @Point.STAsText()
```

The result is as follows:

```
POINT (30 20)
```

The ToString() Method

If you need to return the WKT representation of an object including the z and m coordinate values of each point, you can use the ToString() extended method instead. The syntax for this method is as follows:

```
DECLARE @Point geometry
SET @Point = geometry::STPointFromText('POINT(30 20 10 5)', 0)
SELECT @Point.ToString()
```

This gives the following result:

```
POINT (30 20 10 5)
```

Note The `ToString()` method is defined by the base object class in .NET, from which every specific class of object is derived. When inherited and implemented by the `geometry` and `geography` datatypes, this method returns the corresponding WKT representation of that object.

The AsTextZM() Method

The `AsTextZM()` method is an alternative to the `ToString()` method, which also returns the WKT representation of an object, including the extended coordinate values of z and m. You can use this method as follows:

```
DECLARE @Point geometry
SET @Point = geometry::STPointFromText('POINT(30 20 10 5)', 0)
SELECT @Point.AsTextZM()
```

This gives the following result:

```
POINT (30 20 10 5)
```

Tip You can use the `STAsText()`, `ToString()`, and `AsTextZM()` methods to return the WKT representation of any `geometry` or `geography` column or variable, whatever static method was used to create that data.

Creating Geometries from Well-Known Binary

The WKB format, like the WKT format, is a standardized way of representing spatial data defined by the OGC. In contrast to the text-based WKT format, WKB represents a `geometry` or `geography` object as a contiguous stream of bytes in binary format. Every WKB representation begins with a header section that defines the type of geometry being represented, and the order in which bytes are expressed (for more information on byte ordering, see the upcoming sidebar "Big Endian vs. Little Endian"). Depending on the type of geometry, the header may also contain additional descriptive information such as the number of geometries contained within a multi-element instance, or the number of rings contained in a Polygon geometry. Following the information in the header, a WKB representation lists a stream of 8-byte values representing the coordinates of each point in the geometry. The following code demonstrates the WKB representation of a Point geometry:

0x0000000001401C0000000000040300E000A000000

As you can see, the WKB format is not particularly easy to understand when compared to its sister format, WKT. However, there a number of advantages to the WKB format:

- Creating objects from WKB is faster than using static methods based on either of the text-based (GML or WKT) formats. Each x and y (or latitude and longitude) coordinate value in WKB is stored on 8-byte binary boundaries, as they are in SQL Server's own internal storage representation. The WKB static methods can therefore efficiently process and create the associated instance from WKB, whereas in WKT or GML, the parser must read in the whole text representation first.

- Since it is a binary format, WKB maintains the precision of floating-point coordinate values calculated from binary operations, without the rounding errors introduced in a text-based format.

However, WKB also has the following significant disadvantage:

- Binary values cannot be easily understood by a human reader—it can therefore be hard to detect errors in a WKB representation that could have been easily spotted from examining the equivalent WKT or GML representation.

WKB is most suitable in situations where spatial data must be passed directly between different computer systems, since the speed and precision of this format are beneficial, and the lack of human readability is not significant.

Just as for the WKT format, SQL Server 2008 provides a specific method for creating each type of geometry object from a WKB representation of that object, together with a generalized method STGeomFromWKB() for creating any type of object from valid WKB. Let's look at each method in turn.

BIG ENDIAN VS. LITTLE ENDIAN

A single byte of binary data can store one of 256 different values. When notated in hexadecimal format, these values range from 0x00 (0) to 0xFF (255). In order to describe precise location information, each coordinate value in the WKB format is expressed as a floating-point number stored across 8 bytes of data. In any system that stores multibyte data such as this, there are different accepted ways of ordering the bytes that represent each individual binary value:

- Little-endian binary data stores the least significant, or smallest, byte at the lowest memory address—that is, the "little end" comes first. Little-endian byte order is also known as Network Data Representation (NDR).

- Big-endian binary data stores the most significant byte at the lowest memory address—that is, it is stored "big end" first. Big-endian byte order is also known as External Data Representation (XDR).

A WKB geometry representation, like any binary data, can be stored using either of these formats. For example, the following two binary streams are both valid WKB representations of exactly the same geometry instance, in big-endian and little-endian format, respectively:

```
0x00 00000001 401C000000000000 4030000000000000
0x01 01000000 0000000000001C40 0000000000003040
```

Each byte is represented by two hexadecimal characters, so the little-endian value 0x1234 is equivalent to the big-endian value 0x3412, not 0x4321. In order to know which binary order is being used, the first byte of data in a WKB stream represents a single byte order marker (BOM):

- If the first byte is 0 (0x00), this signifies that the rest of the bytes are expressed in big-endian byte order.

- If the first byte is 1 (0x01), this signifies that the remaining bytes are expressed in little-endian order.

Since the byte order marker itself is only a single byte, it is unaffected by which system is used.

T-SQL binary functions generally use big-endian binary order. However, the STAsBinary() method represents binary data in little-endian order. Therefore, if you use the STGeomFromWKB() method to create an object from big-endian WKB, and then use the STAsBinary() method to retrieve the WKB representation of that object, you will get a different result, representing the same geometry object in little-endian WKB binary.

Creating a Point from WKB

The WKB representation of a Point object is a 21-byte stream of binary data. The elements contained within the representation are as follows:

```
[ByteOrder][Type][X][Y]
```

The first element, [ByteOrder], is a single-byte value that indicates whether the rest of the bytes in the WKB stream are expressed using big-endian (0x00) or little-endian (0x01) byte order. [Type] is a 4-byte unsigned integer indicating the type of geometry. For Point objects, this will always be the value 1, expressed using the relevant byte order as specified in the first byte—that is, 0x00000001 for big endian or 0x01000000 for little endian. [X] and [Y] are both 8-byte floating-point values representing the x and y coordinates of the point, or the longitude and latitude values if describing a point from a geographic coordinate system.

Note As in the WKT format, the WKB representation of a point expressed using geographic coordinates states the longitude coordinate first, then the latitude coordinate. However, unlike WKT, you cannot define additional z and m coordinate values relating to each point in WKB.

Creating a WKB Point Representation

To demonstrate the static method that SQL Server provides to create Point objects from WKB, we need to create a sample WKB Point geometry. In order to do this, first let's declare the parameters that will represent the values of each element contained within the WKB stream:

```
DECLARE @ByteOrder bit
DECLARE @Type int
DECLARE @longitude float
DECLARE @latitude float
```

To build up our WKB representation, we will use the T-SQL CAST function to convert each element into binary format. The result of this function, like all native SQL Server binary functions, uses big-endian binary byte order. Since our WKB representation will be based on the big-endian results of the CAST function, we must state that it too will be in big-endian byte order by setting the value of the BOM to 0:

```
SET @ByteOrder = 0
```

The next element in the WKB binary format defines the type of geometry being represented. In this example, we are describing a Point geometry, which is denoted in WKB as geometry type 1:

```
SET @Type = 1
```

We will create our point based on the following coordinate values, which represent the approximate location of Warsaw, Poland using the EPSG:4326 spatial reference system:

```
SET @longitude = 21.01
SET @latitude = 52.23
```

Having defined the value of each of the components, we can use the CAST function to convert them into binary values of the relevant length, and then append them together in a single binary stream. We will store the result in a varbinary(max) variable called @WKB. This is demonstrated in the following code:

```
DECLARE @WKB varbinary(max)
SET @WKB =
 CAST(@ByteOrder AS binary(1))
 + CAST(@Type AS binary(4))
 + CAST(@longitude AS binary(8))
 + CAST(@latitude AS binary(8))
```

The @WKB parameter now holds a valid WKB representation of our point. To check what it looks like, you can run the following query:

```
SELECT @WKB
```

The result is as follows:

```
0x00000000014035028F5C28F5C3404A1D70A3D70A3D
```

Note that, at this point, we have not yet created a geometry object—the preceding is merely the WKB representation that can be used to create a geometry object in conjunction with one of the static methods that accepts WKB as an input.

Note In practice, you will rarely construct a WKB representation using T-SQL as just shown—it is more likely that you will import existing information in WKB format from an external source. The technique just shown is used primarily as a demonstration of the structure of the WKB format.

Using the STPointFromWKB() Method

Now that we have a WKB representation of a Point, we can use that to create a geometry or geography instance using the STPointFromWKB() method. To create a Point using the geometry datatype, using SRID 0, based on the parameters in the WKB created in the previous section, you would use the following syntax:

```
DECLARE @Warsaw geography
SET @Warsaw = geography::STPointFromWKB(@WKB, 4326)
```

That's it! The @Warsaw variable now holds a geometry Point instance based on the parameters supplied. Here's the code in full:

```
/* Declare variables to hold each element of the WKB */
DECLARE @ByteOrder bit
DECLARE @Type int
DECLARE @longitude float
DECLARE @latitude float

/* Set the byte order marker to indicate big-endian byte order */
SET @ByteOrder = 0

/* Geometry Type 1 denotes a point */
SET @Type = 1

/* Set the values of each coordinate for the point */
SET @longitude = 21.01
SET @latitude = 52.23

/* Declare a new binary variable to hold the WKB */
DECLARE @WKB varbinary(max)

/* Append each of the elements together and store them in @WKB */
SET @WKB =
 CAST(@ByteOrder AS binary(1))
 + CAST(@Type AS binary(4))
 + CAST(@longitude AS binary(8))
 + CAST(@latitude AS binary(8))

/* Declare a new variable to hold the resulting geometry instance */
DECLARE @Warsaw geography

/* Pass the WKB representation created to the STPointFromWKB() method */
SET @Warsaw = geography::STPointFromWKB(@WKB, 4326)
```

To check that the point was created correctly, we can use the STAsText() method to get the WKT representation of our new Point geometry:

```
SELECT @Warsaw.STAsText()
```

The result is as follows:

```
POINT (21.01 52.23)
```

Note Unlike WKT, you cannot use WKB to represent the extended coordinate values of a point, z and m. Each WKB point is defined by exactly two coordinate values, representing x and y or longitude and latitude coordinates.

Creating a LineString from WKB

The WKB representation of a LineString object, containing *n* points, is a stream of bytes as follows:

```
[ByteOrder][Type][NumPoints][X1][Y1][X2][Y2] … [Xn][Yn]
```

As in the WKB representation of a Point, the first byte, [ByteOrder], declares whether the data is represented in little-endian or big-endian format. The next 4 bytes, [Type], represent the type of geometry being created, which is always 2 for a LineString.

Following the declaration of the geometry type is an additional 4-byte value, [NumPoints], which states the number of points in the LineString. Since the WKB structure does not contain delimiters such as the commas and brackets used in WKT, the representation must include this value so that SQL Server knows how many items of coordinate data to expect in the stream, and knows when it has reached the end of the representation. Immediately following the stated number of points in the LineString is a stream consisting of pairs of 8-byte floating-point values representing the coordinate values of each of those points.

Based on this information, we can build a WKB representation of a LineString using the following code:

```
/* Declare the parameters needed to build a WKB representation of a LineString */
DECLARE @ByteOrder bit
DECLARE @Type int
DECLARE @NumPoints int
DECLARE @x1 float
DECLARE @y1 float
DECLARE @x2 float
DECLARE @y2 float

/* We are using CAST to convert the parameters to big-endian byte order */
SET @ByteOrder = 0

/* LineStrings are denoted as geometry type 2 */
SET @Type = 2
```

```
/* This LineString will contain two points */
SET @NumPoints = 2

/* Set the x and y coordinate values of each point */
SET @x1 = 16
SET @y1 = 7
SET @x2 = 23
SET @y2 = 10

/* Declare a new binary parameter to hold the full WKB */
DECLARE @WKB varbinary(max)

/* Append the components together to build the Well-Known Binary representation */
SET @WKB =
 CAST(@ByteOrder AS binary(1))
 + CAST(@Type AS binary(4))
 + CAST(@NumPoints AS binary(4))
 + CAST(@x1 AS binary(8))
 + CAST(@y1 AS binary(8))
 + CAST(@x2 AS binary(8))
 + CAST(@y2 AS binary(8))
```

@WKB now holds the following value:

```
0x000000000200000002403000000000000401C0000000000004037000000000000402400000000000000
```

We can create a LineString geometry from this representation using the STLineFromWKB()
method of either the geography or geometry datatype. For this example, we will use the method
belonging to the geometry datatype, using SRID 0:

```
DECLARE @LineString geometry
SET @LineString = geometry::STLineFromWKB(@WKB, 0)
```

To check that the line was created correctly, once again we can get the WKT representation
of our new LineString using the STAsText() method:

```
SELECT @LineString.STAsText()
```

The result is as follows:

```
LINESTRING (16 7, 23 10)
```

Creating a Polygon from WKB

The WKB representation of a Polygon geometry containing two rings is as follows:

```
[ByteOrder][Type][NumRings][NumPoints][X1][Y1]…[Xn][Yn][NumPoints][X1][Y1]…[Xn][Yn]
                 <--------- Ring 1 --------->< -------- Ring 2 --------->
```

The elements contained within this representation are described in the following list:

- As in all other WKB representations, the stream begins with a single-byte indicator of the byte order of the remaining bytes, [ByteOrder].

- [Type] is a 4-byte unsigned integer representing the type of geometry. The value of [Type] is 3 for all Polygons in WKB.

- The following 4-byte integer value, [NumRings], specifies the total number of rings in the Polygon. This value counts the exterior ring, and any interior rings defined by the geometry.

- The definition of each ring follows, starting with the exterior ring. Because each ring is a closed LineString, each Polygon ring follows the same format as for an individual LineString in WKB—first with a 4-byte binary integer, [NumPoints], which states the number of points in the ring, followed by the x and y coordinates, or longitude and latitude coordinates, of each point.

- Any internal rings contained within the Polygon are listed one after another immediately following the external ring.

To demonstrate this syntax, consider the following code, which builds the WKB representation of a Polygon containing an exterior ring and one interior ring:

```
/* Declare all the elements required */
DECLARE @ByteOrder bit
DECLARE @Type int
DECLARE @NumRings int
DECLARE @Ext_NumPoints int
DECLARE @Ext_x1 float, @Ext_y1 float
DECLARE @Ext_x2 float, @Ext_y2 float
DECLARE @Ext_x3 float, @Ext_y3 float
DECLARE @Ext_x4 float, @Ext_y4 float
DECLARE @Int_NumPoints int
DECLARE @Int_x1 float, @Int_y1 float
DECLARE @Int_x2 float, @Int_y2 float
DECLARE @Int_x3 float, @Int_y3 float
DECLARE @Int_x4 float, @Int_y4 float

/* Set the values */
SET @ByteOrder = 0
SET @Type = 3
SET @NumRings = 2
-- Exterior Ring
SET @Ext_NumPoints = 5
SET @Ext_x1 = -4
SET @Ext_y1 = -5
```

```
SET @Ext_x2 = -4
SET @Ext_y2 = 10
SET @Ext_x3 = 12
SET @Ext_y3 = 10
SET @Ext_x4 = 12
SET @Ext_y4 = -5
-- Interior Ring
SET @Int_NumPoints = 4
SET @Int_x1 = 3
SET @Int_y1 = 1
SET @Int_x2 = 3
SET @Int_y2 = 5
SET @Int_x3 = 7
SET @Int_y3 = 3

/* Build the WKB representation */
DECLARE @WKB varbinary(max)
SET @WKB =
 CAST(@ByteOrder AS binary(1))
 + CAST(@Type AS binary(4))
 + CAST(@NumRings AS binary(4))
-- Exterior Ring
 + CAST(@Ext_NumPoints AS binary(4))
 + CAST(@Ext_x1 AS binary(8)) + CAST(@Ext_y1 AS binary(8))
 + CAST(@Ext_x2 AS binary(8)) + CAST(@Ext_y2 AS binary(8))
 + CAST(@Ext_x3 AS binary(8)) + CAST(@Ext_y3 AS binary(8))
 + CAST(@Ext_x4 AS binary(8)) + CAST(@Ext_y4 AS binary(8))
 + CAST(@Ext_x1 AS binary(8)) + CAST(@Ext_y1 AS binary(8))
-- Interior Ring
 + CAST(@Int_NumPoints AS binary(4))
 + CAST(@Int_x1 AS binary(8)) + CAST(@Int_y1 AS binary(8))
 + CAST(@Int_x2 AS binary(8)) + CAST(@Int_y2 AS binary(8))
 + CAST(@Int_x3 AS binary(8)) + CAST(@Int_y3 AS binary(8))
 + CAST(@Int_x1 AS binary(8)) + CAST(@Int_y1 AS binary(8))
```

We can now use the STPolyFromWKB() method in conjunction with our WKB representation, as follows:

```
DECLARE @polygon geometry
SET @polygon = geometry::STPolyFromWKB(@WKB, 0)
SELECT @polygon.STAsText()
```

The result is as follows:

```
POLYGON ((-4 -5, -4 10, 12 10, 12 -5, -4 -5), (3 1, 3 5, 7 3, 3 1))
```

Creating a Multielement Geometry from WKB

The WKB representations of the multielement geometry types—MultiPoint, MultiLineString, MultiPolygon, and Geometry Collection—are all formed using the same basic structure, as follows:

```
[ByteOrder][Type][NumGeometries]<Geometry1><Geometry2> … <GeometryN>
```

In each case, the elements of the binary stream are as follows:

- `[ByteOrder]` is a single byte that indicates whether the rest of the values describing the multielement instance are expressed using little-endian or big-endian byte order. Note that this only applies to the values specifically relating to the multielement instance itself—that is, the `[Type]` and `[NumGeometries]` values. Each individual geometry contained within the multielement instance must also specify the order in which the bytes of that particular element are stored.

- `[Type]` is a 4-byte unsigned integer indicating the type of multielement geometry being described. This must correspond to one of the following values:

  - 4, representing a MultiPoint

  - 5, representing a MultiLineString

  - 6, representing a MultiPolygon

  - 7, representing a Geometry Collection

- `[NumGeometries]` is a 4-byte unsigned integer representing the number of individual geometry elements contained within each multielement instance.

- `<Geometry1>` … `<GeometryN>` are the *fully formed* WKB representations of the individual geometries contained within the multielement instance, following the same rules that you would use if you were defining them as single geometries. This means that every individual element must explicitly declare all of the elements required in the WKB definition of that type of geometry, including the geometry type and byte order, not just the individual coordinate values.

■**Caution** If you are defining one of the homogenous multielement types, the values of `<Geometry1>` … `<GeometryN>` must all represent instances of individual geometries of the corresponding singular type. For example, the elements contained within a MultiLineString element must all be valid representations of LineString geometries.

The following is an example WKB representation of a Geometry Collection, containing a Point and a LineString:

```
0x0000000007000000020000000001404433333333333C002888A47ECFE9B010200000002000000
9BFEEC478A8802C03333333333333344406666666666F65340B81E85EB51B81B40
```

Table 4-3 breaks down each of the individual elements contained in this representation.

Table 4-3. *Elements Contained Within an Example WKB Multielement Geometry Representation*

Value	Description
0x	Indicates that binary values will be expressed using hexadecimal notation.
00	The following values are expressed in big-endian byte order.
00000007	This geometry is a heterogeneous Geometry Collection, denoted as type 7.
00000002	There are two geometries contained within this collection.
00	The bytes in the first element are expressed in big-endian byte order.
00000001	The first element represents a Point geometry.
4044333333333333	The x coordinate value of this Point.
C002888A47ECFE9B	The y coordinate value of this Point.
01	The bytes in the second element are expressed in little-endian byte order.
02000000	The second element represents a LineString geometry.
02000000	This LineString contains two points.
9BFEEC478A8802C0	The x coordinate of the first point in the LineString.
3333333333334440	The y coordinate of the first point in the LineString.
6666666666F65340	The x coordinate of the second point in the LineString.
B81E85EB51B81B40	The y coordinate of the second point in the LineString.

In order to instantiate a multielement geometry from a multielement WKB representation, you can use the appropriate type-specific method as follows:

- To create a MultiPoint geometry from WKB, use `STMPointFromWKB()`.
- To create a MultiLineString from WKB, use `STMLineFromWKB()`.
- To create a MultiPolygon from WKB, use `STMPolygonFromWKB()`.
- To create a Geometry Collection from WKB, use `STGeomCollFromWKB()`.

In the previous example, the WKB represents a Geometry Collection, so you can use the `STGeomCollFromWKB()` method as follows:

```
SELECT geometry::STGeomCollFromWKB(
0x000000000700000002000000000140443333333333333C002888A47ECFE9B010200000002000000 _
9BFEEC478A8802C033333333333334440666666666F65340B81E85EB51B81B40, 0)
```

Creating Any Type of Geometry from WKB

Just as with WKT, SQL Server 2008 provides a generic method for the geometry and geography datatypes that can be used to create any type of geometry object from well-formed WKB. This generic method is STGeomFromWKB(), which can be used instead of any of the type-specific methods mentioned in this section.

Representing an Existing Geometry As WKB

Although SQL Server 2008 stores spatial data internally as a stream of binary data, it is not the same as the WKB binary data format. As a result, you cannot directly set the value of an item of geography or geometry data from a WKB representation—you must pass that representation to one of the appropriate static methods instead. Likewise, if you directly select the internal binary value that SQL Server uses to store an item of spatial data, it will not be the same as the WKB representation of that feature. One reason for this is that the SQL Server internal binary format stores additional details relating to the spatial reference system that are not present in the original WKB representation.

In order to retrieve the WKB representation of a geometry or geography object, you need to use the STAsBinary() method instead, as follows:

```
SELECT geometry::STPointFromText('POINT(10.572 2.245)', 0).STAsBinary()
```

This returns the WKB representation of the geometry, expressed in little-endian binary format:

```
0x0101000000BE9F1A2FDD242540F6285C8FC2F50140
```

To take this example one stage further, the following code illustrates how to create a Point geometry from WKT using STGeomFromText(), then return the WKB representation of that geometry using STAsBinary(), before creating a second Point using the result of the STAsBinary() method in conjunction with the STGeomFromWKB() method:

```
DECLARE
  @WKT varchar(255) = 'POINT(52 8)',
  @WKB varbinary(max),
  @SRID int = 0,
  @Geometry1 geometry,
  @Geometry2 geometry

SET @Geometry1 = geometry::STGeomFromText(@WKT, @SRID)
SET @WKB = @Geometry1.STAsBinary()
SET @Geometry2 = geometry::STGeomFromWKB(@WKB, @SRID)

SELECT
  @Geometry1.STAsText(),
  @Geometry2.STAsText()
```

The results demonstrate that the geometries created through both methods are identical:

```
POINT (52 8)    POINT (52 8)
```

Tip Although you can create geometries from WKB expressed in either little-endian or big-endian byte order, the representation of a geometry returned by the STAsBinary() method will always be in little-endian byte order.

Creating Geometries from Geography Markup Language

Geography Markup Language is an XML-based language for representing spatial information. When expressed using a GML representation, each property of a geometry is contained within specific element tags within the document structure. This makes GML a very explicit and highly structured format. The following code demonstrates an example of the GML representation of a point:

```
<Point xmlns="http://www.opengis.net/gml">
  <pos>10 30</pos>
</Point>
```

Some advantages of the GML format are the following:

- GML is text based, so it is relatively easy for people to examine and understand the information expressed using it.

- Like all XML-based formats, GML has an explicit, highly structured hierarchical document format. This makes it is easy to understand the structure of complex spatial objects by examining the structure of the associated GML document.

- GML is very verbose, explicitly stating all values within specific elements.

However, GML also has the following disadvantages:

- It is very verbose! Although both GML and WKT are text-based formats, a GML representation of an object occupies substantially more space than the equivalent WKT representation, since it stores all information within associated element tags.

- Since GML is text based like WKT, it too suffers from precision issues caused by rounding when expressing binary floating-point coordinate values.

GML is most commonly used for representing spatial information in an XML-based environment. This includes spatial data syndicated over the Internet, which is discussed in more detail in Chapter 8.

▨**Tip** The GML methods implemented in SQL Server are based on a scaled-down version of the GML 3.1.1 schema. You can view the schema used in SQL Server at `http://schemas.microsoft.com/sqlserver/profiles/gml/`, or you can find the full GML standards on the OGC web site, located at `http://www.opengeospatial.org/standards/gml`.

Structure of a GML Document

The parent element of any GML representation defines the type of geometry being represented. Every GML representation must be contained within one of the following pairs of tags:

- `<Point>` ... `</Point>`
- `<LineString>` ... `</LineString>`
- `<Polygon>` ... `</Polygon>`
- `<MultiPoint>` ... `</MultiPoint>`
- `<MultiCurve>` ... `</MultiCurve>`
- `<MultiSurface>` ... `</MultiSurface>`
- `<MultiGeometry>` ... `</MultiGeometry>`

The names of these element tags, which are used to declare different geometries using GML, are similar to the names of the keywords used to define geometries in WKT. However, there are a few differences:

- Whereas MultiLineString and MultiPolygon geometries may be declared directly in WKT, in GML they are defined as child elements of the abstract *MultiCurve* and *MultiSurface* elements, respectively.

- An element containing multiple, heterogeneous geometries, known as a Geometry Collection in WKT, is called a *MultiGeometry* element in GML.

All of the component properties of a geometry are specified within child elements contained within the top-level parent element, and are enclosed within specific tags describing the property of the geometry that they represent. The particular elements contained within the GML representation of each geometry type will be discussed in more detail later in this section.

Declaring the GML Namespace

Any tag that defines an XML element may include an `xmlns` attribute, which associates the element name with a namespace. The names of elements used in any XML document are only unique within a given namespace, so `<Point xmlns="http://www.opengis.net/gml">` is different from `<Point xmlns="http://www.someothernamespace.com">`, or just `<Point>` with no associated namespace. This allows different XML documents to use the same element name in different situations, by qualifying it with the appropriate namespace.

Every element contained within a GML representation must belong to the GML namespace. This ensures that the names of any elements uniquely identify the appropriate GML element, rather than an element of the same name from any other XML namespace.

To ensure that every element of a GML representation is associated with the correct namespace, you should place a declaration of the GML namespace in the top-level parent element tag. This namespace will then be inherited by each of the child elements nested within that geometry representation. When you define the GML representation of a geometry for use in SQL Server, you should therefore always include the xmlns attribute inside the opening tag of the appropriate parent element, with a value set to a Uniform Resource Identifier (URI) reference to the GML namespace: http://www.opengis.net/gml.

Using the example of a LineString, the GML representation should therefore begin and end with the following tags:

```
<LineString xmlns="http://www.opengis.net/gml"> … </LineString>
```

If you omit the namespace declaration, even though the representation may be well-formed XML, it does not define a valid GML geometry. If you attempt to use such a representation in conjunction with the GeomFromGml() method, you will receive an error, as in the following example:

```
DECLARE @Point geography
SET @Point = geography::GeomFromGml('
  <Point>
    <pos>10 30</pos>
  </Point>',
4326)
```

```
System.FormatException: 24129: The given XML instance is not valid because its
top-level tag was Point. The top-level element of the input Geographic Markup
Language (GML) must be one of Point, LineString, Polygon, MultiPoint,
MultiGeometry, MultiCurve, or MultiSurface.
```

The text of this error message may seem rather confusing—it states that our GML representation is invalid because the top-level tag is Point. Instead, it helpfully gives a list of possible valid elements starting with . . . *Point?*

The reason for the error is that the top-level tag of an XML instance passed to the GeomFromGml() method must be one of the listed element names, *from the GML namespace.* Declaring the namespace in the parent element tag makes the representation valid, as shown in the following example:

```
DECLARE @Point geography
SET @Point = geography::GeomFromGml('
  <Point xmlns="http://www.opengis.net/gml">
    <pos>10 30</pos>
  </Point>',
4326)
```

```
Command(s) completed successfully.
```

Methods to Instantiate Geometry Objects from GML

Unlike the WKT and WKB formats, SQL Server 2008 does not provide different methods for creating specific geometry types from GML; every geometry object, whether Point, LineString, Polygon, or multielement geometry, is created using a single generic method—GeomFromGml(). The type of object created by this method is determined by the structure and content of the GML representation supplied.

The GeomFromGml() method is implemented by both spatial datatypes, so you can use it to create items of geography or geometry data using the appropriate syntax as follows:

```
geometry::GeomFromGml('GML representation', srid)
```

or

```
geography::GeomFromGml('GML representation', srid)
```

Although the GML standard itself is administered by the OGC, support for instantiating geometry objects from GML is not part of the OGC Simple Features for SQL Specification. As a result, the GeomFromGml() static method that creates geometry objects from GML is an extended SQL Server method, which is not prefixed by the letters "ST."

Since every type of geometry may be created from GML using the same syntax of the GeomFromGml() method just shown, in the following examples I won't repeat the method usage in each case. Instead, I will only show you the relevant GML representation of each type of geometry.

Note Don't be misled by the name—although GML stands for *Geography* Markup Language, you can use GeomFromGml() to create objects from a GML representation in both the geography and geometry datatypes.

Creating a Point from GML

An example of the GML representation of a Point is as follows:

```
<Point xmlns="http://www.opengis.net/gml">
  <pos>40.4 -2.31667</pos>
</Point>
```

This example defines a Point geometry, located at coordinates (40.4,–2.31667). The features of this GML representation are as follows:

- The entire representation is contained between the <Point> opening tag and the </Point> closing tag, declared using the GML namespace.

- Within this parent element, the coordinates that define the point's location are contained within the <pos> and </pos> tags.

- Coordinate values themselves are separated by a space, and listed in x–y order for Cartesian coordinates, or latitude–longitude order from a geographic coordinate system.

- Unlike in WKT, the <pos> element of a point in GML must contain exactly two coordinates—x and y, or latitude and longitude. GML does not support z or m coordinates.

■**Caution** GML expresses geographic coordinates in latitude–longitude order, which is the opposite order from that used in the WKT and WKB formats.

To use a GML representation to create a Point geometry, we can use the GeomFromGML() static method of the geometry or geography datatype as follows:

```
DECLARE @gml xml;
SET @gml = '
<Point xmlns="http://www.opengis.net/gml">
  <pos>40.4 -2.31667</pos>
</Point>
';
DECLARE @Point geometry;
SET @Point = geometry::GeomFromGml(@gml, 4326)
```

The variable @Point now holds a geometry Point object based on the parameters we supplied to the GeomFromGml() method—an x coordinate of 40.4 and a y coordinate of –2.31667, based on SRID 4326. Remember that, as in this example, even though the geometry datatype stores flat-earth data, you can still use it to define locations based on geographic coordinate systems, with the values of latitude and longitude mapped directly onto the y and x axes in an equirectan-gular projection.

To test that this object was created correctly, we can now select the WKT representation of this Point by using the STAsText() method, as follows:

```
SELECT @Point.STAsText()
```

This gives the following result:

```
POINT (40.4 -2.31667)
```

Because we have defined a point based on the EPSG:4326 spatial reference system using geographic coordinates of latitude and longitude, perhaps it would be more suitable to create our point using the geography datatype instead of the geometry datatype. In order to do this, we can use the GeomFromGml() method of the geography datatype instead, as follows:

```
DECLARE @gml xml;
SET @gml = '
<Point xmlns="http://www.opengis.net/gml">
  <pos>40.4 -2.31667</pos>
</Point>
';
DECLARE @Point geography;
SET @Point = geography::GeomFromGml(@gml, 4326)
SELECT @Point.STAsText()
```

This gives the following result:

```
POINT (-2.31667 40.4)
```

Notice that the coordinate order is reversed? In the first example, we passed Cartesian coordinates to the GeomFromGml() method of the geometry datatype in x–y order, which is the same order as used by STAsText() to express the result. However, in the second example we stated geographic coordinates required for the geography datatype, but whereas GeomFromGml() accepts those coordinates in latitude–longitude order, STAsText() interprets the result as longitude–latitude.

To create the equivalent Point as the "unprojected" geometry Point in the first example, where longitude is mapped to x and latitude is mapped to y, we need to swap the coordinates passed to GeomFromGml() as follows:

```
<Point xmlns="http://www.opengis.net/gml">
  <pos>-2.31667 40.4</pos>
</Point>
```

When supplied to the GeomFromGml() method of the geography datatype, this gives the following result, represented in WKT:

```
POINT (40.4 -2.31667)
```

Creating a LineString from GML

When defining a single Point in GML, as in the previous example, you specify the coordinate values contained within the <pos> element. When creating a LineString, or any other geometry that requires more than one point in its definition, you use the <posList> element instead. Since every point in GML must contain exactly two coordinates, there is no need to define additional delimiters between each point—within the <posList> element, the coordinate values of each point in the LineString are separated by spaces, with no commas between each coordinate pair.

The following example code demonstrates the GML representation of a LineString instance joining the points at (–6,4) and (3,–5):

```
<LineString xmlns="http://www.opengis.net/gml">
  <posList>-6 4 3 -5</posList>
</LineString>
```

The WKT representation of the LineString created from this GML is as follows:

```
LINESTRING (-6 4, 3 -5)
```

■Caution Unlike the WKT format, there are no commas separating each point in a GML representation.

Creating a Polygon from GML

As in the LineString definition, the coordinate values of each point defining the rings of a Polygon are expressed in space-separated lists contained within a `<posList>` element. In order to specify that the list of points contained within this element represents a closed LineString, each `<posList>` element that defines a Polygon ring is nested within additional `<LinearRing>` ... `</LinearRing>` tags.

Every GML Polygon representation must contain one `<LinearRing>` element within the `<exterior>` element of the Polygon definition, defining the points that form the exterior ring of the polygon. The GML may additionally specify one or more `<LinearRing>` elements contained within the `<interior>` element of the Polygon parent element, defining the interior rings of the polygon.

The following example code demonstrates the GML representation of a Polygon containing three linear rings—one external ring, and two internal rings cut out of the Polygon shape:

```
<Polygon xmlns="http://www.opengis.net/gml">
  <exterior>
    <LinearRing>
      <posList>0 0 100 0 100 100 0 100 0 0</posList>
    </LinearRing>
  </exterior>
  <interior>
    <LinearRing>
      <posList>10 10 20 10 20 20 10 20 10 10</posList>
    </LinearRing>
  </interior>
  <interior>
    <LinearRing>
      <posList>75 10 80 10 80 20 75 20 75 10</posList>
    </LinearRing>
  </interior>
</Polygon>
```

As with all other GML geometry representations, you can instantiate an instance of a geometry based on this representation by passing it to the `GeomFromGml()` method of either the geography or geometry datatype, together with the SRID of the spatial reference system in which the coordinates were obtained.

Creating a MultiPoint from GML

The parent element defining a MultiPoint geometry in GML is defined using the `<MultiPoint>` tag. The individual Point geometries contained within a MultiPoint geometry are defined one after another, using exactly the same syntax as for individual Points, nested within a child element of the parent called `<pointMembers>`.

The following example shows the GML representation of a MultiPoint instance containing two Point geometries:

```
<MultiPoint xmlns="http://www.opengis.net/gml">
  <pointMembers>
    <Point>
      <pos>2 3</pos>
    </Point>
    <Point>
      <pos>4 10</pos>
    </Point>
  </pointMembers>
</MultiPoint>
```

Creating a MultiLineString from GML

When you describe an instance containing more than one LineString geometry in GML, the parent element is actually *MultiCurve* rather than MultiLineString. You may recall that, in Chapter 3, we examined the inheritance hierarchy tree of objects in the geometry and geography datatypes. The hierarchy tree shows that the MultiLineString object is derived from another type of object, called a MultiCurve. A MultiCurve is the generic object type for any geometry that creates a number of paths between different series of points, whereas a MultiLineString is the specific case of the MultiCurve where those paths are calculated by the linear interpolation between the points. Although you cannot create MultiCurve objects directly, the GML representation portrays the fact that MultiLineStrings are descended from MultiCurves. All LineString elements contained within the MultiCurve are expressed in the same format as they would be if they were individual geometries, listed one after another within a child element of the MultiCurve called <curveMembers>.

The following example code demonstrates the GML representation of a MultiLineString geometry containing two LineStrings:

```
<MultiCurve xmlns="http://www.opengis.net/gml">
  <curveMembers>
    <LineString>
      <posList>2 3 4 10</posList>
    </LineString>
    <LineString>
      <posList>4 10 15 40</posList>
    </LineString>
  </curveMembers>
</MultiCurve>
```

Creating a MultiPolygon from GML

The GML element that can contain multiple Polygon elements is called a MultiSurface. Within a MultiSurface, the definition of each member geometry is contained within an element called <surfaceMembers>.

The following example lists the GML representation of a MultiPolygon instance that contains two Polygon geometries, each containing a single exterior ring:

```
<MultiSurface xmlns="http://www.opengis.net/gml">
  <surfaceMembers>
    <Polygon>
      <exterior>
        <LinearRing>
          <posList>2 3 5 3 6 8 2 7 2 3</posList>
        </LinearRing>
      </exterior>
    </Polygon>
    <Polygon>
      <exterior>
        <LinearRing>
          <posList>10 20 20 20 20 30 10 30 10 20</posList>
        </LinearRing>
      </exterior>
    </Polygon>
  </surfaceMembers>
</MultiSurface>
```

Creating a Geometry Collection from GML

The parent element of a Geometry Collection represented in GML is called MultiGeometry. The component elements are therefore contained within opening `<MultiGeometry>` and closing `</MultiGeometry>` tags. As with the other, specific, multielement instances, each of the individual geometries contained within a MultiGeometry is nested within an additional element, in this instance called `<geometryMembers>`.

The following example shows the GML representation of a Geometry Collection containing a Point geometry and a LineString geometry:

```
<MultiGeometry xmlns="http://www.opengis.net/gml">
  <geometryMembers>
    <Point>
      <pos>15 10</pos>
    </Point>
    <LineString>
      <posList>4 10 2 3</posList>
    </LineString>
  </geometryMembers>
</MultiGeometry>
```

As with all other GML geometry representations, you can instantiate an instance of a geometry based on this representation by passing it to the `GeomFromGml()` method of either the geography or geometry datatype.

Representing an Existing Geometry As GML

As with the WKT and WKB formats, SQL Server also provides a method that can be used to represent existing geometry or geography objects in GML format, AsGml(). To demonstrate the use of this method, the following code listing illustrates how to create a LineString geometry from WKT, and then retrieve the GML representation of that geometry using AsGml():

```
DECLARE @Linestring geometry
SET @Linestring  = geometry::STLineFromText('LINESTRING(0 0, 12 10, 15 4)', 0)
SELECT @Linestring.AsGml()
```

The result is as follows:

```
<LineString xmlns="http://www.opengis.net/gml">
  <posList>0 0 12 10 15 4</posList>
</LineString>
```

CREATING POINTS USING POINT()

If you only wish to create Point objects of either the geometry or geography type, there is no need to use a specific language to represent those points—they can be sufficiently described using just three numeric parameters representing the x (or longitude) coordinate, the y (or latitude) coordinate, and the SRID. In this case, you do not need to use one of the dedicated language methods described in this chapter to represent the object—you can use the Point() method instead.

The Point() method accepts three arguments—representing latitude (or x), longitude (or y), and an SRID—to create a Point object of either the geometry or geography datatype. To demonstrate this method, consider the following code:

```
SELECT
geography::Point(41,-87,4269)
```

This example creates a geography Point at latitude 41 degrees, longitude –87 degrees, using the SRID 4269.

Summary

This chapter introduced you to each of the static methods that you can use to instantiate items of geography or geometry data. There are a number of methods available, based on three different standard formats for expressing spatial information: Well-Known Text (WKT), Well-Known Binary (WKB), and Geography Markup Language (GML).

- Each static method requires exactly two parameters—the SRID of the spatial reference system used, and a representation of the geometry expressed using the WKT, WKB, or GML format.

- There are advantages and disadvantages associated with each of the WKT, WKB, and GML representations. Generally speaking, WKT is the simplest to understand, WKB is the fastest, and GML is the most structured.

- When creating items of spatial data from the WKT or WKB format, there are specific methods for each type of geometry (Point, LineString, Polygon, and multielement types), as well as the generic methods STGeomFromText() and STGeomFromWKB(). In contrast, every item of data created from GML uses the same GeomFromGml() method.

- For creating Point geometries, the Point() method can be used, which requires only the two coordinate values and the SRID to which those coordinates relate.

- In addition to the static methods used to create objects from each of these representations, SQL Server also provides instance methods that can be used to express objects in WKT, WKB, and GML formats. This functionality is provided by the STAsText(), STAsBinary(), and AsGml() methods, respectively.

CHAPTER 5

Marking Out Geometries Using Virtual Earth

In the last chapter, we examined a number of static methods that can be used to create items of geometry or geography data. To use any of these methods, you must supply a structured representation of the geometry that you wish to create using the WKT, WKB, or GML format. Therefore, you need to either have this representation already available or create such a representation from known coordinate values of the points in the geometry.

In practice, however, there are many situations in which you simply don't know the coordinates of the spatial features that you wish to describe. You might be able to identify the feature on a map, you might even be able to give someone directions to get there, but you cannot state the latitude and longitude, or x and y values, associated with the points that define its shape. In situations like this, wouldn't it be helpful if you could simply trace the shape of a feature on a virtual map to create a geometry representing that feature?

In this chapter, we will build a small web application that does exactly that, using Microsoft's Virtual Earth web service to create a drawing canvas onto which you can mark out the shapes of geometries. This technique is suitable for creating a small number of discrete features that have fixed spatial properties. For example, it could be used to record the locations of a number of warehouses or distribution depots. Note that in this chapter we are using Virtual Earth as an input device—to help define new items of data in SQL Server. In Chapter 9, I will show you how to retrieve spatial data *from* SQL Server 2008 and output that data graphically on a Virtual Earth or Google Maps control.

Caution The technique discussed in this chapter creates a web application that uses JavaScript to implement the features of the Virtual Earth API. In order to be able to use this application, you must load this page in a JavaScript-enabled web browser, such as Mozilla Firefox or Microsoft Internet Explorer.

Virtual Earth (VE) is an integrated collection of web services from Microsoft that can be used in a range of geospatial applications—plotting routes, finding nearby points of interest (POIs), and examining stunning aerial photography of the world. The main component of the Virtual Earth web service is a JavaScript map control that can be added to a web page to display a two-dimensional map object. In addition to displaying map data and rich imagery of the earth, the

Virtual Earth application programming interface (API) provides a range of methods that you can use to manipulate and overlay spatial data on that map. You have probably already seen the VE map control in use on various sites on the Internet; among other things, it is the main technology used in the Microsoft Live Search Maps service, available at http://maps.live.com.

In this chapter, I am first going to show you how to build your own web page containing a VE map control. Then, I will show you how to use the methods provided by the Virtual Earth API to draw shapes on the map around specific features of interest. Finally, I'll show you how you can extend the API with your own methods to create the appropriate WKT representation of these created shapes, so that they can be used to define items of geography and geometry data in SQL Server.

Creating the Web Application

To create the required HTML and JavaScript code, you can either use a dedicated program, such as Microsoft Visual Studio, or use any text editor, including Windows Notepad. Although it is possible to use a simple text editor, I recommend that you use Visual Studio, since it provides a number of helpful programming aids, such as color coding and syntax highlighting, as well as an integrated ASP.NET development server to preview your site as you develop it. Even if you don't have access to the full version of Visual Studio, you can still get the benefits of these features using Visual Web Developer Express Edition, which is free to download from http://www.microsoft.com/express/vwd/Default.aspx.

In the following section, I assume that you are using Visual Studio 2008. If you choose to use a different tool, then some of the actions may differ, although the underlying code will remain the same.

■**Tip** The full application code is included in the ZIP file accompanying this book, which can be downloaded from the Apress web site. If you'd prefer to download the code rather than create the files, you can skip ahead to the next section.

Creating a New Web Site

Before we can start adding any code, we need to create a new web site project in Visual Studio. To do so, open the Visual Studio application and follow these steps:

1. From the main Microsoft Visual Studio menu bar, select File ➤ New ➤ Web Site.

2. In the New Web Site dialog box, shown in Figure 5-1, highlight the Empty Web Site template and choose a location in which to save the files. For this example, I'll use C:\Spatial\VEDrawingCanvas. You can save the files to any normal drive—they do not need to be published to a web server.

3. Click OK.

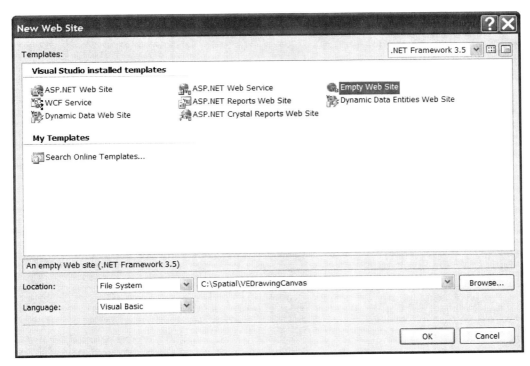

Figure 5-1. *Creating a new web site in Visual Studio 2008*

The new web site will be created, and the status bar at the bottom of the screen will display the following message:

```
Creating Project 'VEDrawingCanvas'… project creation successful.
```

We can now begin adding the elements to our web application.

Adding a Basic Map

To begin, we will add a new HTML page to our web site that will display a basic Virtual Earth map. You can add a new HTML page to the web site as follows:

1. From the Visual Studio menu bar, select Website ➤ Add New Item.

2. In the Add New Item dialog box, shown in Figure 5-2, highlight HTML Page.

3. Choose a name for the page (I am using the default, HTMLPage.htm, but you may change this if you prefer).

4. Click the Add button.

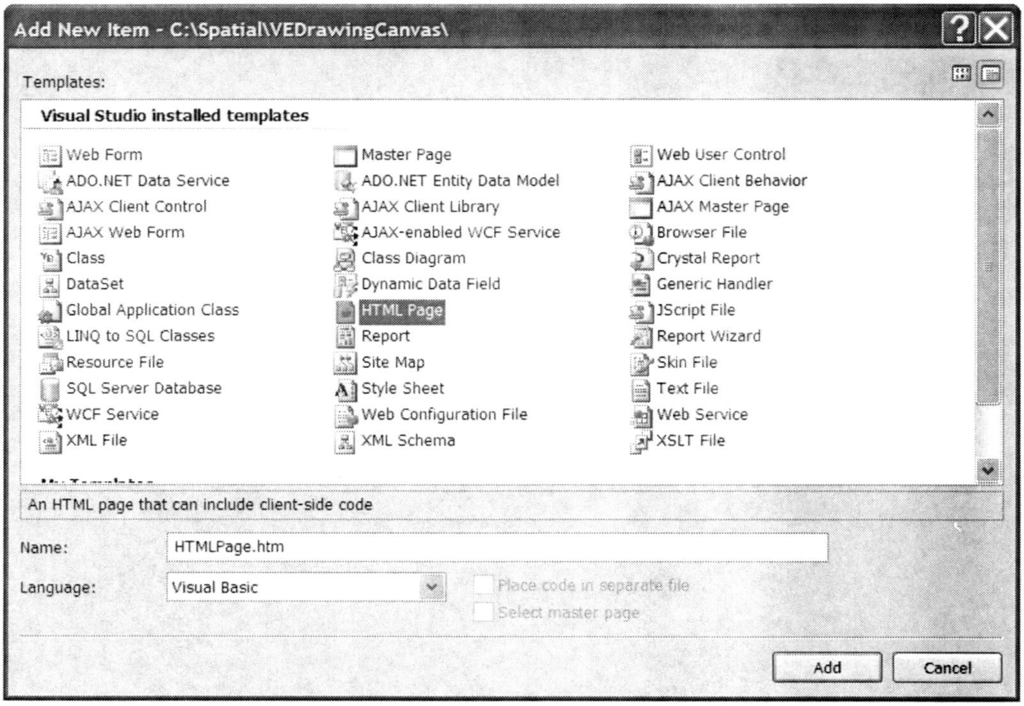

Figure 5-2. *Adding an HTML page to the web site*

The new file is added to the web site project, and the main Visual Studio window changes to show the contents of HTMLPage.htm. This file contains a blank template for a web page, as follows:

```
<!DOCTYPE html PUBLIC "-//W3C//DTD XHTML 1.0 Transitional//EN"
"http://www.w3.org/TR/xhtml1/DTD/xhtml1-transitional.dtd">

<html xmlns="http://www.w3.org/1999/xhtml">
<head>
    <title></title>
</head>
<body>

</body>
</html>
```

Note If you can't see the contents of the HTMLPage.htm page, double-click the HTMLPage.htm file name in the Solution Explorer pane.

To start, we're going to add a few simple lines of code to this template to include a basic VE map control. Change the contents of HTMLPage.htm by inserting the lines highlighted in bold in the following code listing:

```
<!DOCTYPE html PUBLIC "-//W3C//DTD XHTML 1.0 Transitional//EN"
"http://www.w3.org/TR/xhtml1/DTD/xhtml1-transitional.dtd">
<html xmlns="http://www.w3.org/1999/xhtml" >
<head>
  <title>SQL Server 2008 Geospatial Data Generator</title>
  <script src="http://dev.virtualearth.net/mapcontrol/mapcontrol.ashx?v=6.2"
type="text/javascript"></script>
</head>
<body onload="var map = new VEMap('divMap'); map.LoadMap();">
  <div id="divMap" style="position:relative;"> </div>
</body>
</html>
```

That's all that is required! To see what the page looks like, first save the file by selecting File ➤ Save HTMLPage.htm, and then select File ➤ View in Browser (or press Ctrl+Shift+W).

Your default web browser will load, and you should see a page containing a VE control, as shown in Figure 5-3. Even though we haven't added any additional functionality yet, you can already pan around the map, zoom in and out, and change the display style using the controls in the top left corner of the window.

Note Depending on your browser's security settings, you may receive a prompt advising you to enable active content before you can view the page correctly. If prompted whether you would like to view active content, click Yes.

Okay, now let's get back to our code in Visual Studio to see how we created this page. In the <head> section of the document, we added a reference to the following JavaScript script:

```
<script src="http://dev.virtualearth.net/mapcontrol/mapcontrol.ashx?v=6.2"
type="text/javascript"></script>
```

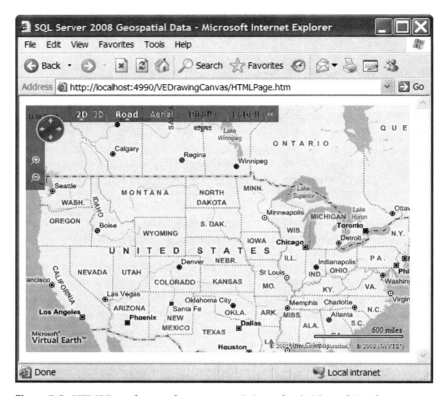

Figure 5-3. *HTMLPage.htm web page containing a basic Virtual Earth map control*

This is the URL of the main Virtual Earth API. At the time of writing, the latest version of the control is version 6.2, which we specified by using the v=6.2 parameter appended to the end of the URL string. Once you have included a reference to this script in a web page, you can then access any of the methods provided by the API. In this example, we used just two such methods:

1. We constructed a new instance of the VE control by using the VEMap() constructor. The single parameter passed to the VEMap() constructor specifies where the created map object should be placed on the page. In our example, we specified 'divMap', which is the ID of the HTML <div> element we added later in the body of the document.

2. We loaded and populated the map, using the LoadMap() method. This method accepts a number of optional parameters that specify the initial properties of the map, such as the center point, zoom level, and map style. However, in this case we called the method without any parameters, which creates the default map, showing a two-dimensional road map of the mainland United States.

We placed the calls to both of these methods in the onload event of the body of the HTML document, which means that the map gets created as soon as the page is loaded by the browser, as follows:

```
<body onload="var map = new VEMap('divMap'); map.LoadMap();">
```

The final addition to our code was to specify a new `<div>` element that would act as a container for our map:

```
<div id="divMap" style="position:relative;">
```

Note that this code simply creates an empty container on the page without specifying any content—it is the `VEMap()` method that dynamically inserts the map control into this element, based on its ID. The `style="position:relative;"` property is a Cascading Style Sheets (CSS) style declaration to ensure that the map appears in the correct position in the browser.

Extending the Map Functionality

Now that we've got our basic Virtual Earth map working, let's add some custom functionality that will enable us to mark out the shape of various geometries that we wish to define in SQL Server. This involves using a range of other methods provided by the main Virtual Earth API, as well as defining some of our own methods. If we were to continue adding these methods into the body of our HTMLPage.htm page, it would quickly become quite complicated and difficult to manage. Instead, we will add a new file to the web site solely to contain our JavaScript functions. To do this, select Website ➤ Add New Item from the Visual Studio menu. In the Add New Item dialog box, shown in Figure 5-4, highlight the JScript File template. The default name is JScript.js, which is fine, so go ahead and create the file by clicking Add.

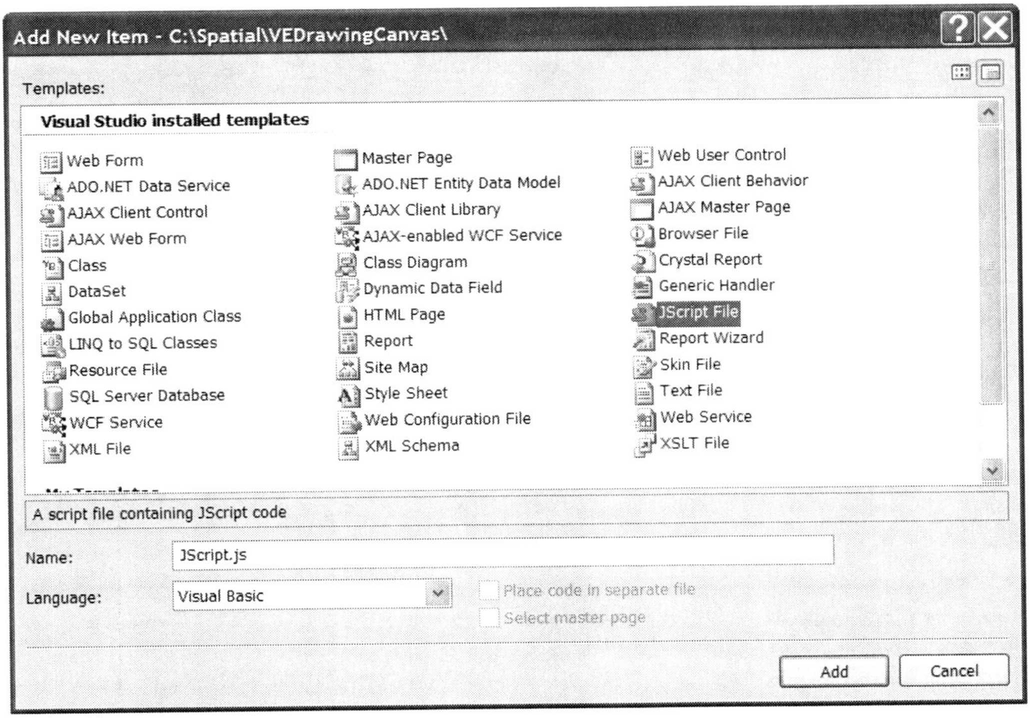

Figure 5-4. *Adding a JavaScript file to the web site*

A new, blank window opens in the main workspace. This is the empty file that we will use to define the JavaScript functions that build on the base functionality provided by the VE control. By keeping these functions separate in the JScript.js file, they are easier to manage than if mixed alongside the main content in the HTMLPage.htm file, which instead will contain the structural elements of the page.

Declaring the Global Variables

We will start by declaring a number of global variables, which we will need to refer to in different functions. Add the following code to the top of the JScript.js file:

```
/**
 * Declare the Global Variables
 */
var map = null; // the map object
var shape = null; // the current shape
var shapeType = null; // the type of the VE shape being created
var shapePoints = new Array(); // the array of points in the shape
```

This code defines four global variables that will be used throughout the web application. map will contain the map object itself, and will be the instance upon which any calls to the Virtual Earth API will act. The shape variable will be used to hold the current shape being drawn on the map. shapeType will store the type of geometry being defined, and shapePoints will store the array of points contained in that geometry's definition.

Initializing the Map

In our first example, we placed the VEMap() constructor and the LoadMap() function directly in the onload event of the HTML <body> tag in HTMLPage.htm. To make our code neater, let's move these separate calls out into a function on their own, which we'll call getMap(). This function will contain all of the necessary information required to set up the map when the page is first loaded. To create the function, add the following code to the JScript.js file:

```
function getMap() {
  // Create a new map instance
  map = new VEMap('divMap');
  // Define the parameters for the map
  map.LoadMap(new VELatLong(51.5, -0.1), 5, VEMapStyle.Road, false);
  // Attach an event handler when you move the mouse across the map
  map.AttachEvent("onmousemove", DisplayCoords);
}
```

The first line contained within the getMap() function, map = new VEMap('divMap'), is exactly the same method as we used in the basic example, to construct a new map and place it in the div element on the page called *divMap*. We then also call the same LoadMap() method as before, to configure the properties of that map. However, whereas in the basic example we used LoadMap() on its own to create a default map, on this occasion we pass a number of parameters to the method to customize different elements of the initial map appearance. Let's take this opportunity to examine the different types of map that you can create using Virtual Earth, denoted by parameters passed to the LoadMap() method:

- The first parameter is a VELatLong() object denoting the center position of the map. In this case, we are centering on a point at 51.5 degrees latitude, –0.1 degrees longitude, which is the approximate location of London, England.

- The second parameter gives the zoom level of the map. Greater numbers indicate that the map should be zoomed in more. For this example, we specify a zoom level of 5.

- The third parameter indicates the display style of the map. The value VEMapStyle.Road denotes that we are creating a road map that displays road and place name data. Virtual Earth supports a number of different map styles, as listed and described in Table 5-1.

Table 5-1. *Virtual Earth Map Styles That Can Be Passed to the LoadMap() Method*

VEMapStyle	Description
Road	Road map style.
Shaded	Shaded map style. This is the same as the Road style but with added shaded relief contours.
Aerial	Aerial map style.
Hybrid	Hybrid style, combining an aerial map with label overlays.
Birdseye	An oblique, bird's-eye view taken from an overhead angle.
Oblique	Same as Birdseye.
BirdseyeHybrid	Oblique imagery as in the Birdseye view, combined with a label overlay.

If you want to mark out real objects that can easily be identified through aerial imagery, such as buildings and other structures, then you should use the Aerial or Hybrid style. If you want to mark out features shown on traditional maps, such as roads, rivers, and state or country boundaries, you may find it easier to set your map to use the Road or Shaded styles.

Note The LoadMap() method also accepts other optional parameters. For a complete list, consult the following web page: http://msdn.microsoft.com/en-us/library/bb412546.aspx.

The final line in getMap() attaches an *event handler* to the map object, using the map.AttachEvent() method. Virtual Earth event handlers can be used to add interactivity to a map by listening for certain events to occur, which then trigger functions to be executed. In this case, map.AttachEvent("onmousemove", DisplayCoords); creates an event handler that causes the DisplayCoords() function to be called every time the mouse is moved over the map. The DisplayCoords() function will be used to calculate and update the current coordinates of the mouse cursor.

Displaying the Coordinates of the Current Mouse Position

To make it easier to mark out exact geometries on the map, we will add a function to our application that will retrieve the coordinates of the mouse cursor as you move across the map, and update the value of two text boxes on the page. To do this, create the DisplayCoords() function by adding the following code to JScript.js:

```
function DisplayCoords(e) {
    // Retrieve the pixel position of the cursor
    var pix = new VEPixel(e.mapX, e.mapY);
    // Convert the pixel location to latitude / longitude
    var pos = map.PixelToLatLong(pix);
    // Update the page to display current cursor latitude / longitude
    document.getElementById("Latitude").value = pos.Latitude;
    document.getElementById("Longitude").value = pos.Longitude;
}
```

The DisplayCoords() function is triggered by the onmousemove event handler added to the getMap() function. Every time the mouse is moved over the map, this handler calls the DisplayCoords() function, passing it a parameter, e, that records details of the event that triggered the handler. In the preceding code, the DisplayCoords() function retrieves the pixel that the mouse cursor is over by using the e.mapX and e.mapY properties. It then uses the PixelToLatLong() method to retrieve the latitude and longitude coordinates of that pixel (based on the position and zoom level of the current map view). Finally, it sets the value of two elements on the HTML page, Latitude, and Longitude (which we will create shortly), to reflect these coordinate values.

Drawing a New Geometry

Our next function, createGeometry(), will be the function called when we click a button to start defining a geometry. The method accepts a parameter, shapetype, which specifies the sort of shape we are creating.

The Virtual Earth API supports three types of shapes that can be created on a map: Pushpins, Polylines, and Polygons. These are equivalent to the SQL Server Point, LineString, and Polygon geometries, respectively, as shown in Table 5-2. Virtual Earth does not support multielement geometry instances.

Table 5-2. *Comparison of Geometric Objects in Virtual Earth and SQL Server 2008*

SQL Server Geometry Type	Virtual Earth Shape Type
Point	Pushpin
LineString	Polyline
Polygon	Polygon

The createGeometry() function prepares to define a new geometry by taking the following steps:

1. It sets the value of the global shapeType variable to the type of geometry being created.

2. It resets the shapePoints array, which is used to record each point in the geometry.

3. It attaches an event handler that calls the addPoint() function every time we click the map. This function will add each point to the shapePoints array of points for the overall geometry.

4. It changes the cursor to a crosshair to show we are in geometry creation mode.

Here's the code for the createGeometry() function, which you should add to the JScript.js file:

```
function createGeometry(shapetype) {
  // Store the type of VEShape we are defining in the global shapeType variable
  shapeType = shapetype;
  // Set the length of the shapePoints array to zero
  shapePoints.length = 0;
  // Attach the addPoint() function to be called every time we click the mouse
  map.AttachEvent("onclick", addPoint);
  // Change the mouse cursor to show we are adding points
  document.getElementById("divMap").childNodes[0].style.cursor = "crosshair";
}
```

Defining Each Point in the Geometry

The addPoint() function is the main function that handles the creation of each point in a geometry shape on the map. Every time you click the mouse button on the map, this function calculates the coordinates of the location where you clicked (using the same e.mapX, e.mapY, and PixelToLatLong() method as used in the DisplayCoords() function), and adds those to the global array of points of the shape currently being created, shapePoints. The switch (shapeType) statement is used to create the correct Virtual Earth shape equivalent to the type of geometry being defined, which is stored in the global variable shape. This shape is then added to the map using the map.AddShape() method. Once you have finished drawing the geometry, the addPoint() method detaches itself from the onclick event handler to prevent any further points from being added to the geometry, and then calls the makeWKT() method to create the WKT representation of the resulting shape. For a Point, this occurs when you click the single position on the map. For a LineString or Polygon, the geometry is ended when you right-click to insert the final point and end the shape definition.

The code for the addPoint() function to be added to JScript.js is as follows:

```
function addPoint(e) {
  // Retrieve the pixel position that we clicked
  var pix = new VEPixel(e.mapX, e.mapY);
  // Convert pixel coordinates to latitude and longitude
  var pos = map.PixelToLatLong(pix);
  // Add these coordinates to the array of points for the current shape
  shapePoints[shapePoints.length] = pos;
  // Handle different geometries
  switch (shapeType) {
```

```
      // We are drawing a VE Pushpin (i.e., a Point)
      case VEShapeType.Pushpin:
        // Create a new Pushpin VEShape based on the point defined
        shape = new VEShape(VEShapeType.Pushpin, shapePoints);
        // Add the pushpin to the map
        map.AddShape(shape);
        break;
     // We are defining a LineString or a Polygon
      case VEShapeType.Polyline:
      case VEShapeType.Polygon:
        // If we have only defined two points for the shape
        if (shapePoints.length == 2) {
          // Create a new Polyline VEShape based on the points defined
          shape = new VEShape(VEShapeType.Polyline, shapePoints);
          // Add the Polyline to the map
          map.AddShape(shape);
        }
        // If we have defined more than two points for the shape
        if (shapePoints.length > 2) {
          // Delete the old shape from the map
          map.DeleteShape(shape);
          // Create a new Polyline or Polygon VEShape based on the points defined
          shape = new VEShape(shapeType, shapePoints);
          // Add the shape to the map
          map.AddShape(shape);
        }
        break;
        // If shapeType is any other value
      default:
        // Stop calling the addPoint() function on every mouseclick
        map.DetachEvent("onclick", addPoint);
        // Throw an error
        throw ("Unexpected shape type");
    }

  // When we have finished the shape definition
  if (shapeType == VEShapeType.Pushpin || e.rightMouseButton == true) {
    // Stop calling the addPoint() function on every mouseclick
    map.DetachEvent("onclick", addPoint);
    // Change the mouse cursor back to normal
    document.getElementById("divMap").childNodes[0].style.cursor = "";
    // Create the WKT representation of this shape
    var WKT = makeWKT(map.GetShapeByID(shape.GetID()))
    // Put the WKT output on the page
    document.getElementById('WKTOutput').innerText = WKT.toString();
  }
}
```

Building the WKT Representation

The makeWKT() function is called by addPoint() when you finish defining a shape on the map (that is, you right-click to insert the final point of a LineString or Polygon geometry, or you click to place a Point geometry), and is used to create the WKT representation of the VE shape just created. The makeWKT() function creates the WKT representation of a geometry by looping through the array of points contained within the shape object, and expressing them in the appropriate syntax required by the equivalent WKT geometry. This involves placing a comma between each coordinate pair, and prefixing the string of point coordinates with the appropriate POINT, LINESTRING, or POLYGON identifier.

To add the makeWKT() function, insert the following code into JScript.js:

```
function makeWKT(shape) {
  // Define a variable to hold what type of WKT shape we are creating
  var wktShapeType = "";
  // Define the WKT type which corresponds to the VEShapeType we have created
  switch (shape.GetType()) {
    // VEShapeType.Pushpin => WKT POINT
    case VEShapeType.Pushpin:
      wktShapeType = 'POINT';
      break;
    // VEShapeType.Polyline => WKT LINESTRING
    case VEShapeType.Polyline:
      wktShapeType = 'LINESTRING';
      break;
    // VEShapeType.Polygon => WKT POLYGON
    case VEShapeType.Polygon:
      wktShapeType = 'POLYGON';
      break;
    default:
      throw ("Unexpected shape type");
  }
  // Define a new string to hold the point list
  var pointsString = ""
  // Retrieve an array of points that make up this shape
  var points = shape.GetPoints();
  // Retrieve the coordinates of the first point
  pointsString = points[0].Longitude + " " + points[0].Latitude;
  // Loop through remaining points in the object definition
  for (var i = 1; i < points.length; i++) {
    // Append the remaining points, with a comma before each coordinate pair
    pointsString += ", " + points[i].Longitude + " " + points[i].Latitude;
  }
  // Build the WKT representation of the shape
  var WKT = null
  if (wktShapeType == 'POLYGON')
    // Polygons require double brackets around the points of the exterior ring
    WKT = wktShapeType + "((" + pointsString + "))";
```

```
  else
    // Other WKT geometry types have single brackets
    WKT = wktShapeType + "(" + pointsString + ")";
  // Return the final WKT representation
  return WKT;
}
```

Starting Again

Finally, to complete our JScript.js file, we use the StartAgain() function to remove any existing shapes from the map and reset any variables, so that we can begin creating a new shape. Here's the code:

```
function StartAgain() {
  // Delete all shapes from the map
  map.DeleteAllShapes();
  // Reset the cursor to default style
  document.getElementById('divMap').childNodes[0].style.cursor = "";
  // Reset the text
  document.getElementById('WKTOutput').innerText =
    'The WKT representation of the geometry will appear here.';
}
```

Adding Controls to HTMLPage.htm

Now that we've added the extra JavaScript functions required to draw geometries on our map into the JScript.js file, we need to revisit and update our HTMLPage.htm file—the main page viewed by the browser.

Open HTMLPage.htm (if it's not already visible) by double-clicking the HTMLPage.htm file name in the Solution Explorer pane. Change the code to be as follows:

```
<!DOCTYPE html PUBLIC "-//W3C//DTD XHTML 1.0 Transitional//EN"
"http://www.w3.org/TR/xhtml1/DTD/xhtml1-transitional.dtd">
<html xmlns="http://www.w3.org/1999/xhtml" >
<head>
  <title>SQL Server 2008 Geospatial Data Generator</title>
  <script src="http://dev.virtualearth.net/mapcontrol/mapcontrol.ashx?v=6.2"
        type="text/javascript"></script>
  <script src="JScript.js" type="text/javascript"></script>
</head>
<body onload="getMap();">
  <form action="">
```

```
  <h2>1.) Select the type of geometry to create</h2>
    <input id="DrawPoint" title="Mark a point"
onclick="createGeometry(VEShapeType.Pushpin)" type="button" value="Point" />
    <input id="DrawLineString" title="Draw a line"
onclick="createGeometry(VEShapeType.Polyline)" type="button" value="LineString" />
    <input id="DrawPolygon" title="Draw a polygon."
onclick="createGeometry(VEShapeType.Polygon)" type="button" value="Polygon" />

  <h2>2.) Click the map to define point(s) of this geometry</h2>
  <div id="divMap" style=" position: relative;">
    <!-- The Virtual Earth Map control will be automatically inserted here -->
  </div>
  <label for="Latitude">Latitude</label><input id="Latitude" />
  <label for="Longitude">Longitude</label><input id="Longitude" />

  <h2>3.) Well-Known Text</h2>
  <div id="WKTOutput">The WKT representation of the geometry will appear here.</div>

  <hr />
  <p>
    <input id="Reset" onclick="StartAgain();" type="button" value="Start Again" />
  </p>
  </form>
</body>
</html>
```

The new HTMLPage.htm contains some important functional changes, as well as some additional structural and descriptive elements to help you use the page. Let's review some of the key amendments, highlighted in bold in the preceding code listing.

In the head section of the HTML page, we include a reference to the JScript.js file we created earlier, using the line `<script src="JScript.js" type="text/javascript"></script>`. Then, we change the `onload` event of the body of the page to call the `getMap()` function defined in the JScript.js file, which performs all the necessary operations to create and configure the map on the page.

Within the body of the HTML page itself, we add a number of elements that will provide the user interface that enables us to use the application. First, we define a number of form input buttons, which we will use to start drawing on the map. These call the `createGeometry()` function, passing a parameter to state what kind of geometry we are about to define, as in this example:

```
<input id="DrawPoint" title="Mark a point"
onclick="createGeometry(VEShapeType.Pushpin)" type="button" value="Point" />
```

We also add form elements that the DisplayCoords() function will update to display the latitude and longitude values of the current mouse position:

```
<label for="Latitude">Latitude</label><input id="Latitude" />
<label for="Longitude">Longitude</label><input id="Longitude" />
```

And finally, we add a <div> element that will contain the WKT representation of our geometry when it is complete:

```
<div id="WKTOutput">The WKT representation of the geometry will appear here.</div>
```

Once all of these elements have been added, our page is now ready to use!

Using the Web Application

Once you have made all the changes described in the previous section, save the HTMLPage.htm and JScript.js files by selecting File ➤ Save All (Ctrl+Shift+S). Next, right-click HTMLPage.htm in the Solution Explorer pane and select the View in Browser menu option (Ctrl+Shift+W). The page appears as shown in Figure 5-5.

Tip If you choose to create your files using a text editor rather than in Visual Studio, you should create the HTMLPage.htm and JScript.js files separately based on the code in the preceding section and save them in the same directory. Then load the HTMLPage.htm page in your browser (you can normally do this by locating the file in Windows Explorer and double-clicking its icon).

To use the page, follow these steps:

1. Before you begin, adjust the map view so that it clearly shows the feature you will be tracing. You can click and drag the map (or use the cursor keys) to pan, and zoom in and out using the plus (+) and minus (–) keys. Alternatively, you can use the controls in the top left corner of the map.

2. Click the relevant button to select the type of geometry you want to create: Point, LineString, or Polygon.

3. Click each point of that geometry on the map. Remember that, when defining the exterior ring of a Polygon for use in the geography datatype, you should enter the points of the ring in a counterclockwise direction, so that the area enclosed within the Polygon lies on the left side of the line drawn between the points.

4. When you are done, right-click to insert the final point and finish the geometry.

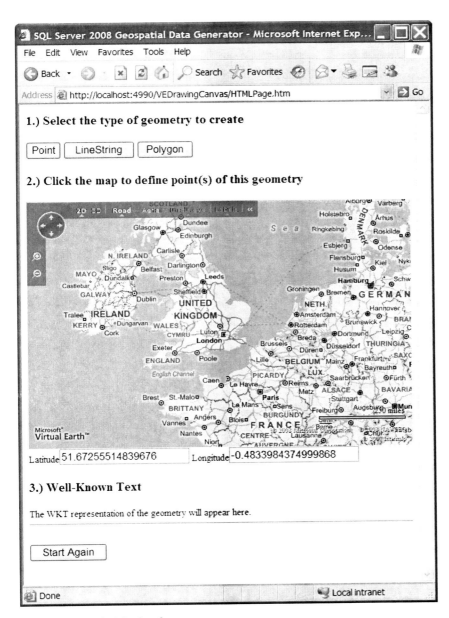

Figure 5-5. *The finished web page*

Caution The application described in this chapter cannot create Polygons containing interior rings, or multielement geometries.

The WKT representation of the geometry will appear at the bottom of the screen. Figure 5-6 illustrates an example of how the page can be used to create a Polygon geometry representing the state of Colorado.

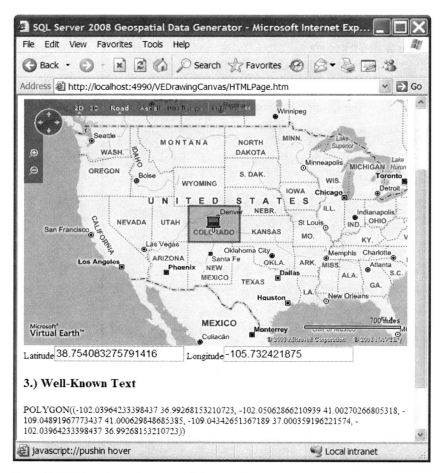

Figure 5-6. *Using the web page to define a Polygon representing the state of Colorado*

Creating a Geometry from the WKT Output

Every static method that can be used to create spatial data in SQL Server requires two parameters—a representation describing the coordinates of each point of the geometry, and the SRID denoting the spatial reference system in which those coordinates were defined. Now that we have obtained the WKT representation of a geometry, before we can use that to create an item of spatial data in SQL Server, we need to know the spatial reference system used by Virtual Earth. Since the VE control presents a two-dimensional map of the earth, we know that the spatial information portrayed must be from a projected spatial reference system. The WKT representation of the particular spatial reference system used by the VE map display, denoted by EPSG:3785, is as follows:

```
PROJCS[
  "Popular Visualisation CRS / Mercator",
  GEOGCS[
    "Popular Visualisation CRS",
    DATUM[
      "Popular Visualisation Datum",
      SPHEROID[
        "Popular Visualisation Sphere",
        6378137,
        0,
        AUTHORITY["EPSG",7059]
        ],
      TOWGS84[0, 0, 0, 0, 0, 0, 0],
      AUTHORITY["EPSG",6055]
      ],
    PRIMEM["Greenwich", 0, AUTHORITY["EPSG", "8901"]],
    UNIT["degree", 0.0174532925199433, AUTHORITY["EPSG", "9102"]],
    AXIS["E", EAST],
    AXIS["N", NORTH],
    AUTHORITY["EPSG",4055]
    ],
  PROJECTION["Mercator"],
  PARAMETER["False_Easting", 0],
  PARAMETER["False_Northing", 0],
  PARAMETER["Central_Meridian", 0],
  PARAMETER["Latitude_of_origin", 0],
  UNIT["metre", 1, AUTHORITY["EPSG", "9001"]],
  AXIS["East", EAST],
  AXIS["North", NORTH],
  AUTHORITY["EPSG",3785]
]
```

The SPHEROID parameter of this system specifies a reference ellipsoid with a semimajor axis of 6,378,137 m, which is the same as the WGS 84 ellipsoid. However, the second parameter, 0, indicates that no flattening should be applied—EPSG:3785 is based on the same WGS 84 datum used by GPS systems, but is applied to a perfectly spherical model of the earth rather than to a spheroid or an ellipsoid. The WKT representation also shows that EPSG:3785 is based on the Mercator projection, and uses the Greenwich Prime Meridian.

Note EPSG:3785 is the same spatial reference system used by both Microsoft Virtual Earth (http://maps.live.com) and Google Maps (http://maps.google.com).

From what you've learned so far, you might well be thinking that, since we have defined geometries by creating points on this *projected* map, any coordinates obtained using this technique will be projected coordinate values that will have to be stored using the geometry datatype and defined using SRID 3785. If so, well done! You have a good understanding of the geometry and geography datatypes. However, in this instance, this isn't the case. You see, Virtual Earth actually uses two spatial reference systems:

- When *displaying* projected data, Virtual Earth uses a Mercator projection based on the WGS 84 datum, but applied to a sphere. This is the EPSG:3785 spatial reference just described.

- When *accessing* data programmatically through the API methods, however, Virtual Earth uses (unprojected) geographic longitude/latitude coordinates based on the standard WGS 84 system—that is, EPSG:4326.

Since our web page uses methods provided by the API to obtain results based on a geographic coordinate system, we can therefore store the results in either the geography or geometry datatype, using the familiar SRID 4326.

Having used the map to create a new geometry, and obtained the relevant WKT representation, we can then use that representation in conjunction with the STGeomFromText() method, as follows:

```
geography::STGeomFromText(@WKT, 4326)
```

@WKT is substituted here for the WKT representation displayed by the web page after you define a geometry.

To demonstrate this method, the following code creates an item of geography data from the Polygon representation of Colorado illustrated in Figure 5-6:

```
DECLARE @Colorado geography
SET @Colorado = geography::STGeomFromText(
  'POLYGON((-102.03964233398437 36.99268153210723, -102.05062866210939
   41.00270266805318, -109.04891967773437 41.000629848685385, -109.04342651367189
   37.000359196221574, -102.03964233398437 36.99268153210723)) '
  ,4326
)
```

Summary

In this chapter I showed you a technique that enables you to use Microsoft Virtual Earth as a drawing canvas, on which you can mark out the shape of geometries representing features on the earth. You learned the following:

- Virtual Earth is a web-based mapping service that you can embed within your own web applications.

- The Virtual Earth API allows you to define Pushpins, Polylines, and Polygons on the map display. Using JavaScript, these shapes can be converted to the equivalent WKT Point, LineString, or Polygon representation.

- The map image displayed by Virtual Earth is projected using the EPSG:3785 spatial reference system. This system uses a Mercator projection of the WGS 84 datum, applied to a spherical model of the earth. However, coordinates accessed through the API are geographic coordinates defined using EPSG:4326.

- The Well-Known Text representation of a Virtual Earth shape, as created by the application described in this chapter, can be supplied to the STGeomFromText() method to create an item of geography or geometry data representing that feature in SQL Server.

CHAPTER 6

■ ■ ■

Importing Spatial Data

Many spatial applications combine custom-defined spatial features, such as the location of a set of customers, with spatial data representing widely accepted, generic features on the earth, such as the boundaries of countries and states, the locations of major world cities, and the paths of main roads and railways. Rather than having to create this information yourself, there are a number of alternative sources from which you can obtain commonly used spatial data on which to base your spatial applications.

In this chapter, I will introduce you to some of the sources from which you can obtain publicly available spatial information, the formats in which that data is commonly supplied, and the techniques you can use to import that information into SQL Server.

Sources of Spatial Data

There is a wealth of existing spatial information, which you can obtain from a variety of commercial data vendors as well as from educational institutions and government agencies who make the information available for free. Table 6-1 gives details of a few possible Internet sources from which spatial data is free to download.

Table 6-1. *Sources of Freely Downloadable Spatial Information*

Source[a]	Description
http://www.census.gov/	The US Census Bureau Geography Division has lots of high-quality spatial information, including a US Gazetteer, Zip Code Tabulation Areas (ZCTAs), and the TIGER database of streets, rivers, railroads, and many other geographic entities (United States only).
http://geodata.grid.unep.ch/	The United Nations Geo Data Portal includes global, national, regional, and subregional statistics and spatial data, covering themes such as covering themes such as freshwater, population, forests, emissions, climate, disasters, health, and GDP.
http://biogeo.berkeley.edu/gadm/	The global administrative areas database (GADM) contains the boundaries of countries, states, counties, provinces, and their equivalents covering the whole world, and is available as a single ZIP file hosted at the University of California, Berkeley.

Table 6-1. *Sources of Freely Downloadable Spatial Information (Continued)*

Source[a]	Description
`http://earth-info.nga.mil/gns/html/`	The US National Geospatial-Intelligence Agency (NGA) GEOnet Names Server (GNS) is the official repository of all foreign place names, containing information about location, administrative division, and quality.
`http://geodata.gov/wps/portal/gos`	The US government "Geospatial One Stop" web page of geographic data contains classified links to a variety of sources covering areas including ecology, geology, health, transportation, and demographics.

[a] *There may be restrictions on the use of data obtained from these sources. Please refer to the respective providers for specific details.*

As demonstrated in Chapter 4, each of the SQL Server 2008 static spatial methods can only create a single item of spatial data at a time, from either a WKT, WKB, or GML representation. However, sources of spatial data such as those listed in Table 6-1 may be stored in a variety of other spatial formats, and may describe many thousands of individual items in a single document. You therefore cannot directly create geography or geometry data from these sources using any static methods.

In the remainder of this chapter, I discuss some of the common alternative formats of spatial data that are available, and explain techniques that you can use to import this data into SQL Server 2008.

Importing Tabular Spatial Data

Although arguably not a spatial data format, the most abundant (and also the simplest) source of freely available geographic information generally takes the form of a list of place names, together with a single pair of latitude and longitude coordinates describing the location of each place. These sources may also contain other columns of associated information, such as demographic or economic measures. Information presented in this format is commonly known as a *gazetteer*, a dictionary of geographic information.

If you want to import spatial information from a structured table of data containing columns of latitude and longitude (or northing and easting coordinate values from a projected coordinate system) such as a gazetteer, you can use one of the available static methods to create a geography or geometry Point object based on the coordinate values representing each item of data. This involves the following steps:

1. Import the structured data into a new table by using one of the bulk import methods provided by SQL Server 2008: the OPENROWSET and BULK INSERT T-SQL statements, the BCP utility, or the Import and Export Wizard.

2. Use the ALTER TABLE statement to add to the table a new geography or geometry column that will hold the derived spatial data.

3. Use the T-SQL UPDATE statement in conjunction with a static method to populate the new column, based on the values of the coordinate columns in the imported data.

To demonstrate this approach, let me show you an example using a file of earthquake data provided by the United States Geological Survey (USGS). The USGS makes a number of datasets freely available, which you can download from their web site at http://www.usgs.gov. One such dataset lists real-time, worldwide earthquake lists in the past 7 days, which you can download directly from http://earthquake.usgs.gov/eqcenter/catalogs/eqs7day-M1.txt. This file is a comma-separated list of data, containing various attributes of each earthquake in columnar format, as listed and described in Table 6-2.

Table 6-2. *Columns of Data in the eqs7day-M1.txt File*

Column	Description
Src	The two-character identifier of the source network that contributed the data
Eqid	The unique identifier for this earthquake
Version	The version number
Datetime	A text string describing the date at which the recording was made
Lat	The latitude of the epicenter, stated in the EPSG:4326 spatial reference system
Lon	The longitude of the epicenter, stated in the EPSG:4326 spatial reference system
Magnitude	The magnitude of the earthquake, determined by the strength of the seismic waves detected at each station
Depth	The depth of the earthquake's center, measured in kilometers
NST	The number of reporting stations
Region	A text string description of the area in which the earthquake occurred

To obtain a copy of this data, follow these steps:

1. Load your web browser and, in the address bar, type the following URL address: http://earthquake.usgs.gov/eqcenter/catalogs/eqs7day-M1.txt. The browser will show the contents of the latest feed, as demonstrated in the example in Figure 6-1.

2. Save this file to an accessible location by choosing File ➤ Save As (or Save Page As, depending on your browser). You will be prompted for a file name and location. For this example, I will assume that you name the file eqs7day-M1.txt and save it to the C:\Spatial folder.

Note Because the eqs7day-M1.txt file contains a constantly updated feed of data from the last 7 days, the actual content of this file will be different from that demonstrated in this chapter.

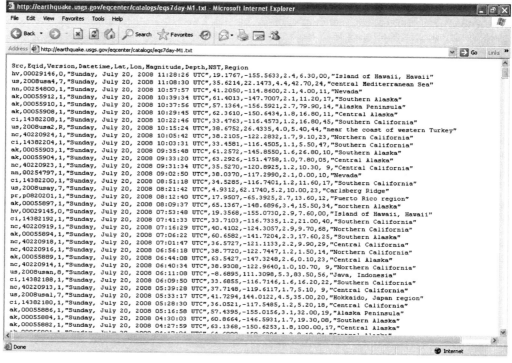

Figure 6-1. *The USGS earthquake data file*

Importing the Text File

There are a number of different ways to import data into SQL Server 2008. This example uses the Import and Export Wizard, which allows you to step through the creation of a simple package to move data from a source to a destination. The steps follow:

1. From the Object Explorer pane in Microsoft SQL Server Management Studio, right-click the name of the database into which you would like to import the data, and select Tasks ➤ Import Data.

2. The Import and Export Wizard appears. Click Next to begin.

3. The first page of the wizard prompts you to choose a data source. Select Flat File Source from the Data Source drop-down list at the top of the screen.

4. Click the Browse button and navigate to the eqs7day-M1.txt text file that you saved earlier. Highlight the file and click Open.

5. By default, the Text Qualifier field for the connection is set to <none>. The text strings within the eqs7day-M1.txt file are contained within double quotes, so change this value to be a double quote character (") instead.

6. The eqs7day-M1.txt text file contains headings, so check the Column Names in the First Data Row check box.

7. Click the Advanced option in the left pane. Click each column in turn and, from the properties pane on the right side, amend the values of the DataType and OutputColumnWidth fields to match the values shown in Table 6-3.

8. Once you have made the appropriate changes, click the Next button. The wizard prompts you to choose a destination.

9. Enter the details of your SQL Server 2008 instance and database, and then click Next.

10. The wizard prompts you to select source tables and views. By default, the wizard automatically creates a destination table called eqs7day-M1, so click Next.

11. On the Save and Run Package screen, click Finish (if you are using SQL Server 2008 Express, Web, or Workgroup Edition, this screen is called Run Package). The package summary appears, and you are prompted to verify the details.

12. Click Finish again to execute the package.

■**Note** In SQL Server 2008 Express, Web, or Workgroup Edition, you can use the Import and Export Wizard to create a package for immediate execution only. To save packages created by the wizard, you must use SQL Server Standard, Developer, or Enterprise Edition.

Table 6-3. *Column Properties for the USGS Earthquake Text File Connection*

Name	DataType	OutputColumnWidth
Src	string [DT_STR]	2
Eqid	string [DT_STR]	8
Version	string [DT_STR]	1
Datetime	string [DT_STR]	50
Lat	double-precision float [DT_R8]	
Lon	double-precision float [DT_R8]	
Magnitude	float [DT_R4]	
Depth	float [DT_R4]	
NST	two-byte signed integer [DT_I2]	
Region	string [DT_STR]	255

You will receive a message informing you that the execution was successful, and stating the number of rows transferred from the text file into the destination table. You may now close the wizard by clicking the Close button.

Let's check the contents of our new table. You can do this by opening a new query window and issuing the following command:

```
SELECT * FROM [eqs7day-M1]
```

You will see the data inserted from the text file, as shown in Figure 6-2.

	Src	Eqid	Version	Datetime	Lat	Lon	Magnitude	Depth	NST	Region
1	ci	14394700	1	Wednesday, September 24, 2008 19:21:55 UTC	36.0864	-117.8501	1.4	4.5	21	Central California
2	ci	14394696	1	Wednesday, September 24, 2008 18:57:31 UTC	34.4421	-118.0073	2.2	7.7	45	Southern California
3	ak	00069720	1	Wednesday, September 24, 2008 18:29:11 UTC	60.1676	-153.8832	2.8	100	53	Southern Alaska
4	ci	14394684	1	Wednesday, September 24, 2008 18:25:28 UTC	32.6763	-115.9196	1.7	6.2	27	Southern California
5	ci	14394680	1	Wednesday, September 24, 2008 18:23:54 UTC	33.027	-116.4236	1.3	7.4	33	Southern California
6	ci	14394676	1	Wednesday, September 24, 2008 18:23:11 UTC	33.2503	-116.2673	1.2	13.2	32	Southern California
7	ak	00069718	1	Wednesday, September 24, 2008 18:14:45 UTC	60.2384	-141.3107	1.6	0	11	Southern Alaska
8	ci	14394672	1	Wednesday, September 24, 2008 18:07:01 UTC	33.2283	-116.7318	1.3	11.4	60	Southern California

Figure 6-2. *The data inserted from the eqs7day-M1.txt file*

Adding the geography Column

The location of each earthquake is currently described in the eqs7day-M1 table using the latitude and longitude coordinate values stored in the Lat and Lon columns. In order to use any of the spatial methods provided by SQL Server, we need to use these coordinates to create a representation of each earthquake using the geography or geometry datatype instead. Since the Lat and Lon columns contain geographic coordinates describing an exact location, we will create a Point object representing each earthquake using the geography datatype. To add to the table a new column of the geography datatype called Location, execute the following T-SQL query:

```
ALTER TABLE [eqs7day-M1]
ADD Location geography
GO
```

Populating the Spatial Column

Having added a new geography column to the eqs7day-M1 table, we now need to populate it with Point geometries representing each individual earthquake. We can do this by using the Point() method of the geography datatype, supplying the values contained within the Lat and Lon columns, together with the SRID 4326 on which they are based. We will then set the value of the Location column to the result of this method by using a SQL UPDATE statement. To populate the Location column, execute the following code:

```
UPDATE [eqs7day-M1]
SET Location =
  geography::Point(Lat, Lon, 4326)
```

You receive a message stating the number of rows affected, as shown here (the number of rows affected differs depending on the number of earthquakes in the dataset you downloaded):

```
843 row(s) affected.
```

To test the contents of the Location column, you can now run the following query:

```
SELECT TOP 5
  Eqid,
  Location.STAsText() AS Epicenter
FROM
  [eqs7day-M1]
```

The results are as follows:

Eqid	Epicenter
14394700	POINT (-117.85 36.0864)
14394696	POINT (-118.007 34.4421)
00069720	POINT (-153.883 60.1676)
14394684	POINT (-115.92 32.6763)
14394680	POINT (-116.424 33.027)

Using the Point() method, we have been able to populate the Location column with Point geometries representing the latitude and longitude of each earthquake's epicenter, which lies on the surface of the earth. However, the point of origin of an earthquake (its *hypocenter*) normally lies deep within the earth, tens or hundreds of miles underground. In the eqs7day-M1 dataset, the depth of the hypocenter, in kilometers, is recorded in the Depth column. To be able to represent the position of the hypocenter of each earthquake instead, we need to define each Point in the Location column with an additional z coordinate based on the value of the Depth column. Although we cannot use the Point() method to do this, because it only accepts two coordinate values, we can use the static methods based on the WKT syntax, which *do* support z coordinates.

The following code illustrates how to update the Location column using the STPointFromText() method instead, by creating the WKT representation of a Point based on the latitude, longitude, and depth of each earthquake. Since the Depth column represents a distance *beneath* the earth's surface, the z coordinate of each Point is set based on the negative value of the Depth column.

```
UPDATE [eqs7day-M1]
SET Location =
  geography::STPointFromText(
    'POINT('
      + CAST(Lon AS varchar(255)) + ' '
      + CAST(Lat AS varchar(255)) + ' '
      + CAST (-Depth AS varchar(255)) + ')',
    4326)
```

You can now select the data contained in the eqs7day-M1 table, including the Point representation of the hypocenter of each earthquake, as follows:

```
SELECT
  Eqid,
  Location.AsTextZM() AS Hypocenter
FROM
  [eqs7day-M1]
```

The results follow:

Eqid	Hypocenter
14394700	POINT (-117.85 36.0864 -4.5)
14394696	POINT (-118.007 34.4421 -7.7)
00069720	POINT (-153.883 60.1676 -100)
14394684	POINT (-115.92 32.6763 -6.2)
14394680	POINT (-116.424 33.027 -7.4)

■**Tip** Once you have populated the Location column with Points representing the location of each earthquake, you can delete the original Lat, Lon, and Depth columns from which they were derived. If you ever need to retrieve the original coordinate values, you can do so using the Lat, Long, and Z properties (explained in more detail in Chapter 11).

Importing Data from Keyhole Markup Language

KML is an XML-based language originally developed by Keyhole, Inc., for use in its EarthViewer application. In 2004, Google acquired Keyhole, together with EarthViewer, which Google used as the foundation on which to develop its popular Google Earth platform (http://earth.google.com). Although the KML format has undergone some revisions since then (at the time of writing, the latest version is KML 2.2), it continues to be the native format for storing spatial information used in Google Earth. In 2008, KML was adopted by the Open Geospatial Consortium as a standard format for spatial information, and you can now find the latest implementation of the KML specification at the OGC web site, at the following address: http://www.opengeospatial.org/standards/kml/.

While KML has always been used within the Google Earth community to share user-created spatial data, the popularity and accessibility of the Google Earth platform among the wider Internet community means that KML is becoming increasingly used for educational and research purposes, as well as in critical applications such as emergency and disaster services. Coupled with its adoption as a standard by the OGC, KML is becoming an increasingly important format for the interchange of spatial data.

Comparing KML to GML

Like GML, a KML file may contain different types of geometric instances to describe spatial features: Points, Paths (which are equivalent to LineStrings), and Polygons. However, whereas the GML format (like WKT and WKB) is purely used to describe the shape and location of geographic features, a KML file additionally specifies how those features should be styled and presented in a graphical display.

To demonstrate the KML document format, Listing 6-1 shows the KML representation of a Path, taken from the sample code available at http://code.google.com/apis/kml/documentation/kml_tut.html.

Listing 6-1. *An Example KML Document*

```
<?xml version="1.0" encoding="UTF-8"?>
<kml xmlns="http://www.opengis.net/kml/2.2">
  <Document>
    <name>Paths</name>
    <description>Examples of paths. Note that the tessellate tag is by default
      set to 0. If you want to create tessellated lines, they must be authored
      (or edited) directly in KML.</description>
    <Style id="yellowLineGreenPoly">
      <LineStyle>
        <color>7f00ffff</color>
        <width>4</width>
      </LineStyle>
      <PolyStyle>
        <color>7f00ff00</color>
      </PolyStyle>
    </Style>
    <Placemark>
      <name>Absolute Extruded</name>
      <description>Transparent green wall with yellow outlines</description>
      <styleUrl>#yellowLineGreenPoly</styleUrl>
      <LineString>
        <extrude>1</extrude>
        <tessellate>1</tessellate>
        <altitudeMode>absolute</altitudeMode>
        <coordinates> -112.2550785337791,36.07954952145647,2357
          -112.2549277039738,36.08117083492122,2357
          -112.2552505069063,36.08260761307279,2357
          -112.2564540158376,36.08395660588506,2357
          -112.2580238976449,36.08511401044813,2357
          -112.2595218489022,36.08584355239394,2357
          -112.2608216347552,36.08612634548589,2357
          -112.262073428656,36.08626019085147,2357
          -112.2633204928495,36.08621519860091,2357
          -112.2644963846444,36.08627897945274,2357
          -112.2656969554589,36.08649599090644,2357
        </coordinates>
      </LineString>
    </Placemark>
  </Document>
</kml>
```

Notice that this KML representation contains a lot more information than is needed to describe the purely geometric properties of the LineString in question: there are also many different styling and descriptive elements. If we were to describe this same feature using the GML format, which only contains elements relating to the shape of the features, we would only require the code listing shown in Listing 6-2.

Listing 6-2. *Equivalent GML LineString Representation*

```
<LineString xmlns="http://www.opengis.net/gml">
  <posList>
    36.079549521456471 -112.2550785337791
    36.081170834921217 -112.25492770397381
    36.082607613072788 -112.25525050690629
    36.083956605885056 -112.25645401583761
    36.08511401044813 -112.2580238976449
    36.085843552393939 -112.2595218489022
    36.086126345485887 -112.2608216347552
    36.086260190851469 -112.262073428656
    36.086215198600911 -112.2633204928495
    36.086278979452743 -112.2644963846444
    36.086495990906442 -112.26569695545891
  </posList>
</LineString>
```

Transforming KML to GML

One of the advantages of the highly structured nature of XML is that specifying explicit transformations to convert from one XML dialect into another is relatively easy. By creating and applying the necessary transformation(s), it is therefore possible to convert from KML (such as shown in Listing 6-1) into the GML (shown in Listing 6-2), which can then be imported into SQL Server using the GeomFromGml() method of the geography or geometry datatype. In order to convert from KML to GML, the following transformations must occur:

1. Remove any KML elements that purely describe styling or descriptive properties, which are not relevant in the GML file. These elements include <LookAt>, <visibility>, <styleUrl>, <Style>, and <name>.

2. Retrieve the contents of those elements that do relate to geometric properties, and replace them with the equivalent GML elements, as shown in Table 6-4.

Table 6-4. *Geographic KML Elements and Their GML Equivalents*

KML	GML	Description
<GeometryCollection>	<GeometryCollection>	Denotes a Geometry Collection element
<Polygon>	<Polygon>	Denotes a Polygon geometry
<LineString>	<LineString>	Denotes a LineString geometry
<Point>	<Point>	Denotes a Point geometry

Table 6-4. *Geographic KML Elements and Their GML Equivalents*

KML	GML	Description
<outerBoundaryIs>	<exterior>	Denotes the exterior boundary of a Polygon
<innerBoundaryIs>	<interior>	Denotes the interior boundary of a Polygon
<coordinates>[a]	<pos> <posList>	The element containing the coordinate list of a geometry

[a] *In GML, there is a distinction between the <pos> element, which is used to contain a single coordinate tuple (such as used to define a single Point geometry), and the <posList> element, which contains multiple coordinate tuples (as in a LineString or Polygon). In KML, there is no such distinction, and the <coordinates> element is used in every case.*

■ **Note** All multielement geometries in KML are represented as a Geometry Collection—there are no specific homogenous element types that are equivalent to MultiPoint, MultiLineString, or MultiPolygon geometries.

3. Manipulate the contents of the <coordinates> element, which contains the coordinates of each point in a KML geometry, into the appropriate format for the equivalent GML <posList> or <pos> element, as follows:

   a. Replace the comma separator used between each value of a coordinate tuple with a space.

   b. If the coordinates within the KML <coordinates> element are stated with an altitude (z) coordinate, then disregard this value.

   c. Reverse the coordinate order of the two remaining coordinates to state them in latitude–longitude order.

■ **Note** Early versions of the GML standard contain a <coordinates> element very similar to that used in KML. However, this was deprecated in GML version 3.1.0 and is not supported by SQL Server 2008. You must use the <posList> or <pos> element instead.

There are several methods that you could use to make the changes necessary to transform from the KML document shown in Listing 6-1 into the equivalent GML representation shown in Listing 6-2. For example, you could use XQuery, or you could apply an Extensible Stylesheet Language Transformation (XSLT). XQuery and XSLT are both approved standards administered by the World Wide Web Consortium (W3C). For more information on how they can be used to transform and query XML data, refer to the W3C web pages at http://www.w3.org/TR/xslt.html and http://www.w3.org/TR/xquery/, respectively. However, note that any method that performs a simple conversion between the two XML formats may suffer from the following limitations:

- You will lose any additional descriptive and styling elements contained in the original KML file, which may contain useful additional information about each geometry instance that cannot be represented in GML.

- There is no validation or error checking performed on the source document to check whether it would create a valid geometry. For instance, remember that in order to create a geography Polygon instance from the converted GML representation, the points of a ring containing an area of space must be listed in counterclockwise order, which might not be as they were listed in the original KML <coordinates> element.

If you want to use a more robust method to import KML data into SQL Server 2008, you might want to investigate the range of third-party tools available designed specifically for this purpose, some of which are listed at the end of this chapter.

Note KML and GML are not the only XML-based spatial data formats. For example, GPS Exchange Format (GPX) is an XML format used to store and share data between many different types of handheld GPS devices.

Importing Data from ESRI Shapefile Format

The shapefile format was designed and is maintained by Environmental Systems Research Institute, Inc. (ESRI). Originally developed for use in its ARC/INFO suite of GIS software, the shapefile is now a very common format used for exchanging spatial information between all kinds of systems, and is the format in which most commercial spatial data is supplied. Over time, a large body of spatial datasets has been created in ESRI shapefile format.

Although a set of data provided in shapefile format is commonly referred to as "*a* shapefile" (singular), this is a slight misnomer, since a single shapefile actually consists of several files. Each file relating to a given shapefile dataset shares the same file name, with one of the following file extensions:

.shp: The SHP file contains the raw geometrical shape data. Each SHP file can contain items of only one kind of geometry shape: Points, LineStrings, or Polygons.

.shx: The SHX file maintains the shapefile index, which holds one index entry for every shape in the shapefile document. Each index entry describes the start position and length of the associated shape record in the SHP file.

.dbf: The DBF file contains additional, nonspatial attributes of each shape. For instance, in a shapefile containing Polygons representing the states of America, the DBF file might contain the name of each state, its population, or the name of its state capital.

.prj: The PRJ file gives details about the projection in which the coordinates of the geometry data are represented, in the same format as used in the well_known_text column of the sys.spatial_reference_systems table. When importing a shapefile into SQL Server, this file contains the information that is required to determine the correct spatial reference identifier (SRID).

▨**Note** Any valid document in shapefile format must contain a SHP file and an associated SHX file. Files with the .dbf and .prj extensions are optional files that contain additional information about the data.

Obtaining Sample Shapefile Data

To demonstrate how to import data from the shapefile format into SQL Server, we'll use data from the US Census Bureau representing the Zip Code Tabulation Areas (ZCTAs) of the state of California. ZCTAs were defined by the US Census Bureau during the US 2000 census, and are approximately equivalent to the delivery area for a five-digit ZIP code as used by the US Postal Service. You can download the ESRI shapefile of the ZCTA areas in the state of California directly from the US Census web site at the following URL: http://www.census.gov/geo/cob/bdy/zt/z500shp/zt06_d00_shp.zip.

Download this ZIP file and extract its contents. You will find that it contains the following files:

- zt06_d00.shp

- zt06_d00.shx

- zt06_d00.dbf

The SHP file, which is the largest of these files (4,560KB), contains the raw data that defines the Polygon shapes representing each ZCTA. The SHX file is the index file that records the start position and length of each shape in the shapefile. In addition to these two files, which contain the purely spatial information representing each ZCTA, the DBF file includes additional associated columns of data, as listed and described in Table 6-5.

Table 6-5. *Columns of Data in the zt06_d00.dbf File*

Column	Description
Area	The internal area of each ZCTA, in square kilometers.
Perimeter	The length of the perimeter of each ZCTA, in kilometers.
ZT06 D00	An automatically generated sequential feature number.
ZT06 D00 ID	A user-defined feature number.
ZCTA	The ZCTA reference five-digit number.
NAME	Same as ZCTA.
LSAD	The Legal/Statistical Area Description (LSAD) code. This is a two-character field that corresponds to a legal or statistical type of entity. For a five-digit ZCTA code, this is always Z5.
LSAD_TRANS	The description associated with the LSAD of each shape. For ZCTAs, this is '5-Digit ZCTA'.

Tip For more information on any of the data contained in zt06_d00.dbf, consult http://www.vcgi.org/ metadata/BoundaryOther_ZCTA2000.txt.

Notice that there is no included PRJ file relating to the ZCTA shapefile. How then do we know what spatial reference system has been used to define the coordinates? A further search of the US Census Bureau web site reveals a page of metadata, located at http://www.census.gov/geo/ www/cob/zt_metadata.html. This page states that the ZCTA data is defined using a geographic coordinate system based on the NAD 83 datum. We know that the WKT representation of any spatial reference system must begin with a keyword representing the type of coordinates used—for geographic coordinates as used in the ZCTA data, this is GEOGCS. We also know that the WKT representation of a spatial reference system must state the name of the datum on which it is based—NAD83. With this information, we can search for the correct identifier for this spatial reference system in the sys.spatial_reference_systems table using the following query:

```
SELECT
  spatial_reference_id
FROM
  sys.spatial_reference_systems
WHERE
  well_known_text LIKE 'GEOGCS%"NAD83"%'
```

The single result returned is as follows:

---

spatial_reference_id
4269

So, when importing the data contained in this shapefile, we should use SRID 4269. Since this is a spatial reference system based on geographic coordinates, we will choose the geography datatype to store the spatial data. With this information, we are now ready to import the data from the shapefile into SQL Server.

Importing Shapefile Data with Shape2SQL

Shape2SQL is a popular, simple application specifically designed to load shapefile data into SQL Server 2008. You can download it as part of the SQL Spatial Tools package, freely available from http://www.sharpgis.net/page/SQL-Server-2008-Spatial-Tools.aspx.

Once you have downloaded and unzipped the SqlSpatialTools.zip archive, load the application by double-clicking the Shape2Sql.exe file. When you run the Shape2SQL application for the first time, you are prompted to enter details of the database connection, as shown in Figure 6-3.

In the Database Configuration dialog box, enter the name of the SQL Server 2008 instance that you want to import shapefile data to, and provide any authentication details required to connect to that server. Then select the appropriate database from the drop-down list, and click OK.

Once the database configuration is complete, the main window of the Shape2SQL application appears. Figure 6-4 illustrates the Shape2SQL application, showing the settings required to import the California ZCTA data that you saved earlier.

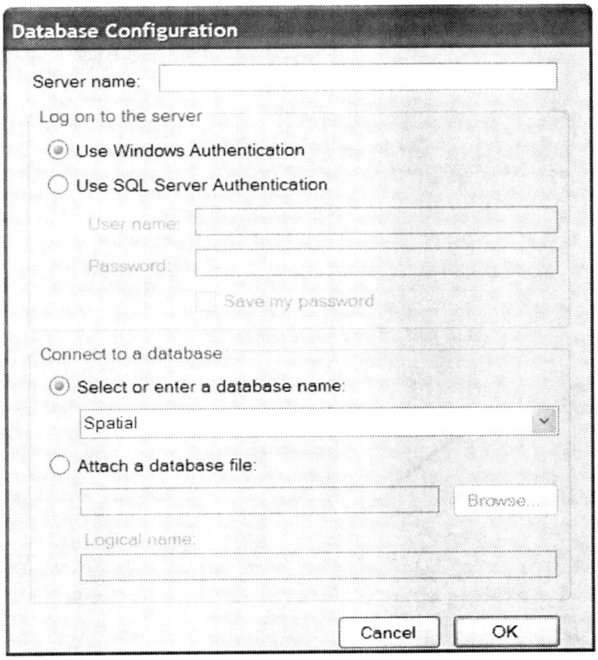

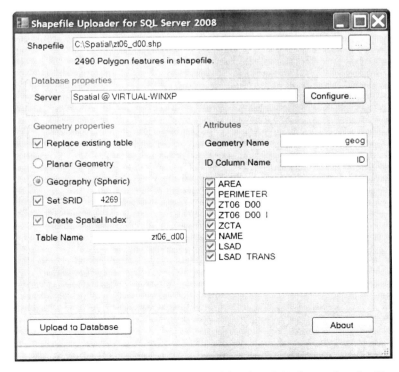

Figure 6-3. *Configuring the database connection for the Shape2SQL application*

Figure 6-4. *Setting options to import California ZCTA data using the Shape2SQL application*

To set the appropriate options to import the California ZCTA data, follow these steps:

1. Click the ... button in the top right corner of the screen to select a shapefile. Browse to the location where you saved the zt06_d00.shp file, highlight it, and click Open. The application will display the number and type of features found in the shapefile. The zt06_d00.shp file used in this case contains 2490 Polygon features.

▪**Caution**   In order to import a shapefile using Shape2SQL, the SHP file and associated SHX file must be saved in the same location. If an associated DBF file is also saved in the same location, you will be able to import additional, nonspatial attributes associated with each shape from the DBF file into columns of the SQL Server table.

---

2. Check that the Server field in the Database Properties section specifies the correct server and database into which to insert the data. If you need to change any of the details, click the Configure button to open the Database Configuration dialog box.

3. In the Geometry Properties section at the bottom left of the application window, use the following options to specify how you want the data to be imported into SQL Server:

   • *Replace existing table*: When this option is checked, if a table already exists with the specified name, it is dropped and replaced with a new table containing the imported shape data. If unchecked, data instead is appended to an existing table. If no table exists with the name specified, a new table is created regardless of whether this option is checked or not.

   • *Planar Geometry/Geography (Spheric)*: Choosing one of these options determines the datatype of the column into which imported data will be inserted. Selecting the Planar Geometry option leads to the geometry datatype being used, whereas choosing Geography (Spheric) leads to the geography datatype being used. The California ZCTA data is defined using geographic coordinates, so choose Geography (Spheric).

   • *Set SRID*: You must enter the integer value that identifies the spatial reference system in which the coordinates of the shape have been defined. If your shapefile came with an associated file with a .prj extension, you should choose the SRID corresponding to the details contained in that file. If you do not have a PRJ file, you must instead find out the correct spatial reference system by examining other metadata related to the file, or by contacting the supplier of the data. Note that this option specifies the spatial reference in which the coordinates have *already* been defined—it does not specify a spatial reference system into which you want to reproject the data. For the California ZCTA data, enter the SRID **4269**.

▪**Caution**   Even if the shapefile contains a file with the .prj extension, you must manually specify the corresponding SRID to match the details contained in the PRJ file. Shape2SQL does not automatically set the correct SRID.

- *Create Spatial Index*: When this option is checked, Shape2SQL automatically creates a spatial index on the table into which the data is inserted. Spatial indexes improve the speed of certain types of spatial query, and are discussed in more detail in Chapter 14.

- *Table Name*: This value determines the name of the table into which the data will be imported. If the table already exists and the Replace Existing Table option is not checked, then the schema of this table must match the columns of data imported from the shapefile. The default table name is the same as the file name of the source shapefile, in this case **zt06_d00**.

4. In the Attributes section at the bottom right of the window, use the following options to identify which columns of data should be imported:

- *Geometry Name*: This field determines the name of the geography or geometry column that will hold the imported spatial data. The default column name is geom. Since we are importing data based on geographic coordinates, change this to **geog**.

- *ID Column Name*: If you want to include a unique identifier for each row of data in the table, then you should enter a name for that column here. This value is used to create an IDENTITY(1,1) integer column in the created table, and to create a primary key based on that column. If you do not want to include an identity column, leave this value blank.

- *Optional attributes box*: The final box in the Attributes section allows you to select optional additional columns of data to import relating to each feature in the shapefile. Click the check box next to each attribute that you want to import—each attribute will correspond to its own column in the created SQL Server table.

▓Note Additional columns of data are based on the data contained in the associated DBF file. If no DBF file is present, you will not be able to import any additional information relating to each shape.

---

5. When you are satisfied with the options that you've chosen, click the Upload to Database button.

Once the Shape2SQL application has finished importing, you can test the data by running the following query in SQL Server Management Studio:

```
SELECT TOP 5
  ZCTA,
  geog.STAsText() AS WKT
FROM
  zt06_d00
```

The results are as follows:

```
ZCTA   WKT
96044  POLYGON ((-122.681733 41.914761, -122.682284 41…
960XX  POLYGON ((-122.681733 41.914761, -122.677967 41…
96023  POLYGON ((-122.161862 41.881475, -122.156022 41…
96064  POLYGON ((-122.161862 41.881475, -122.156785 41…
96086  POLYGON ((-123.136812 41.866187, -123.1422 41…
```

Using Third-Party Conversion Tools

The techniques discussed in this chapter have demonstrated a variety of simple methods that allow you to import spatial data in various formats into SQL Server 2008. However, if you are planning to import a large amount of existing spatial data, you may find it more suitable to use a dedicated application designed for this purpose. There are a number of tools available to facilitate the conversion of spatial data into one of the formats that SQL Server 2008 can use. Some of these tools are freely available, whereas others are commercial tools developed by Microsoft's spatial partners. This section introduces some tools that you might find useful.

Commercial Tools

Many of the commercial tools that support SQL Server 2008 provide full extract, transform, load (ETL) processes for spatial data, enabling you to convert between many different types of spatial datatypes. In addition to simply changing the format of the file itself, these tools may also provide reprojection of the data contained within the file, as well as methods to visualize and analyze data. Some commercial tools are as follows:

- ArcGIS (version 9.3 and above), the long-established industry-standard GIS platform from ESRI (http://www.esri.com), integrates with SQL Server 2008 to provide a wealth of tools supporting all aspects of spatial data manipulation, analysis, and presentation.

- Manifold (http://www.manifold.net) also uses a direct connection to SQL Server 2008 to provide conversion and importing between hundreds of spatial data formats, reprojection, and visual editing of spatial data.

- Feature Manipulation Engine (FME) from Safe Software (http://www.safe.com) is designed to facilitate the conversion of spatial data between a huge range of supported formats, although it also includes basic visualization tools. In addition to providing a stand-alone desktop application, FME is able to extend SQL Server Integration Services (SSIS) so that you can perform complex spatial ETL as part of your existing SSIS projects.

Free Tools

If you only require a simple conversion between two formats, then you might want to investigate one of the following sources for free tools instead:

- FWTools (http://fwtools.maptools.org) is a package that contains a variety of useful open source GIS components, including the OGR library and command-line tools for converting between spatial data formats.

- Spatial Order (http://spatialorder.com/downloads.htm) is a consultancy firm that provides free of charge a number of basic command-line conversion tools, including tools to convert from ESRI shapefile format to GML (shp2gml) and from ESRI shapefile format to WKT (shp2wkt).

- Zonum Solutions (http://www.zonums.com) provides utilities for converting between ESRI shapefile format and Google Earth KML files, including the ability to convert between a range of coordinate systems and datums.

Summary

In this chapter, you learned about a variety of data formats in which existing spatial data may be provided, and how you can import that data into SQL Server 2008. Specifically, this chapter covered the following:

- There are many alternative file formats in which spatial information is commonly stored and shared, including tabular geographic information, the ESRI shapefile format, and the KML file format used by Google Earth.

- There are a number of sources from which you can obtain freely available spatial data over the Internet. The data obtained from these sources ranges in quality and in coverage. If you are downloading spatial data for use in a critical application, be sure to check the accuracy of that data first!

- Simple spatial information provided in tabular format can be imported using the Import and Export Wizard. The T-SQL UPDATE statement can then be used, in conjunction with the geography datatype Point() static method, to populate a geography or geometry column from the coordinate values of each row of data in the table.

- To construct an item of spatial data from tabular information containing a z coordinate, you can manually construct the WKT representation of a geometry based on each coordinate value, and then pass that representation to the relevant WKT static method, such as STPointFromText().

- KML is an XML-based spatial data format, like GML. There are several similarities between the two formats, but there are also important differences. In order to convert from KML into GML that can be used in SQL Server, you must transform each KML element into the equivalent GML element.

- The Shape2SQL tool is a free tool designed to import data from ESRI shapefile format into SQL Server 2008.

- There are a number of other tools capable of converting and importing spatial data that are compatible with SQL Server.

CHAPTER 7

■■■

Geocoding

Over the last few chapters, I have introduced a number of different techniques that you can use to add spatial data into SQL Server 2008: use a `geometry` or `geography` static method directly with a WKT, WKB, or GML representation of a geometry; visually mark out a spatial feature by plotting points on a Virtual Earth web control; or import existing spatial data from a variety of other spatial data formats. In this chapter, we will consider one last technique that can help us to define items of spatial information in SQL Server: geocoding.

> **Caution** The geocoding method discussed in this chapter requires the creation of a .NET assembly that extends the functionality of the SQLCLR to interact with an external web service. In order to follow the example code used to create this assembly, you must use Microsoft Visual Studio. There are many different editions of Visual Studio available—the code in this chapter works with both Microsoft Visual Basic 2008 Express Edition and Microsoft Visual C# 2008 Express Edition, both of which you can download for free from `http://www.microsoft.com/express/download/`.

What Is Geocoding?

Even though the spatial datatypes, `geometry` and `geography`, are a new feature introduced in SQL Server 2008, almost every existing SQL Server database already contains some form of spatial information—that is, information that allows us to describe the location of an item of data. This spatial information might be addresses of customers or suppliers, postal codes, delivery routes, or the names of cities or regions for which a sales manager is responsible. Wouldn't it be useful if you could conduct spatial analysis based on this existing information? That is exactly what geocoding enables you to do.

The practice of geocoding involves taking textual information describing the location of an item or a place, such as a street address, the name of a landmark, or a postal code, and deriving a structured spatial representation associated with that information so that it can be used in spatial applications. Most geocoding methods return the coordinates of a single Point geometry representing a precise location relating to each item of data. For example, the address of the White House, the official residence of the President of the United States, is 1600 Pennsylvania Avenue NW, Washington DC, 20500. If you were to geocode this address, you might obtain the coordinates of 38.8980 degrees latitude, -77.0365 degrees longitude, corresponding to a Point located at this address. The geocoding process is illustrated in Figure 7-1.

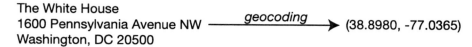

The White House
1600 Pennsylvania Avenue NW ——— *geocoding* ——→ (38.8980, -77.0365)
Washington, DC 20500

Figure 7-1. *The geocoding process*

    There are a number of different ways to provide geocoding functionality: some **geocoding** tools are desktop-based applications, whereas others are services that you access over the **Web**. In this chapter, I will demonstrate how you can integrate the Find service from Microsoft's MapPoint Web Service to add geocoding functionality directly into SQL Server 2008.

Tip If you want to geocode only a single address, you can do this online using Microsoft Live Search Maps at http://maps.live.com. Use the search field at the top of the page to search for an address, and when the map is centered on the location in question, click Share (located above the top right corner of the map). In the URL link that is generated, look for the two numbers immediately following &cp=. These are the latitude and longitude coordinates of the address on which the map is centered (using the EPSG:4326 spatial reference system).

MapPoint Web Service

According to the Microsoft web site, the MapPoint Web Service is

> *. . . an XML Web service with a SOAP API hosted by Microsoft and used by enterprises and independent software developers to integrate location-based services into Web applications.*

—http://www.microsoft.com/virtualearth/platform/mappoint.aspx

---

Note The MapPoint Web Service is not the same as Microsoft MapPoint, which is a desktop application for Microsoft Windows, primarily used for journey planning.

    Like the Virtual Earth web service introduced in Chapter 5, the MapPoint Web Service provides a range of services relating to geospatial content, including the capability to find nearby points of interest and to find routes between locations. Whereas the Virtual Earth API allows developers to issue requests via JavaScript to control a map object on a web page, the MapPoint Web Service is accessed over the Internet by Simple Object Access Protocol (SOAP) methods. The particular MapPoint service we will use to perform geocoding in this chapter is the *Find* service, which provides the following set of methods:

- Find()

- FindAddress()

- FindByID()

- `FindByProperty()`

- `FindNearby()`

- `FindPolygon()`

- `FindNearRoute()`

- `GetLocationInfo()`

- `ParseAddress()`

These methods all offer various functionality related to geocoding features on the earth. The most common application of geocoding is to obtain the coordinates of a point from a street address, such as that of a customer or supplier. This is what the `FindAddress()` method of the Find service does, and it is the method that we will use in this chapter.

■ **Tip** You can find full details for the `FindServiceSoap` class in the MSDN Library at `http://msdn.microsoft.com/en-us/library/aa502416.aspx`.

Accessing the MapPoint Web Service from SQL Server

To add geocoding functionality to SQL Server, we will once again be taking advantage of the methods offered by the .NET Base Class Library (BCL), by creating a .NET method that connects to the MapPoint Web Service over the Internet. Since SQL Server 2008 hosts the .NET Framework CLR, we will be able to interact with the MapPoint Web Service using T-SQL from SQL Server, supplying the details of an address and receiving the geocoded result from the CLR directly in SQL Server. This is another example of how .NET can be used in SQL Server to provide functionality not possible using T-SQL alone.

In this chapter, I'll take you through the following steps to add geocoding functionality to your SQL Server 2008 database:

1. Create a .NET class that connects to the MapPoint Find service and sends a set of parameters describing a postal address: the house name/number, street address, town, region, postal code, and country. The class will receive the response from the service and turn it into the WKT syntax for a Point representing the given address.

2. Build the .NET class into a compiled assembly file (`.dll` extension) that can be imported into SQL Server 2008.

3. Define a new T-SQL function in SQL Server that can be used to call the .NET method defined in the assembly.

4. Create a new geography Point instance from an address, by supplying the WKT representation of the geocoded result to the `STPointFromText()` method.

Before we can begin, you need to sign up for an account to use the MapPoint Web Service. This is explained in the following section.

Signing Up for the MapPoint Web Service

The MapPoint Web Service provides two different environments:

> The *staging* environment provides a free-to-use service that can be used when developing applications for trial, demonstration, or proof-of-concept purposes.

> The *production* environment is the commercial environment that must be used when you deploy any live applications using the MapPoint services.

Before you can use either environment, you must first register for a Virtual Earth Platform developer account (the MapPoint Web Service and Virtual Earth are sister services, and are accessed using the same account details). In the example code in this chapter, I will use the staging environment, for which you can use a free developer account. If you are registered to use the production environment, you may substitute the appropriate references and credentials as required. To request an evaluation developer account, follow the instructions at the following web page: https://mappoint-css.live.com/MwsSignUp/Default.aspx.

Once you have completed the necessary registration form, you will receive an e-mail containing your developer account ID and instructions on how to set up your password. Once you have completed the registration process, be sure to keep your account ID and password in a safe yet handy place, as you will need to supply them later to allow SQL Server to access the MapPoint Web Service in order to perform the geocoding function.

▨**Caution** The account you use to access the MapPoint Web Service is linked to your Windows Live ID, but it does not use the same username and password.

Creating the .NET Assembly

To perform geocoding from within SQL Server, we will build a custom .NET assembly that will be executed by the SQLCLR. This is another example of the benefits of using the .NET CLR within SQL Server—it allows you to extend the functionality of SQL Server using any of the methods contained in the BCL. The easiest way to create and compile a .NET assembly is to use an integrated development environment (IDE), such as Microsoft Visual Studio. There are several editions of Visual Studio available—the code listed in this chapter will work with Visual Studio 2005 or Visual Studio 2008, including the freely available Visual Basic and Visual C# Express Editions that you can download from http://www.microsoft.com/express/download/.

Remember that .NET applications are language neutral. That is to say that .NET code may be written in any one of a number of supported Common Language Specification (CLS) languages, including C# and VB .NET. Code written in any CLS language may access the BCL and may be executed using the common language runtime (CLR) engine. I will give you the required code for both C# and Visual Basic, so you may choose the language that you are most comfortable with.

Note The following steps describe the creation of a .NET assembly using Visual Basic 2008 Express Edition. If you create the assembly using a different version of Visual Studio, you may find that some of the menu items appear under different headings.

Creating a New Project

Your first task is to create a new class library project, by following these steps:

1. From the Visual Studio menu bar, select File ➤ New Project (or press Ctrl+N).

2. In the New Project dialog box, shown in Figure 7-2, select the Class Library template and type a name for the new project. For this example, I named the project Geocoder.

3. Click OK.

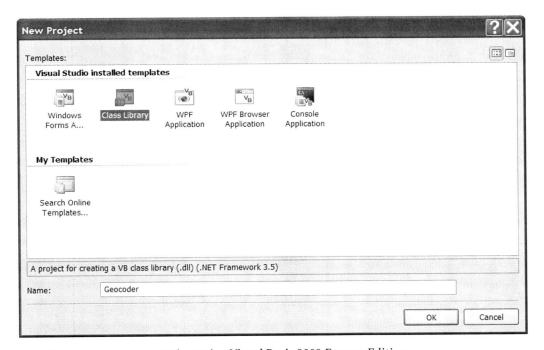

Figure 7-2. *Creating a new project using Visual Basic 2008 Express Edition*

Tip If you are using Visual Studio Professional Edition or Team System, you can use the SQL Server Project template rather than the generic Class Library template, which allows you to automatically deploy the assembly to SQL Server from Visual Studio, and provides additional debugging facilities.

The status bar at the bottom of the application window will display the following message:

```
Creating project 'Geocoder'… project creation successful.
```

Once the project has been created, the project workspace will appear, and the main window will show the contents of the Class1.vb file within the project (or Class1.cs, if using Visual C#).

Configuring the Project

Before we add the geocoding function, we need to make a number of changes to configure the project. To make these changes, first open the project properties page by selecting Project ➤ Geocoder Properties from the main menu bar.

Setting the Output Directory

You need to specify the output directory where Visual Studio will generate the compiled assembly (DLL) file. It is important to choose a memorable and easily accessible location, since you will need to manually specify the file path later when importing the compiled assembly into SQL Server. To set a location where the compiled assembly will be placed:

1. Select Compile from the list of tabs on the left side of the project properties page, as shown in Figure 7-3 (if using Visual C# 2008 Express Edition, this tab is called Build instead).

2. In the Build Output Path field, enter the location in which you want the compiled assembly to be created. In this example, I set the path to C:\Spatial\ (in Visual C# 2008 Express Edition, this field is simply called Output Path).

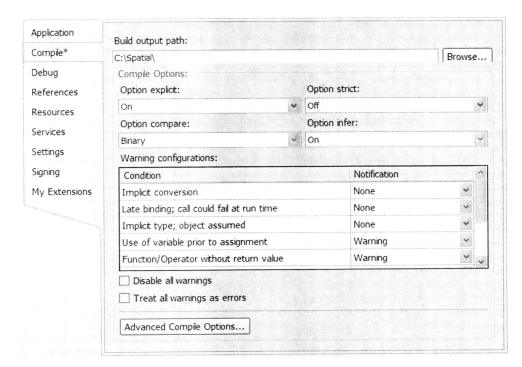

Figure 7-3. *Setting the compile options for the Geocoder project using Visual Basic 2008 Express Edition*

Generating Serialization Assemblies

Our geocoding function relies on sending and receiving data to the MapPoint Web Service over an Internet connection. Whenever a .NET assembly accesses resources over the Web in this way, all the data must first be *serialized*—converted into a series of bytes for transmission across a network. Visual Studio can generate an additional assembly containing the required serialization methods for our code, so to ensure that this additional assembly is created, perform the following steps:

1. Ensure that you are on the Compile (VB .NET) or Build (C#) tab of the project properties page.

2. Click the Advanced Compile Options button at the bottom of the page. The Advanced Compiler Settings dialog box will appear. If using Visual C#, simply scroll down to the bottom of the Build tab instead.

3. Choose On in the Generate Serialization Assemblies drop-down list.

4. Click OK to close the Advanced Compiler Settings window (Visual Basic 2008 Express Edition only).

Adding the Web Reference

Before we can use the Find service methods within our geocoding function, we need some way of providing the .NET project with details of the MapPoint Web Service and the methods it provides. This is done by adding a *web reference*. A web reference gives the URL of a Web Services Description Language (WDSL) file—a type of XML file that describes a web service and the methods it provides. We can add a web reference to the MapPoint Web Service by following these steps:

1. From the main menu bar, choose Project ➤ Add Service Reference.

2. In the Add Service Reference dialog box, click the Advanced button in the bottom left corner.

3. In the Service Reference Settings dialog box, click the Add Web Reference button at the bottom of the window to open the Add Web Reference dialog box, a completed example of which is shown in Figure 7-4 for your reference as you work through these steps.

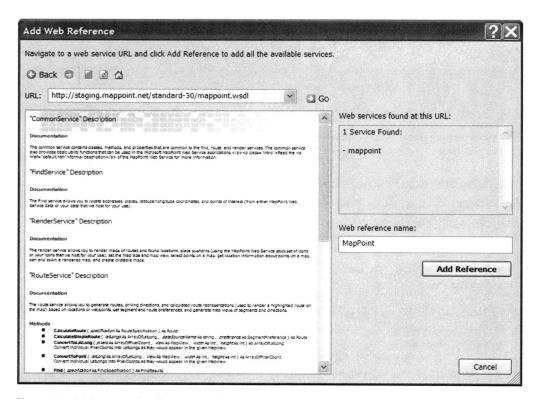

Figure 7-4. *Adding a web reference to the MapPoint Web Service*

4. In the URL box, enter the URL of the MapPoint Web Service staging environment, which is `http://staging.mappoint.net/standard-30/mappoint.wsdl`. If you want to use the production environment (and have a license to do so), you should specify the following URL instead: `http://service.mappoint.net/standard-30/mappoint.wsdl`.

5. Click the Go button to the right of the URL box. After a short while, you should receive a message stating that one service, *mappoint*, was found at this URL, and the main window will present documentation about the service, including the following description of the Find service:

   The Find service allows you to locate addresses, places, latitude/longitude coordinates, and points of interest (from either MapPoint Web Service data or your data that we host for your use).

6. The Web Reference Name field gives the name that will be used to refer to the service in our project. It currently states net.mappoint.staging, which is not very convenient, so change it to something simpler instead, such as *MapPoint*.

7. Click the Add Reference button.

The newly created MapPoint reference will appear in the Solution Explorer pane, contained within the Web References folder of the Geocoder project. That's it for the project configuration, so now we can get on with writing the code for our geocoding function.

Note The Solution Explorer pane is normally visible on the right side of the workspace. If you can't see it, choose View ➤ Solution Explorer or press Ctrl+Alt+L.

Adding the Geocoding Function

Now that the project is set up, it's time to add the user-defined function that will perform the geocoding itself. We will add the necessary code into the Class1.vb/Class1.cs file that was added when the project was first created, so load this file into the main Visual Studio window by double-clicking the relevant file name in the Solution Explorer pane. If you are writing your code using VB .NET (using either Visual Studio 2005/2008 or Visual Basic 2008 Express Edition), then edit the contents of the Class1.vb file to read as shown in Listing 7-1. If you are developing your code using C# instead (using Visual Studio 2005/2008 or Visual C# 2008 Express Edition), then edit the Class1.cs file to read as shown in Listing 7-2.

In either case, you need to insert your account ID and password in place of `ENTER YOUR ACCOUNT ID HERE` and `ENTER YOUR PASSWORD HERE`. (The username is the numeric Virtual Earth Platform developer account ID that you were sent after registering for the MapPoint Web Service, and the password is the password you chose when confirming the account.) You may also need to change the value of the `findAddressSpec.DataSourceName` property, depending on the country that contains the addresses that you want to geocode (explained in more detail following the code).

Listing 7-1. *The Class1.vb File (for VB .NET Project Types)*

```vbnet
'Import references for generic .NET functionality
Imports System
Imports System.Data
Imports System.Data.SqlClient
Imports System.Data.SqlTypes
Imports Microsoft.SqlServer.Server
Imports System.Net

'Import the web reference to the MapPoint Web Service
Imports Geocoder.MapPoint

Partial Public Class UserDefinedFunctions
  <Microsoft.SqlServer.Server.SqlFunction()> _
  Public Shared Function geocode( _
    ByVal AddressLine As SqlString, _
    ByVal PrimaryCity As SqlString, _
    ByVal Subdivision As SqlString, _
    ByVal PostalCode As SqlString, _
    ByVal CountryRegion As SqlString) As SqlString

    'Initialize the MapPoint Find service
    Dim findService As New FindServiceSoap

    'Provide the logon credentials
    Dim UserName As String = "ENTER YOUR ACCOUNT ID HERE"
    Dim Password As String = "ENTER YOUR PASSWORD HERE"
    findService.Credentials = New Net.NetworkCredential(Username, Password)

    'FindAddressSpecification contains the details passed to the Find service
    Dim findAddressSpec As New FindAddressSpecification

    'Build a new address object from the parameters provided to the function
    Dim address As New Address
    address.AddressLine = AddressLine.ToString
    address.PrimaryCity = PrimaryCity.ToString
    address.Subdivision = Subdivision.ToString
    address.PostalCode = PostalCode.ToString
    address.CountryRegion = CountryRegion.ToString
    findAddressSpec.InputAddress = address

    'Specify the data source in which to search for the address
    findAddressSpec.DataSourceName = "MapPoint.NA"

    'Create the options to limit the result set
    Dim findOptions As New FindOptions
```

```
    'Filter the results to only show LatLong information
    findOptions.ResultMask = FindResultMask.LatLongFlag

    'Only return the first matching result
    Dim findRange As New FindRange
    findRange.StartIndex = 0
    findRange.Count = 1
    findOptions.Range = findRange

    'Apply the options to the specification
    findAddressSpec.Options = findOptions

    'Call the MapPoint Web Service and retrieve the results
    Dim findResults As FindResults
    findResults = findService.FindAddress(findAddressSpec)

    'Create the WKT representation of the geocoded result
    Dim WKT As New SqlString
    If findResults.Results.Length > 0 Then
      WKT = "POINT(" & _
        findResults.Results(0).FoundLocation.LatLong.Longitude & " " & _
        findResults.Results(0).FoundLocation.LatLong.Latitude & ")"
    Else
      WKT = "POINT EMPTY"
    End If

    'Return the result to SQL Server
    Return WKT

  End Function
End Class
```

Listing 7-2. *The Class1.cs File (for C# Project Types)*

```csharp
// Import references for generic .NET functionality
using System;
using System.Data;
using System.Data.SqlClient;
using System.Data.SqlTypes;
using Microsoft.SqlServer.Server;
using System.Net;

// Import the web reference to the MapPoint Web Service
using Geocoder.MapPoint;

namespace Geocoder
{
```

```
public partial class UserDefinedFunctions
{
  [Microsoft.SqlServer.Server.SqlFunction()]
  public static SqlString geocode(
    SqlString AddressLine,
    SqlString PrimaryCity,
    SqlString Subdivision,
    SqlString PostalCode,
    SqlString CountryRegion)
  {

    // Initialize the MapPoint Find service
    FindServiceSoap findService = new FindServiceSoap();

    // Provide the logon credentials
    string UserName = "ENTER YOUR ACCOUNT ID HERE";
    string Password = "ENTER YOUR PASSWORD HERE";
    findService.Credentials = new NetworkCredential(UserName, Password);

    // FindAddressSpecification contains the details passed to the Find service
    FindAddressSpecification findAddressSpec = new FindAddressSpecification();

    // Build a new address object from the parameters provided to the function
    Address address = new Address();
    address.AddressLine = AddressLine.ToString();
    address.PrimaryCity = PrimaryCity.ToString();
    address.Subdivision = Subdivision.ToString();
    address.PostalCode = PostalCode.ToString();
    address.CountryRegion = CountryRegion.ToString();

    // Add the address to search for to the specification
    findAddressSpec.InputAddress = address;

    // Specify the data source in which to search for the address
    findAddressSpec.DataSourceName = "MapPoint.NA";

    // Create the options to limit the result set
    FindOptions findOptions = new FindOptions();

    // Filter the results to only show LatLong information
    findOptions.ResultMask = FindResultMask.LatLongFlag;

    // Only return the first matching result
    FindRange findRange = new FindRange();
    findRange.StartIndex = 0;
    findRange.Count = 1;
    findOptions.Range = findRange;
```

```
// Apply the options to the specification
findAddressSpec.Options = findOptions;

// Call the MapPoint Web Service and retrieve the results
FindResults findResults;
findResults = findService.FindAddress(findAddressSpec);

// Create the WKT representation of the geocoded result
SqlString WKT = new SqlString();
if (findResults.Results.Length > 0)
{
  WKT = "POINT(" +
    findResults.Results[0].FoundLocation.LatLong.Longitude + " " +
    findResults.Results[0].FoundLocation.LatLong.Latitude + ")";
}
else
{
  WKT = "POINT EMPTY";
}

// Return the result to SQL Server
return WKT;

  }
}
}
```

Whether you create the code based on the Visual Basic or C# example, both sets of code perform the same functionality:

- Create a new instance of the FindServiceSoap class.

- Set the credentials to use the connection, using the findService.Credentials property. As previously noted, you must change the values of UserName (ENTER YOUR ACCOUNT ID HERE) and Password (ENTER YOUR PASSWORD HERE) to reflect the account information you obtained when you signed up for the MapPoint Web Service.

- Set the InputAddress (the address to be geocoded) of the class based on string parameters supplied to the method.

- Select the data source in which to search for the address, using the findAddressSpec.DataSourceName property. The information used by the MapPoint Web Service is split across several data sources, which hold information about different countries. You must specify the name of the data source containing the data of the country in which the address you want to geocode is located. Table 7-1 lists the data sources available within the MapPoint Web Service, and the countries whose address details they contain. The example code in Listings 7-1 and 7-2 specifies the MapPoint.NA data source, which can be used in the United States, Canada, and Puerto Rico. If you wanted to geocode addresses from the United Kingdom instead, for instance, you should change the code to read findAddressSpec.DataSourceName = "MapPoint.EU".

Table 7-1. *MapPoint Web Service Data Sources that Support Address-Geocoding*

DataSourceName	Description
MapPoint.AP	The Asia-Pacific data source supports address-level geocoding in Australia, New Zealand, Hong Kong, Singapore, and Taiwan.
MapPoint.BR	The Brazilian data source can be used to find addresses in Brazil.
MapPoint.EU	The European data source supports address-level geocoding in Austria, Belgium, Denmark, Finland (Helsinki only), France, Germany, Italy, Luxembourg, The Netherlands, Norway (Oslo only), Portugal, Spain, Sweden, Switzerland, Greece (Athens only), and the United Kingdom.
MapPoint.NA	The North America data source supports address-level geocoding in the United States, Canada, and Puerto Rico.

- Specify options to limit the result set returned by the Find service. These reduce the size of the response, thereby conserving bandwidth and increasing the efficiency of the function. First, a filter is applied to only return the latitude and longitude information in the results, not any additional associated information provided by the Find service. Second, a range option is specified to only return a single result. By default, the SOAP response from the Find service contains an array of all similar matching addresses to that supplied, but for geocoding purposes, you only want to return the top, exact match.

- Call the FindAddress() method with the full specification of the item to be geocoded, and retrieve the response.

- Construct the WKT representation of the Point geometry represented by the geocoded response, and return that result to SQL Server. If no matching address can be found, then an empty Point is returned.

For more information on any of these methods, please consult the MapPoint Web Service SDK at http://msdn.microsoft.com/en-us/library/bb507684.aspx.

Compiling the Assembly

You can now compile the assembly by selecting Build ➤ Build Geocoder (Build ➤ Build Solution in Visual C# 2008 Express Edition), or press Ctrl+Shift+B. You should see the following message appear in the output window (if you cannot see the output window, select it from the View menu or press Ctrl+Alt+O):

```
------ Build started: Project: Geocoder, Configuration: Release Any CPU ------
Geocoder -> C:\Spatial\Geocoder.dll
========== Build: 1 succeeded or up-to-date, 0 failed, 0 skipped ==========
```

That's all that is required from Visual Studio, so you can go back to SQL Server now.

Configuring the Database

Before we can use the geocoding function, we need to make a few configuration changes to the SQL Server Database Engine to allow it to use the new Geocoder assembly correctly.

Enabling CLR Support

We know that SQL Server runs the .NET CLR process—that's how the geometry and geography datatypes work. However, although the system-defined CLR datatypes require no additional CLR configuration, we cannot normally import and run *user-defined* CLR functions in SQL Server, since this feature is disabled by default. This is a deliberate safety mechanism to ensure that a database administrator has allowed the use of these powerful (although potentially dangerous) features. In order to use the custom .NET geocoding function, we first need to configure the database to enable CLR support. We can do this by running the following T-SQL code in a Query Editor window:

```
EXEC sp_configure 'clr enabled' , '1'
GO
```

If the configuration change is successful, we receive the following result:

```
Configuration option 'clr enabled' changed from 0 to 1.
Run the RECONFIGURE statement to install.
```

To complete the change, we need to reconfigure the server to reflect the changed value, by issuing a T-SQL query with the RECONFIGURE statement as follows:

```
RECONFIGURE
GO
```

The SQL Server configuration settings are updated to allow us to run user-defined CLR code, and we receive the following message:

```
Command(s) completed successfully.
```

Setting Security Permissions

Secondly, since we will be using our .NET managed code to access information from a web service, we need to set the appropriate security permissions on the database to enable us to access external data. This can be done by running the following T-SQL code (note that you should change the name Spatial to match the name of your database):

```
ALTER DATABASE Spatial SET TRUSTWORTHY ON
GO
```

We receive the following message:

```
Command(s) completed successfully.
```

The database is now configured and ready to import the geocoding assembly.

Importing the Assembly

Having created and compiled our .NET assembly and made the necessary configuration changes to our server, we can now import the assembly into the database. We can do this by executing the following T-SQL script:

```
CREATE ASSEMBLY Geocoder
FROM 'C:\Spatial\Geocoder.dll'
WITH PERMISSION_SET = EXTERNAL_ACCESS;
GO
```

This creates an assembly in the database called `Geocoder`, from the Geocoder.dll output file compiled by Visual Studio. You will need to change the file path specified from C:\Spatial\ Geocoder.dll to match the build output location that you set in Visual Studio earlier. The `PERMISSION_SET` argument specifies the permission level granted to this assembly. By default, new SQL Server assemblies are marked as `SAFE`, which means that they can only access restricted, local resources. This is a security feature to ensure that any code cannot access external (potentially dangerous) resources to which it is not permitted access. For our geocoding function to work, we need to explicitly allow our code to access external resources, by specifying `PERMISSION_SET = EXTERNAL_ACCESS`.

Remember that, when we configured our Visual Studio project, we specified that XML serialization assemblies should also be automatically created, so that data transferred to the web service could be serialized. This created a second DLL file containing the necessary serialization functions, which must also be imported as follows:

```
CREATE ASSEMBLY [Geocoder.XmlSerializers]
FROM 'C:\Spatial\Geocoder.XmlSerializers.dll'
WITH PERMISSION_SET = SAFE;
GO
```

Again, you should change the specified file path as necessary to match the output location set in Visual Studio. The `Geocoder.XmlSerializers` assembly does not access outside data, so granting the default `PERMISSION_SET = SAFE` is sufficient.

When the assemblies have been created, you should see them appear in the SQL Server Management Studio Object Explorer, listed under Assemblies within the Programmability node of the database into which they were imported, as shown in Figure 7-5 (you may need to refresh the Object Explorer view before the assemblies become visible, by right-clicking the Assemblies node and selecting Refresh).

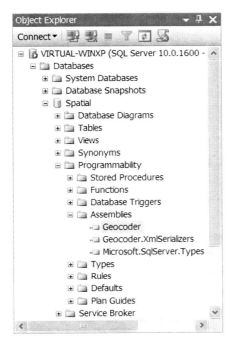

Figure 7-5. *The Geocoder and Geocoder.XmlSerializers assemblies listed in SQL Server Management Studio Object Explorer*

Tip If you execute the code to create these assemblies in the Model database, they will be available for all future databases created on that SQL Server instance.

Creating the Function

Now that we have imported our assembly, we need to define a function so that we can use the geocoding method from within T-SQL code. The function will specify a number of input parameters representing different descriptive fields of an address, and return an output string containing the WKT representation of a Point geometry representing that address, constructed from the coordinates returned from the MapPoint Find service. To create the function, execute the following T-SQL code:

```
CREATE FUNCTION dbo.Geocode(
  @AddressLine nvarchar(255),
  @PrimaryCity nvarchar(32),
  @Subdivision nvarchar(32),
  @Postcode nvarchar(10),
  @CountryRegion nvarchar(20))
RETURNS nvarchar(255)
AS EXTERNAL NAME
Geocoder.[Geocoder.UserDefinedFunctions].geocode
GO
```

This code creates a T-SQL function called dbo.Geocode that provides an interface to the geocode method contained within the Geocoder assembly. It specifies that the parameters that must be provided when using the Geocode function are an address, the name of a city, the subdivision (i.e., county/state), the postal code (e.g., ZIP code), and the country or region. These parameters correspond exactly to the parameters passed by the .NET method to the MapPoint Web Service. The return value of the function is an nvarchar(255) string containing the WKT representation of the Point associated with that address.

■**Note** When creating a function from a .NET assembly, the syntax for the AS EXTERNAL NAME clause is as follows: AssemblyName.[Namespace.ClassName].FunctionName.

That's it! Congratulations—you've just used .NET to add a new function to SQL Server, extending the existing spatial functionality by allowing you to geocode address data.

Using the Geocode Function

Finally, we get to use our geocoding function. To test it out, let's try retrieving the latitude and longitude of a Point at the Apress office, located at the following postal address:

Apress, Inc.
2855 Telegraph Avenue, Suite 600
Berkeley,
CA 94705

The syntax for the dbo.Geocode function is

dbo.Geocode('@Address','@PrimaryCity','@Subdivision','@Postcode','@CountryRegion')

To retrieve the WKT representation of a Point based on the Apress office address, we can therefore execute the following query in SQL Server Management Studio:

SELECT dbo.Geocode('2855 Telegraph Avenue','Berkeley','CA','94705','USA')

The result is as follows:

POINT(-122.259388935609 37.8584886447626)

■**Note** It might take a few seconds to retrieve the results from the MapPoint Web Service staging environment. The production environment is faster.

You can now use the dbo.Geocode function from within your SQL Server database in any situations where you want to be able to retrieve the WKT representation of a Point representing an address. Since the coordinate values returned by the MapPoint Web Service use the EPSG:4326 spatial reference system, you can use the resulting representation with the STPointFromText() method of the geography or geometry datatype as follows:

```
SELECT geography::STPointFromText(
  dbo.Geocode('2855 Telegraph Avenue','Berkeley','CA','94705','USA'),
  4326).STAsText()
```

The result is as follows:

```
POINT(-122.259388935609 37.8584886447626)
```

To extend this example further, suppose that you have a typical table of address data, with columns containing a street address, town, county, and postal code. The following example code creates such a table, containing the details of five outlets of the Barnes and Noble bookstore in the United States:

```
CREATE TABLE #BarnesAndNoble (
  Address varchar(255),
  City varchar(255),
  State char(2),
  ZIP int
  )
GO

INSERT INTO #BarnesAndNoble (Address, City, State, ZIP) VALUES
('100 Quinn Drive', 'Pittsburgh', 'PA', 15275),
('10180 South State Street', '', 'UT', 84070),
('2800 South Rochester Road', 'Rochester Hills', 'MI', ''),
('3909 Sth Cooper St', 'Arlington','TX', 76015),
('555 Fifth Avenue', 'New York', 'NY', '')
GO
```

Note Even if you have incomplete data, such as missing postal codes or partial addresses as shown in the preceding example, the MapPoint Web Service will still return the best match for the information supplied.

You can now add a geography column to this table and populate it with Points representing each row of data, by using the dbo.Geocode function:

```
ALTER TABLE #BarnesAndNoble
  ADD Location geography
GO
```

```
UPDATE #BarnesAndNoble
  SET Location =
  geography::STPointFromText(dbo.Geocode(Address,City,State,Zip,'USA'), 4326)
GO
```

■**Note** In this example, since all the rows of data relate to addresses in the United States, the final parameter passed to the geocodingc function is always 'USA'.

SQL Server Management Studio displays a message showing the number of geocoded results:

```
(5 row(s) affected)
```

You can now examine the table to see that it contains Point instances based on the geocoded address of each store, which you can use as you would any other item of spatial data, such as in the following query:

```
SELECT
  Address,
  City,
  Location.STAsText() AS WKT
FROM
  #BarnesAndNoble
```

Here are the results:

Address	City	WKT
100 Quinn Drive	Pittsburgh	POINT (-80.178599 40.44773)
10180 South State Street		POINT (-111.890829 40.566146)
2800 South Rochester Road	Rochester Hills	POINT (-83.131891 42.639297)
3909 South Cooper Street	Arlington	POINT (-97.133609 32.682357)
555 Fifth Avenue	New York	POINT (-73.979266 40.755704)

■**Tip** To automatically geocode any new addresses as they are entered into the table, you can create a *trigger* on the table that fires the dbo.Geocode function every time a new row is inserted. For more information on triggers, please see http://msdn.microsoft.com/en-us/library/ms187834(SQL.90).aspx.

Summary

In this chapter, you learned how to extend the functionality of SQL Server to include the ability to create spatial information from geocoded data. Specifically, you learned the following:

- Geocoding can be used to derive a structured spatial representation of a feature on the earth from descriptive information about that feature.

- The MapPoint Web Service provides a number of methods that can be used to geocode data, accessible via a SOAP interface over the Web.

- By creating a .NET assembly and importing it into SQL Server, you can define a new function that connects to the MapPoint Web Service and constructs the WKT representation of a geocoded street address.

- A geocoding function, such as the one created in this chapter, can be used in combination with the T-SQL UPDATE statement to populate a column of the geography datatype with Points representing each row of data in a table of addresses.

Geocoding has many useful applications, and the dbo.Geocode function is one way of adding this functionality to the spatial capabilities provided in SQL Server 2008. Remember that if you want to perform geocoding using the MapPoint Web Service in a production environment, you must first upgrade your account to register for the commercial service.

PART 3

Presenting Spatial Data

Describing a geometry or geography object as a string of coordinates may be a convenient way of storing spatial data, but it is not very helpful for presenting that data or visually analyzing it. Because geospatial data represents the physical properties of features on the earth, we typically want to be able to "see" those objects—on a map or on a globe, for example.

This part of the book demonstrates several techniques for presenting spatial data in an accessible, graphical format. Chapter 8 shows how you can create a syndicated feed of spatial information using GeoRSS, and how to import and display that information on a web site. Chapter 9 explains how to use Virtual Earth and Google Maps to create a dynamic spatial interface to SQL Server, which you can use to build exciting spatial applications. Finally, Chapter 10 introduces the new Spatial Results tab in SQL Server Management Studio, which you can use to quickly and easily visualize the results of ad hoc spatial queries.

CHAPTER 8

■■■

Syndicating Spatial Data

Syndication (in the context of the Internet) refers to the publication of information in a standard format so that it can be easily accessed, aggregated, and shared by many potential users. Although syndication is most commonly associated with the publication of news headlines, many different types of information are now syndicated on the Internet, including blog entries, sports results, product listings, new web site content, and traffic and weather reports. One feature common to many of these types of information is that, frequently, an item of syndicated content relates to a particular thing or place on the earth, or can be described at a particular location. As such, spatial data is ideally suited for syndication alongside other syndicated content.

In this chapter, I'll show you how to create a syndicated feed containing spatial information from SQL Server 2008 using the GeoRSS format, and how to make that feed available over the Internet so that users can view and subscribe to it. I'll also show you how to build a page that consumes the GeoRSS feed, and displays it on a map using Microsoft Virtual Earth or Google Maps.

Note Syndicated content is commonly referred to as a *feed*, which users *consume* using a feed reader.

Why Syndicate Spatial Information?

Much of the content already syndicated on the Internet relates to a place on the earth—for instance, the location of an item of news, the area affected by a traffic or weather report, or the venue at which an event is being held. All of these examples naturally lend themselves to syndication in a feed that also contains associated spatial information. However, even if you are dealing with data related to a subject that is not traditionally thought of as being suitable for syndication, there are a number of reasons why you still might want to consider providing a feed as a method to output spatial information from SQL Server 2008:

- A large amount of spatial information also has a time-based element (remember that the "ST" prefix, used by all methods in SQL Server that are compliant with the OGC specifications, stands for spatio-*temporal*)—for instance, consider flight information, weather forecasts, or the position of any moving vehicle. Creating and updating a feed of information is an effective way to ensure that your users always have access to the latest position of changing spatial data.

- A number of web mapping services (including Google Maps, Virtual Earth, and Yahoo! Maps) allow for the direct importation and visualization of spatial feeds. Instead of creating your own tool to plot spatial data on a map, you can simply load a spatial feed and the web mapping service will automatically display and style each of the items contained within it.

- Even without using a map viewer, spatial information delivered via a feed in XML format is text based and hence understandable by an end user. This means that it can be accessed on a wide range of compatible devices, including PDAs and mobile phones.

- Users can subscribe to a spatial feed using their existing feed reader, and read the feed alongside other, nonspatial feeds.

- By making spatial information available in a standardized, syndicated format, you enable users to easily combine it with other (spatial and nonspatial) data to create their own "mashup" applications.

To give you an example of the practical benefits that are possible from syndicating spatial data, suppose that you were using SQL Server to provide the back-end database for an online magazine. By combining spatial information with syndicated content, you could offer the following features to your users:

- See the latest news items plotted on a map of the world, as they are reported.

- Subscribe to a report that shows any traffic incidents affecting a journey between two named places, displaying alternative, suggested routes instead.

- Be alerted when an item is offered for sale within a given distance of their home location.

Now that you've seen some of the benefits that syndication has to offer, we'll look in detail at how you can create a syndicated feed of spatial data from SQL Server 2008.

Syndication Formats

Web feeds are simply text documents that contain the necessary information to be syndicated, structured in an appropriate (usually XML) format, and placed on a server that is accessible over the Internet. Users read, or subscribe to, the feed by entering the URL of the document into their web browser or feed reader. The feed document itself normally contains a number of items, each containing a title, a short summary, a link to further details, and a publication date or date on which the item was last modified so that readers can tell when the feed has been updated.

The two most commonly used formats for providing web feeds are Really Simple Syndication (RSS) and Atom format. Many sites containing a large amount of syndicated content offer feeds in both formats; for instance, Google News provides RSS and Atom feeds, available at http://news.google.com/?output=rss and http://news.google.com/?output=atom, respectively. Although there are some semantic differences between the two formats, they both have broadly the same structure, and most feed readers will consume feeds provided in either format.

To illustrate these two syndication formats, compare how the same sample news item would appear in both RSS format and in Atom format, shown in Listings 8-1 and 8-2, respectively.

Listing 8-1. *Example News Feed in RSS Format*

```
<?xml version="1.0" encoding="UTF-8"?>
<rss version="2.0">
<channel>
  <title>Example RSS News Feed</title>
  <link>http://www.example.com/</link>
  <language>en</language>
  <webMaster>feedback@example.com</webMaster>
  <copyright>&amp;copy;2008</copyright>
  <pubDate>Tue, 05 Aug 2008 12:00:00 GMT</pubDate>
  <lastBuildDate>Tue, 05 Aug 2008 12:00:00 GMT</lastBuildDate>
  <item>
    <title>World's Smallest Snake Discovered</title>
    <link>http://www.example.com/smallestsnake</link>
    <pubDate>Tue, 05 Aug 2008 12:00:00 GMT</pubDate>
    <description>The world's smallest species of snake, at just 10cm in length,
      has been discovered on the Caribbean island of Barbados.</description>
  </item>
  <item>
    …
  </item>
</channel>
</rss>
```

Listing 8-2. *Example News Feed in Atom Format*

```
<?xml version="1.0" encoding="UTF-8"?>
<feed xmlns="http://w3.org/2005/Atom" xml:lang="en">
  <title>Example Atom News Feed</title>
  <link href="http://www.example.com/"/>
  <author>
    <name>John Smith</name>
    <email>feedback@example.com</email>
  </author>
  <copyright>©2008</copyright>
  <modified>2008-08-05T12:00:00Z</modified>
  <entry>
    <title>World's Smallest Snake Discovered</title>
    <link href="http://www.example.com/smallestsnake "/>
    <issued>2008-08-05T12:00:00Z</issued>
    <modified>2008-08-05T12:00:00Z</modified>
    <summary>The world's smallest species of snake, at just 10cm in length,
      has been discovered on the Caribbean island of Barbados.
    </summary>
  </entry>
  <entry>
    …
  </entry>
</feed>
```

Listings 8-1 and 8-2 both demonstrate simple feeds containing just one item, and each feed only specifies the minimum core elements required to describe that item. The key elements of each format, and their equivalent elements, are listed and described in Table 8-1.

Table 8-1. *Comparison of Key RSS and Atom Feed Elements*

RSS	Atom	Description
<rss>		The top-level element of any RSS feed. It must contain a required attribute, version, which states the version of RSS to which the document conforms. There is no equivalent Atom element.
<channel>	<feed>	The parent element of the feed, which contains all of the individual feed entries, together with metadata about the feed itself.
<title>	<title>	The name of the feed, or the name of a particular item within the feed.
<link>	<link>	A URL reference relating to the feed or an individual item contained in the feed.
<language>	xml:lang	The language code in which the feed is written. In RSS, <language> is a separate element, whereas the equivalent Atom xml:lang is expressed as a property of the <feed> element.
<webMaster>	<author>	The contact name/e-mail address related to the feed.
<pubDate>	<issued>	The date/time on which an item was added to the feed. Notice the different date/time formats used by RSS (RFC 822) and Atom (ISO 8601).
<lastBuildDate>	<modified>	The date/time on which an item was last updated.
<item>	<entry>	An individual item of information in the feed (e.g., a particular story in a news feed).
<description>	<summary>	A synopsis of an individual item.

Note that some of the elements listed in Table 8-1 describe properties of individual items within a feed, whereas others apply to the feed as a whole. Some elements, such as <title>, can be used in either case, depending on the context in which they are used.

■**Note** In addition to the elements shown in Table 8-1, RSS and Atom feeds may contain many more, optional elements. You can view a full list of tags available for each format at http://validator.w3.org/feed/docs/rss2.html (RSS) and http://tools.ietf.org/html/rfc5023 (Atom).

The GeoRSS Format

Both Atom and RSS specify a number of different required, and optional, elements that can be used to describe different sorts of syndicated information. However, neither the RSS nor the Atom format is designed to express spatial information—that's where *GeoRSS* comes in. GeoRSS is an XML-based language that allows geographic information in XML format to be attached to other XML content. Since RSS and Atom are both XML based, this means that the GeoRSS format can be used to syndicate spatial information in XML format as part of a web feed. For more information on the GeoRSS specification, please visit `http://georss.org/`.

■**Note** GeoRSS does not replace RSS or Atom as a format for syndication, but allows you to extend either of these formats to include spatial information as part of a feed.

To examine the GeoRSS format in more detail, the following two sections introduce the various ways in which spatial information can be encoded in GeoRSS, and how that spatial information can be related to specific items of data.

GeoRSS Spatial Encodings

The GeoRSS format, in itself, does not specify the format in which the main XML content or the associated spatial information must be provided—it simply provides a mechanism for joining the two together. You can think of it as the "glue" that enables you to stick spatial information to other XML content. The content to which geographic information is attached can be one of the RSS or Atom formats already discussed, but can also be any other structured XML document, such as an Extensible HTML (XHTML) or a Resource Description Framework (RDF) document. The spatial information itself may also be encoded in one of several XML-based formats, as follows:

W3C Geo (`http://www.w3.org/2003/01/geo/`): The Geo standard, proposed by the World Wide Web Consortium (W3C), defines a class of points, each of which may contain a `lat` (latitude), `long` (longitude), and `alt` (altitude) element. The location of Paris, as represented by a Point at 48.87°N latitude and 2.33°E longitude, may be described in W3C Geo using the following two element tags: `<geo:lat>48.87</geo:lat>` and `<geo:long>2.33</geo:long>`. The W3C Geo format can only be used to define Point locations—not more complicated LineString or Polygon geometry types. Furthermore, the coordinate values of any Points can only be expressed using the WGS 84 spatial reference system. W3C Geo is therefore of limited use, and is now largely deprecated, although existing GeoRSS feeds based on this format are still accepted and used by many sites, including Yahoo! Maps (`http://developer.yahoo.com/maps/georss/`).

Simple (`http://georss.org/simple`): As its title suggests, the Simple GeoRSS encoding provides a simple XML vocabulary that allows you to describe each of the three basic geometry types—Point, LineString, and Polygon (together with an additional element, *Box*, which defines a rectangular region). The Simple format is very concise—each type of geometry can be represented using only a single tag. For instance, the location of Paris could be represented as `<georss:point>48.87 2.33</georss:point>`. For LineStrings and Polygons, the appropriate `<georss:line>` or `<georss:polygon>` element contains a space-separated list of each coordinate in the geometry, similar to the `<posList>` element used in GML. The GeoRSS Simple encoding is more sophisticated than the W3C Geo encoding, but is still limited to using coordinates expressed in the WGS 84 spatial reference system, and lacks the flexibility or compatibility of the full GML standard.

GML (http://www.opengeospatial.org/standards/gml): GeoRSS supports the Geography Markup Language encoding of spatial information, which we have already examined in Chapter 4. It is this feature of GeoRSS that is of particular interest to us, since we know that SQL Server 2008 can natively output spatial information in GML format using the AsGml() method. Not only does this format offer the most powerful and complete set of features possible to describe elements of the GeoRSS feed, but it is the one that is most easy to create from SQL Server. We will therefore focus on the GML encoding of GeoRSS for the rest of this chapter.

Relating Spatial Information with the GeoRSS <where> Tag

The main component of the GeoRSS vocabulary is the <where> element. You can add the <where> tag within any XML element to associate the content of that element with a geometry encoded in any of the supported spatial encodings previously described. Figure 8-1 illustrates how the GeoRSS <where> element can be used to attach spatial content to other XML content.

Note When you use any GeoRSS elements, such as <where>, in an XML document, you must prefix them with the GeoRSS namespace, which is http://www.georss.org/georss.

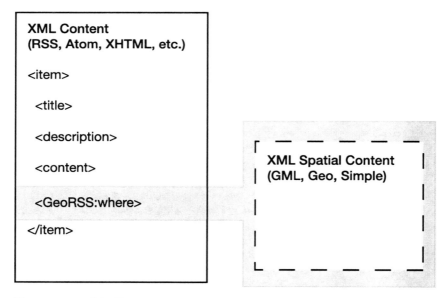

Figure 8-1. *Model of how GeoRSS may be used to attach XML spatial information to other XML content*

Note that <GeoRSS:where> may be used to attach associated spatial information to the XML document as a whole, or to any individual element within that document.

Attaching Spatial Information to a Feed

To illustrate how GeoRSS can be used to syndicate spatial information, let's reconsider the example Atom and RSS feeds shown in Listings 8-1 and 8-2. The news item contained in these feeds describes the discovery of a snake on the island of Barbados. According to the US Central Intelligence Agency's online edition of *The World Factbook* (https://www.cia.gov/library/publications/the-world-factbook/geos/bb.html), the island of Barbados is situated at a latitude of 13°10'N and a longitude of 59°32'W. When expressed as a Point geometry in the GML representation, this is as follows:

```
<gml:Point>
  <gml:pos>13.16 -59.53</gml:pos>
</gml:Point>
```

We can attach this GML representation to describe the location of the news item, by using the GeoRSS `<where>` element to embed it in the RSS and Atom feeds, as shown in Listings 8-3 and 8-4, respectively. Those lines that have been added from the original RSS and Atom feeds given in Listings 8-1 and 8-2 are highlighted in bold.

Listing 8-3. *GeoRSS Feed Using GML Encoding in an RSS Feed*

```
<?xml version="1.0" encoding="UTF-8"?>
<rss version="2.0"
  xmlns:georss="http://www.georss.org/georss"
  xmlns:gml="http://www.opengis.net/gml">
<channel>
  <title>Example RSS News Feed</title>
  <link>http://www.example.com/</link>
  <item>
    <title>World's Smallest Snake Discovered</title>
    <link>http://www.example.com/smallestsnake</link>
    <pubDate>Tue, 05 Aug 2008 12:00:00 GMT</pubDate>
    <description>The world's smallest species of snake, at just 10cm in length,
     has been discovered on the Caribbean island of Barbados.</description>
    <georss:where>
      <gml:Point>
        <gml:pos>13.16 -59.53</gml:pos>
      </gml:Point>
    </georss:where>
  </item>
</channel>
</rss>
```

Listing 8-4. *GeoRSS Feed Using GML Encoding in an Atom Feed*

```
<?xml version="1.0" encoding="UTF-8"?>
<feed xmlns="http://w3.org/2005/Atom" xml:lang="en"
  xmlns:georss="http://www.georss.org/georss"
  xmlns:gml="http://www.opengis.net/gml">
```

```
<title>Example Atom News Feed</title>
<link href="http://www.example.com/"/>
<entry>
  <title>World's Smallest Snake Discovered</title>
  <link href="http://www.example.com/smallestsnake"/>
  <issued>2008-08-05T12:00:00Z</issued>
  <modified>2008-08-05T12:00:00Z</modified>
  <summary>The world's smallest species of snake, at just 10cm in length,
   has been discovered on the Caribbean island of Barbados.</summary>
  <georss:where>
    <gml:Point>
      <gml:pos>13.16 -59.53</gml:pos>
    </gml:Point>
  </georss:where>
</entry>
</feed>
```

Since we are including the `<where>` element from GeoRSS and the `<Point>` and `<posList>` elements from GML, we must include declarations of the appropriate GeoRSS and GML namespaces. This is done by adding the `xmlns` attribute to the parent element of the feed (`<rss>` and `<feed>`, for RSS and Atom, respectively), as follows:

```
xmlns:georss="http://www.georss.org/georss"
xmlns:gml="http://www.opengis.net/gml"
```

Within the news item that we want to attach the locational information to, we add the `<where>` tag (prefixed with the GeoRSS namespace), which contains the GML representation of a Point describing the location of the item:

```
<georss:where>
  <gml:Point>
    <gml:pos>13.16 -59.53</gml:pos>
  </gml:Point>
</georss:where>
```

By simply adding the GeoRSS and GML namespace declarations and inserting the GML representation of the item within the GeoRSS `<where>` tag as just shown, you can turn a regular Atom or RSS feed into a fully formed GeoRSS spatial feed.

Creating a GeoRSS Feed

Having illustrated how GeoRSS can be used to spatially enable syndicated feeds of information, let's now look at how we can create a feed of spatial information directly from SQL Server. For this example, suppose that you work for a real-estate agency that advertises a number of properties for sale. The information about properties available for purchase will change regularly, so providing a feed to your customers would be a useful service. By using GeoRSS, you can also include spatial information as part of that feed, describing the position and shape of each property.

For this example, we will use the GML encoding contained in an RSS feed. In order to construct the elements of a GeoRSS feed in the necessary format, we will use three different features of SQL Server 2008:

AsGml(): We will use the AsGml() method to create the geographic representation of each property in the GML format. For this example, these will be Point representations of each property for sale, but if we were advertising a property with a substantial amount of associated land, we could use Polygon geometries instead. There will be one GML geometry attached to each item of content in the RSS feed.

SELECT ... FOR XML: The SELECT T-SQL statement normally returns the results of the query as a tabular dataset, with rows and columns of data. However, when used with the FOR XML modifier, a SELECT statement returns the results as elements of an XML document. We will use this to create the main content of the RSS feed, such as the elements containing the title, description, address, and price of each property.

XQuery: XQuery is a method of formatting, selecting, and querying elements of an XML document. We will use XQuery to combine the GML representation created by the AsGml() method with the other elements of the feed from the SELECT ... FOR XML statement, and style them to form the appropriate GeoRSS output.

■**Tip** For more information on the SELECT ... FOR XML syntax and the implementation of XQuery in SQL Server, refer to http://msdn.microsoft.com/en-us/library/ms178107.aspx and http://msdn.microsoft.com/en-us/library/ms189075.aspx, respectively.

Creating the Sample Data

Let's start by creating a new table that will hold the information about properties for sale—that is, the information that we will syndicate in our feed. The table will contain basic information such as the address, description, and price of each property, together with the date it was listed and a column of the geography datatype to record the location of the property. To create the table, execute the following T-SQL statement in SQL Server Management Studio:

```
CREATE TABLE PropertiesForSale (
  id int,
  address varchar(255),
  location geography,
  price money,
  description varchar(max),
  listdate datetime
)
```

Now let's insert some sample data into the table:

```
INSERT INTO PropertiesForSale VALUES
(1,
'Pilgrims Way, Chew Stoke, Somerset',
geography::Point(51.354940,-2.635765,4326),
750000,
'Grade II Listed former Rectory, with magnificent architectural features and
stunning gardens.',
'2008-08-01 17:00:00'
),
(2,
'Moulsford, Wallingford, Oxfordshire',
geography::Point(51.549963,-1.149013,4326),
1650000,
'Situated on the River Thames, this period house features landscaped gardens
extending up to 240ft, and private mooring.',
'2008-07-25 14:30:00'
),
(3,
'Pantings Lane, Highclere, Newbury',
geography::Point(51.347206,-1.375828,4326),
965000,
'A newly developed 5-bedroom house on the edge of Highclere, with very high build
specifications used throughout.',
'2008-07-25 12:00:00'
)
```

To create the GeoRSS document to syndicate the information from this table, we will create a stored procedure. Stored procedures are sections of T-SQL code that perform a particular function. Stored procedures can be stored in the database and used again and again, so rather than having to pass a long and complicated T-SQL statement to SQL Server each time we want to access the feed, we can simply call the stored procedure. Stored procedures have a number of performance advantages over regular inline T-SQL statements, since SQL Server can create and reuse the same execution plan each time the query is executed.

Stored procedures also offer better security. Our GeoRSS feed will be requested from the Internet, and sending SQL statements directly from a web page to SQL Server is not good practice because, if intercepted, they might give away important information such as the names of tables, databases, or login information. By using a stored procedure, all of the T-SQL code remains secure on the SQL Server, and only the name of the necessary stored procedure that creates the GeoRSS feed needs to be sent for the relevant results to be sent back to the calling web page.

Listing 8-5 shows the T-SQL code required to create a stored procedure called uspGeoRssFeeder, which selects data from the PropertiesForSale table and returns the output in the GeoRSS format, using GML encoding within an RSS feed.

Listing 8-5. *A Stored Procedure to Create a GeoRSS Feed*

```
CREATE PROCEDURE [dbo].[uspGeoRSSFeeder]
AS
BEGIN
-- SET NOCOUNT ON added to prevent extra result sets from
-- interfering with SELECT statements.
SET NOCOUNT ON;

-- Declare an XML variable to hold the GeoRSS output
DECLARE @GeoRSS xml;

/**
 * Create the elements of the feed using SELECT … FOR XML and AsGml()
**/
WITH XMLNAMESPACES (
  'http://www.opengis.net/gml' AS gml,
  'http://www.georss.org/georss' AS georss
)
SELECT @GeoRSS =
  (SELECT
    [address] AS title,
    [description] + ' £' + CAST([price] AS varchar(32)) AS description,
    'http://www.beginningspatial.com/' + CAST([id] AS varchar(8)) AS link,
    LEFT(DATENAME(dw, [listdate]),3) + ', '
    + STUFF(CONVERT(nvarchar,[listdate],113),21,4,' GMT') AS pubDate,
    location.AsGml() AS [georss:where]
  FROM
    PropertiesForSale
  FOR XML PATH('item'), ROOT('channel')
)

/**
 * Style the results using XQuery
**/
SELECT @GeoRSS.query('
<rss version="2.0"
  xmlns:georss="http://www.georss.org/georss"
  xmlns:gml="http://www.opengis.net/gml">
<channel>
  <title>SQL Server GeoRSS Feed</title>
  <description>This feed contains information about some fictional properties for
sale in order to demonstrate how to syndicate spatial data using the GeoRSS format.
</description>
  <link>http://www.beginningspatial.com</link>
  {
```

```
            for $e in channel/item
            return
            <item>
            <title> { $e/title/text() }</title>
            <description> { $e/description/text() }</description>
            <link> { $e/link/text() }</link>
            <pubDate>  { $e/pubDate/text() }</pubDate>
            <georss:where>
              {
                for $child in $e/georss:where/*
                return
                if (fn:local-name($child) = "Point")
                then  <gml:Point> { $child/* } </gml:Point>
                else  if (fn:local-name($child) = "LineString")
                then  <gml:LineString> { $child/* } </gml:LineString>
                else  if (fn:local-name($child) = "Polygon")
                then  <gml:Polygon> { $child/* } </gml:Polygon>
                else  if (fn:local-name($child) = "MultiPoint")
                then  <gml:MultiPoint> { $child/* } </gml:MultiPoint>
                else  if (fn:local-name($child) = "MultiCurve")
                then  <gml:MultiCurve> { $child/* } </gml:MultiCurve>
                else  if (fn:local-name($child) = "MultiSurface")
                then  <gml:MultiSurface> { $child/* } </gml:MultiSurface>
                else  if (fn:local-name($child) = "MultiGeometry")
                then  <gml:MultiGeometry> { $child/* } </gml:MultiGeometry>
                else  ()
              }
            </georss:where>
          </item>
          }
        </channel>
        </rss>
        ') AS GeoRSSFeed
END
```

There are a number of different steps involved in this procedure, so let's look at some of the important points in more detail:

- The WITH XMLNAMESPACES clause preceding the SELECT statement specifies the GML and GeoRSS namespaces so that they can be used to identify items in the following result set.

- The publication date of each item is set using the function LEFT(DATENAME(dw, [listdate]),3) + ', ' + STUFF(CONVERT(nvarchar,[listdate],113),21,4,' GMT'). This function is used to express a SQL Server datetime value in a string format such as Fri, 01 Aug 2008 17:00:00 GMT. This is the date format dictated by the RFC 822 standard, which is the format required in RSS feeds.

- The value of `link`, the URL hyperlink associated with each item, is created by appending the `id` value onto the end of the string `http://www.beginningspatial.com/`. For example, property number 3 links to the URL `http://www.beginningspatial.com/3`.

- The `FOR XML PATH('item'), ROOT('channel')` modifier after the `SELECT` statement means that the columns for each row of data will be represented as subelements of an `<item>` element, contained within a `<channel>` root element. This mirrors the structure used in the RSS format.

- The results of the `SELECT` query are stored in an XML variable called `@GeoRSS`. XQuery is then used to modify the structure of the XML contained by this variable using the `@GeoRSS.query` syntax. XQuery is used to add in the channel elements to the feed, such as the title and link, as well as to wrap the results in the root `<rss>` element.

- XQuery is also used to manually append the `gml` namespace prefix to the beginning of each element contained within the GML representation. When you use the `AsGml()` method to return results from SQL Server, the GML namespace is only declared on the top-level item—that is, `<Point xmlns="http://www.opengis.net/gml"><pos>51.35494 -2.635765</pos></Point>`. Since it is the default namespace, the GML namespace does not need to be restated for every child element—by default, any child elements inherit the namespace from their parent item. However, when this GML representation is placed as a fragment into a larger GeoRSS feed, the GML namespace is no longer the default namespace, and some RSS parsers will not correctly assign the `<pos>` element to the GML namespace, thus failing to read the feed. In order to ensure that the GeoRSS feed is correctly parsed, the stored procedure explicitly states the `gml` namespace prefix for every element in the GML representation—that is, `<gml:Point><gml:pos>51.35494 -2.635765</gml:pos></gml:Point>`.

When you create the stored procedure by running the Listing 8-5 code in SQL Server Management Studio, you should receive the following message:

```
Command(s) completed successfully.
```

To demonstrate the GeoRSS output created by this stored procedure, you can now execute the stored procedure using the exec T-SQL command, as follows:

```
exec uspGeoRSSFeeder
```

The result is a GeoRSS-formatted syndicated feed of spatial data:

```
<rss xmlns:georss="http://www.georss.org/georss"
    xmlns:gml="http://www.opengis.net/gml" version="2.0">
<channel>
  <title>SQL Server GeoRSS Feed</title>
    <description>This feed contains information about some fictional properties
      for sale in order to demonstrate how to syndicate spatial data using the
      GeoRSS format. </description>
```

```
      <link>http://www.beginningspatial.com</link>
      <item>
        <title>Pilgrims Way, Chew Stoke, Somerset</title>
        <description>Grade II Listed former Rectory, with magnificent architectural
          features and stunning gardens. £750000.00</description>
        <link>http://www.beginningspatial.com/1</link>
        <pubDate>Fri, 01 Aug 2008 17:00:00 GMT</pubDate>
        <georss:where>
          <gml:Point>
            <gml:pos>51.35494 -2.635765</gml:pos>
          </gml:Point>
        </georss:where>
      </item>
      <item>
        <title>Moulsford, Wallingford, Oxfordshire</title>
        <description>Situated on the River Thames, this period house features
landscaped gardens extending up to 240ft, and private mooring. £1650000.00</description>
        <link>http://www.beginningspatial.com/2</link>
        <pubDate>Fri, 25 Jul 2008 14:30:00 GMT</pubDate>
        <georss:where>
          <gml:Point>
            <gml:pos>51.549963 -1.149013</gml:pos>
          </gml:Point>
        </georss:where>
      </item>
      <item>
        <title>Pantings Lane, Highclere, Newbury</title>
        <description>A newly developed 5-bedroom house on the edge of Highclere, with
          very high build specifications used throughout. £965000.00</description>
        <link>http://www.beginningspatial.com/3</link>
        <pubDate>Fri, 25 Jul 2008 12:00:00 GMT</pubDate>
        <georss:where>
          <gml:Point>
            <gml:pos>51.347206 -1.375828</gml:pos>
          </gml:Point>
        </georss:where>
      </item>
    </channel>
</rss>
```

Serving the GeoRSS Feed

Having created a stored procedure that outputs spatial data in an appropriately structured GeoRSS format, we could now save the resulting output as a static XML file onto a web server from which users could access it. However, it would be better if we were able to create that GeoRSS feed dynamically, to ensure that it always reflected the latest data in the database. To do this, we will create a *web handler*, a web page that connects to the SQL Server, executes the uspGeoRssFeeder stored procedure, and returns the results to the user's browser (or feed reader) whenever it is called. We will create the handler using .NET, and make use of the library of existing methods provided by the Base Class Library to retrieve and serve the GeoRSS feed.

The following steps describe how to create the handler using Visual Studio 2008, but you can use instead the freely available Visual Studio Web Developer Express Edition, downloadable from http://www.microsoft.com/express/vwd/:

1. From the main Visual Studio menu bar, select File ➤ New ➤ Website.

2. Select ASP.NET Web Site.

3. As with all the .NET examples introduced in this book, choose whether to create the handler using Visual Basic or C# by selecting the appropriate option from the Language drop-down list.

4. Choose a location in which to save the handler.

5. Click OK.

Once the project has been created, we need to add a new handler. To do so, follow these steps:

1. Select Website ➤ Add New Item (or press Ctrl+Shift+A).

2. In the Add New Item dialog box, select Generic Handler from the list of templates.

3. Choose a name for the handler (the default is Handler.ashx). This will be the name of the file that your users request in order to read the GeoRSS feed, so perhaps change it to **GeoRSSFeed.ashx**.

4. Click Add.

When the new handler is created, the main window will show a template for a generic web handler. Change this code so that it appears as shown in Listing 8-6 or Listing 8-7, depending on whether you chose to base your project on the VB .NET or C# language, respectively. Be sure to change the properties of the connection string (highlighted in bold) to reflect the credentials required to connect to your SQL Server instance.

Listing 8-6. *The GeoRSS Feed Handler (VB .NET)*

```vbnet
<%@ WebHandler Language="VB" Class="Handler" %>

Imports System
Imports System.Web
Imports System.Data.SqlClient
Imports System.Configuration
Imports System.Text

Public Class Handler : Implements IHttpHandler

  Public Sub ProcessRequest(ByVal context As HttpContext) _
  Implements IHttpHandler.ProcessRequest

    'Set the response headers
    context.Response.ContentType = "text/xml"
    context.Response.Charset = "iso-8859-1"
    context.Response.CacheControl = "no-cache"
    context.Response.Expires = 0

    'Define the connection to the database
    Dim myConn = New SqlConnection( _
      "server=ENTERYOURSERVERNAMEHERE;" & _
      "Trusted_Connection=yes;" & _
      "database=Spatial")

    'Open the connection
    myConn.Open()

    'Define the query to execute
    Dim myQuery As String = "exec dbo.uspGeoRSSFeeder"

    'Set the query to run against the connection
    Dim myCMD As New SqlCommand(myQuery, myConn)

    'Create a reader for the results
    Dim myReader As SqlDataReader = myCMD.ExecuteReader()

    'Read through the results
    While myReader.Read()

      'Write the GeoRSS response back to the client
      context.Response.Write(myReader("GeoRSSFeed").ToString)

    End While
```

```
    'Close the reader
    myReader.Close()

    'Close the connection
    myConn.Close()

  End Sub

  Public ReadOnly Property IsReusable() As Boolean _
  Implements IHttpHandler.IsReusable
    Get
      Return False
    End Get
  End Property

End Class
```

Listing 8-7. *The GeoRSS Feed Handler (C#)*

```
<%@ WebHandler Language="C#" Class="Handler" %>

using System;
using System.Web;
using System.Data.SqlClient;
using System.Configuration;
using System.Text;

public class Handler : IHttpHandler
{

  public void ProcessRequest(HttpContext context)
  {
    // Set the response headers
    context.Response.ContentType = "text/xml";
    context.Response.Charset = "iso-8859-1";
    context.Response.CacheControl = "no-cache";
    context.Response.Expires = 0;

    // Define a connection to the database
    SqlConnection myConn = new SqlConnection(
      @"server=ENTERYOURSERVERNAMEHERE;
      Trusted_Connection=yes;
      database=Spatial");

    // Open the connection
    myConn.Open();
```

```
    // Define the query to execute
    string myQuery = "exec dbo.uspGeoRSSFeeder";

    // Set the query to run against the connection
    SqlCommand myCMD = new SqlCommand(myQuery, myConn);

    // Create a reader for the results
    SqlDataReader myReader = myCMD.ExecuteReader();

    // Read through the results
    while (myReader.Read())
    {
      // Write the GeoRSS response back to the client
      context.Response.Write(myReader["GeoRSSFeed"].ToString());
    }

    // Close the reader
    myReader.Close();

    // Close the connection
    myConn.Close();
  }

  public bool IsReusable
  {
    get { return false; }
  }
}
```

Having entered the appropriate code, you should save the new handler by selecting File ➤ Save GeoRSSFeed.ashx (Ctrl+S).

Testing the Feed Handler

To test the GeoRSSFeed.ashx handler, you can preview the feed by selecting File ➤ View in Browser (Ctrl+Shift+W). The output will appear in your browser, as shown in Figure 8-2.

Note To subscribe to or view your GeoRSS feed, your users must be able to access the server on which it resides. Thus, if you want to make your feed available to the public, you must deploy the GeoRSSFeed.ashx file to a public web server connected to the Internet.

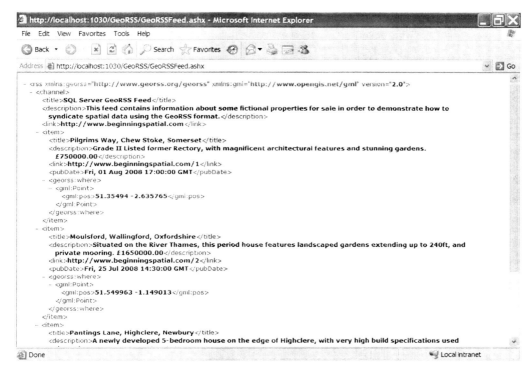

Figure 8-2. *The GeoRSS feed as viewed directly in a browser*

Consuming the GeoRSS Feed

Viewing the text-based GeoRSS feed in a browser is all very well, but spatial information can be presented much more clearly on a map. One of the advantages of syndicating spatial information over the Internet is that a number of web mapping services can directly consume GeoRSS feeds, including Yahoo! Maps, Google Maps, and Virtual Earth. In this section, I will show you how to use these services to take the GeoRSS output created by the web handler and present it directly on a map.

Using the Google Maps Web Site

Google provides a very easy way of consuming GeoRSS feeds directly from the main Google Maps web site, as follows:

1. Point your browser to the Google Maps site, `http://maps.google.com`.

2. Enter the URL of a GeoRSS feed in the search box at the top of the screen; for instance, `http://www.beginningspatial.com/georss_example.xml`. (If you deployed your GeoRSSFeed.ashx handler to a public web server, you may enter its URL here instead.)

3. Click Search Maps.

After the contents of the GeoRSS feed are parsed, a map displaying markers representing the location of each item appears on the right side of the window, with a list to its left of all the items corresponding to the markers. This is illustrated in Figure 8-3. Clicking any of the markers displays a pop-up window containing the information associated with that item.

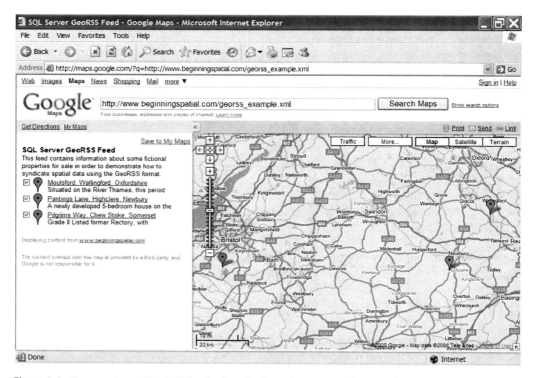

Figure 8-3. *Consuming a GeoRSS feed using the http://maps.google.com site*

▓**Tip** You can consume a GeoRSS feed directly into a Google Maps map by appending the URL of the GeoRSS feed as a query parameter to the end of the URL string, as follows: `http://maps.google.com/maps?q=http://www.beginningspatial.com/georss_example.xml`.

Using Embedded Google Maps

Rather than visiting the `http://maps.google.com` site, it is also possible to embed a map from Google Maps directly into your own web page. One advantage of this approach is that it allows you to show on your own web site the locations of all properties for sale, for example, alongside other related information.

▓**Caution** When using Google Maps to plot a GeoRSS feed, the URL of the feed must reside on a publicly accessible server, since the content must be fetched and processed by the Google servers before being returned to the browser. You therefore cannot use Google Maps to plot a GeoRSS feed located on a local server, such as `http://localhost/GeoRSSFeed.ashx`, or any server on a private network. If you do not have access to a publicly accessible web server to deploy your feed, you can still use Google Maps to plot other publicly available GeoRSS feeds on the Internet, such as the example feed at `www.beginningspatial.com/georss_example.xml`. In order to plot locally held GeoRSS feeds, you can use Microsoft's Virtual Earth control, which is discussed in the next section of this chapter.

Signing Up for a Google Maps API Key

Before you can access the methods required to plot a GeoRSS feed on a Google Map in your own web page, you first need to sign up for a Google Maps API key. You can obtain a free key at the following web site: `http://code.google.com/apis/maps/signup.html`. When requesting a new key, you will be prompted to enter a web domain (e.g., `http://www.example.com`). The API key is linked to the domain that you submit when the key is created, so that a key generated for `http://www.example.com` can be used to add Google Maps to any pages hosted on that web domain, but not on `http://www.anotherexample.com`. You can request multiple API keys for different domains, if you require.

The Google Maps API key is an alphanumeric string of about 86 characters, and looks something like this (this is the API key required for the site `http://localhost`):

```
ABQIAAAAv5k471fvcyRtqrTpWxn55BT2yXp_ZAY8_ufC3CFXhHIE1NvwkxQrYahVeo3mdN6v2qqO
zLdpgJiwlA
```

Having obtained an API key, make a note of it, because you will need to specify it in the code you use to add a map to your web page.

Embedding Google Maps in a Web Page

To create a web page containing an embedded Google Map to consume the GeoRSS feed, we will add a new item to the existing web site project we created earlier:

1. Select Website ➤ Add New Item (Ctrl+Shift+A).

2. In the Add New Item dialog box, select HTML Page from the list of templates.

3. Type a name for the new page, such as **GoogleMaps.htm**.

4. Click Add.

5. When the default HTML page template appears, enter the following code. Edit the lines highlighted in bold to include your Google Maps API key, and the URL of the GeoRSS feed handler you created earlier (if it resides on a public server). If you have access only to a local development server, you may still test the map functionality by using the URL of the example feed shown.

```html
<!DOCTYPE html PUBLIC "-//W3C//DTD XHTML 1.0 Strict//EN"
    "http://www.w3.org/TR/xhtml1/DTD/xhtml1-strict.dtd">
<html xmlns="http://www.w3.org/1999/xhtml"
      xmlns:v="urn:schemas-microsoft-com:vml">
  <head>
    <meta http-equiv="content-type" content="text/html; charset=utf-8"/>
    <title>Google Maps GeoRSS Overlay</title>
    <script
      src="http://www.google.com/jsapi?key=INSERTYOURAPIKEYHERE"
      type="text/javascript"></script>
    <script type="text/javascript">

      // Load the Google Maps API
      google.load("maps", "2");

      // When the API is loaded, call the initialize() function
      google.setOnLoadCallback(initialize);

      function initialize() {

        // Create a new map object on the page
        var map = new google.maps.Map2(document.getElementById("divMap"));

        // Set the center point and zoom level of the map
        map.setCenter(new google.maps.LatLng(52, -0.8), 6);

        // Specify the GeoRSS feed URL
        var url = "http://www.beginningspatial.com/georss_example.xml";
        var geoXml = new google.maps.GeoXml(url);

        // Add navigational control to the map
        map.addControl(new google.maps.LargeMapControl());

        // Load the GeoRSS onto the map
        map.addOverlay(geoXml);
        }
    </script>
  </head>
  <body onload="initialize()">
    <div id="divMap" style="width: 640px; height: 480px; float:left;"></div>
  </body>
</html>
```

6. Save the new page by selecting File ➤ Save GoogleMaps.htm (Ctrl+S).

7. View the new page by selecting File ➤ View in Browser (Ctrl+Shift+W).

Figure 8-4 illustrates the web page created, which displays the items from the GeoRSS feed on an embedded Google Maps object. The map already includes basic functionality, such as panning and zooming. Clicking any of the markers on the map opens an information window with the content relating to that item in the feed.

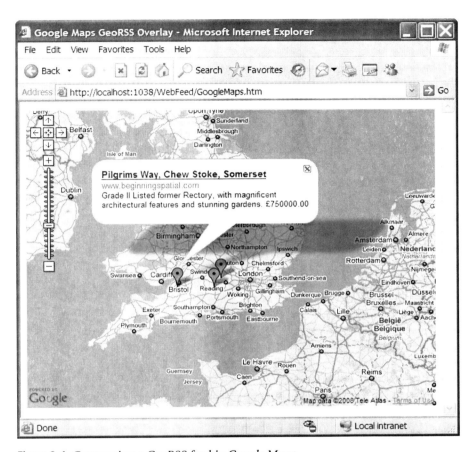

Figure 8-4. *Consuming a GeoRSS feed in Google Maps*

Using Microsoft Virtual Earth

To create a web page that displays the GeoRSS feed on an embedded Virtual Earth map instead, follow these steps:

1. Select Website ➤ Add New Item (Ctrl+Shift+A).

2. In the Add New Item dialog box, select HTML Page from the list of templates.

3. Type a name for the new page, such as **VirtualEarth.htm**.

4. Click Add.

5. When the default HTML page template appears, enter the following code. If necessary, change the line highlighted in bold to reflect the URL of the GeoRSS feed handler that you created earlier.

```
<!DOCTYPE html PUBLIC "-//W3C//DTD XHTML 1.0 Transitional//EN"
  "http://www.w3.org/TR/xhtml1/DTD/xhtml1-transitional.dtd">
<html xmlns="http://www.w3.org/1999/xhtml">
<head>
  <title>Virtual Earth GeoRSS Overlay</title>
  <script src="http://dev.virtualearth.net/mapcontrol/mapcontrol.ashx?v=6.2"
        type="text/javascript"></script>
  <script type="text/javascript" language="javascript">
    function initialize () {

      // Insert a new map object into the page
      map = new VEMap('divMap');

      // Specify the map options
      map.LoadMap(new VELatLong(52, -0.8), 6 ,'r' ,false);

      // Specify a new shape layer on the map
      var geoRSSLayer = new VEShapeLayer();
      var url = "./GeoRSSFeed.ashx";
      var geoRSSLayerSpec = new VEShapeSourceSpecification(VEDataType.GeoRSS, url,
        geoRSSLayer);
      map.ImportShapeLayerData(geoRSSLayerSpec, null, false);
    }
  </script>
</head>
<body onload=" initialize ();">
  <div id="divMap" style="position: relative; width: 640px; height: 480px;">
  </div>
</body>
</html>
```

6. Save the new page by selecting File ➤ Save VirtualEarth.htm (Ctrl+S).

7. View the new page by selecting File ➤ View in Browser (Ctrl+Shift+W).

The page will appear in your default browser, as illustrated in Figure 8-5. Like the Google Map, the Virtual Earth control implements a number of default functions for viewing GeoRSS files—you can pan and zoom around the map, and hover the mouse cursor over any feature to open an information window containing the detailed description of that item from the feed, together with a link to the URL contained in the <link> element.

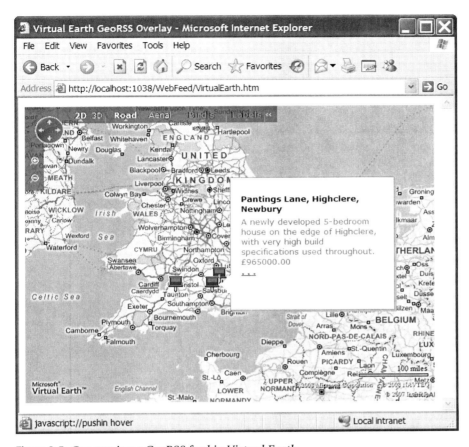

Figure 8-5. *Consuming a GeoRSS feed in Virtual Earth*

Using Yahoo! Maps

Although Yahoo! Maps does support plotting information from GeoRSS feeds, at the time of writing it only supports the W3C Geo encoding. You therefore cannot plot the GeoRSS feed created in this chapter using Yahoo! Maps, since that feed is based on the GML encoding.

■**Tip** Although the example in this chapter only uses Point geometries, you may also use GML LineString or Polygon geometries to represent items in a GeoRSS feed.

Summary

In this chapter I demonstrated how to syndicate spatial information from SQL Server. Topics covered included the following:

- Syndication is a useful way of presenting up-to-date spatial information in a standard format that may be easily accessed by many users.

- The GeoRSS format can be used to attach spatial information in XML format to any other XML-based content, which can be used to create feeds of spatial information based on the RSS or Atom formats.

- By combining AsGml(), SELECT ... FOR XML, and XQuery, you can create a stored procedure that produces a GeoRSS feed of spatial information directly from SQL Server 2008.

- To enable users to access, view, and subscribe to the GeoRSS feed, you can create a .NET web handler that executes the stored procedure and returns the resulting GeoRSS feed to a browser or feed reader.

- Google Maps and Microsoft Virtual Earth provide methods that consume a GeoRSS feed and automatically display the contents on a map.

Presenting Spatial Data Using Web Mapping Services

In the last chapter I showed you how to build a web page that consumes a GeoRSS feed from SQL Server and displays the data contained in that feed on a Microsoft Virtual Earth or Google Maps control. While this provides a convenient way to present spatial information, it does not allow for much flexibility in how that information should be displayed—both Google Maps and Virtual Earth present GeoRSS feeds by applying a standard template that is difficult to customize. Displaying GeoRSS information has other limitations—each item in the feed can be represented by only a single geometry, and the information that can be displayed on the map is limited to those columns of data specified by the Atom or RSS format on which the GeoRSS feed is based. For instance, you cannot easily display different colored markers based on the price of each property in a GeoRSS feed, since the RSS and Atom syndication formats do not contain any elements that record a monetary value relating to each item.

In this chapter, I'll show you how to take control over exactly how your spatial data is displayed, by using .NET and JavaScript to integrate the output from SQL Server directly into the Virtual Earth and Google Maps application programming interfaces (APIs). By programmatically creating each element on the map through the relevant API, you can take advantage of the many rich features offered by these tools to customize exactly how spatial data should be presented.

Tip You can see the full API reference documentation for Google Maps and Virtual Earth online at `http://code.google.com/apis/maps/documentation/` and `http://msdn.microsoft.com/en-us/library/bb429619.aspx`, respectively.

The Application

Online web mapping services are not limited to being a static display medium for spatial data. Rather, they provide a powerful interface that enables the user to interact and explore spatial data in a way that is not possible using traditional presentation methods. In this chapter, I'll show you a simple example that touches the surface of some of the features that you can implement using this approach. We will use a SQL Server database that contains a table of information about airports in the United States. This table has columns of data containing the name and address of each airport, its elevation above sea level, and a geography Point instance representing its location. We will then build an application that presents the user with a web page displaying a map of the United States, using either the Google Maps or Virtual Earth web mapping providers. When the user clicks any point on the map, the application will trigger a stored procedure in SQL Server that selects all airports within a 50 km radius of the location chosen, displays them on the map, and provides an information box containing the details of each airport. This type of application is commonly used to provide "store locator" functionality, but can easily be adapted for many different purposes.

Tip You can download the dataset used in this chapter, together with the full application code, from the Source Code/Download area of the Apress web site, `http://www.apress.com`.

Process Overview

Virtual Earth and Google Maps are both controlled by JavaScript APIs—that is, a library of methods that are imported and executed by the client's browser. These methods perform various tasks, including requesting the raster tiles that make up the map view from the web mapping service provider, creating the vector data representing features on the map, and processing the results to render the map object in the browser. However, JavaScript code running on the client's machine cannot interact *directly* with spatial data held in SQL Server. In order to integrate spatial data from SQL Server 2008, we must introduce a middle layer that acts as a go-between to allow information to be passed between the two platforms. For this example, we will once again use .NET to provide this middle layer, although you could use any one of many different programming languages to provide this functionality.

In data architecture terminology, the application described in this chapter is an example of a three-tier model:

Presentation tier. This is the user map interface, which is provided by JavaScript and HTML code executed within a web browser on the client machine.

Logic tier. The logic tier is provided by a .NET handler running on a web server. All information passed between the presentation tier and the data tier passes through the logic tier, which processes the data accordingly and controls the application logic.

Data tier. SQL Server 2008 provides the data tier, responsible for the storage and retrieval of data.

The specific interactions between each tier in the example described in this chapter are as follows:

- When the user clicks on the map, a JavaScript function creates an HTTP request to the .NET handler, passing details of the map position on which the user clicked (presentation tier ➤ logic tier).

- The .NET handler connects to SQL Server and executes a query to select data based on the parameters supplied (logic tier ➤ data tier).

- The .NET handler retrieves the results from SQL Server and dynamically constructs the JavaScript functions that create the elements on the map relating to each item of data (data tier ➤ logic tier).

- The JavaScript output of the .NET handler is passed back to the browser for execution on the client's machine, which causes the map display to be updated with the new features (logic tier ➤ presentation tier).

This is a relatively complex technique, which involves the interaction of a number of different technologies. A simplified overview of the process is illustrated in Figure 9-1.

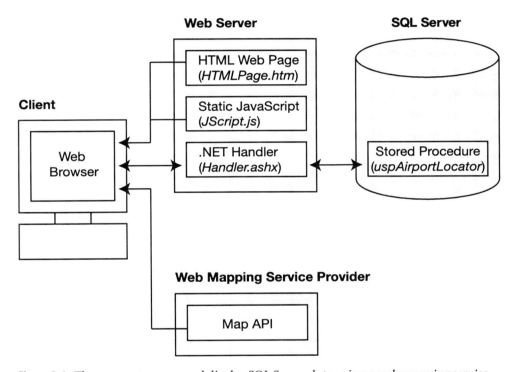

Figure 9-1. *The process to query and display SQL Server data using a web mapping service*

To implement this approach, we must create four separate elements, shown in italic text in Figure 9-1. The function performed by each of these elements is as follows:

HTMLPage.htm: The HTML file defines the overall structure of the web page, including the size, position, and style of the container in which the map object will be embedded. It also specifies a reference to the relevant mapping service API and the custom JavaScript code (JScript.js) required to provide the functionality of the page.

JScript.js: The JavaScript code provided by the web server uses the methods of the relevant API to create the map object on the page and configure the initial map settings. This is described as *static* code, since the same code will be reused every time the application is used, whatever features are shown on the map.

uspAirportLocator. The stored procedure uses the latitude and longitude of the point clicked on the map as parameters in a SELECT query, to return those airports lying within a 50 km radius of the selected location. The query uses the STAsText() method to return the Well-Known Text (WKT) representation of the geometry of each airport in the result set, together with other columns of descriptive information.

Handler.ashx: The .NET handler is triggered every time the user clicks on the map, which creates a connection to SQL Server and executes the stored procedure. It then retrieves the results and manipulates them to create the function calls required by the API of the relevant mapping provider. Like the JScript.js file, the .NET handler returns JavaScript code that is executed by the client's browser. However, whereas JScript.js contains static JavaScript functions, the JavaScript created by the .NET handler is *dynamic*—it changes based on the results returned by the SQL Server stored procedure. The resulting JavaScript code is returned to the browser, where it is executed to add the relevant features to the map.

For each of these components, I'll show you the code required to create the same functionality using either Google Maps or Virtual Earth. Now that I've explained the approach we'll be taking, let's begin!

Note In November 2008, Microsoft released a Community Technology Preview (CTP) of Windows Live Tools for Visual Studio 2008. One of the tools included in this package allows you to control a Virtual Earth map using server-side .NET code, rather than having to write client-side JavaScript (as used in the technique described in this chapter). At the time of writing, Windows Live Tools has not been fully released, but you can find out more information, or download a preview copy of the software, by visiting http://dev.live.com/tools/.

CHOOSING A WEB MAPPING SERVICE PROVIDER

There are many online web mapping service providers, including OpenLayers (http://openlayers.org), Multimap (http://www.multimap.com), Yahoo! Maps (http://maps.yahoo.com/), Virtual Earth (http://maps.live.com), and Google Maps (http://maps.google.com). All of these services offer similar functionality, and they can all be controlled programmatically through their own API methods, allowing you to integrate your own data alongside the rich datasets provided. In addition to simply overlaying data on a map, some of these APIs also offer a range of other spatial functionality, including geocoding, route mapping, coordinate transformations, and spatial operations.

In this chapter, I'll show you the code necessary to integrate SQL Server 2008 with two of the most popular web mapping services: Google Maps and Virtual Earth. There is little to choose between the two providers in terms of functionality, so the decision is largely one of personal preference. Both offer a very complete, well-documented set of API methods, are very well supported, and offer an astounding quality of base map data (both road and aerial) on which you can overlay your spatial features.

Obtaining the Source Data

The source data used in this example is based on information from the National Transportation Atlas Database published by the US Department of Transportation, available online at http://www.bts.gov/publications/national_transportation_atlas_database/2008/. This site contains many useful and interesting datasets—we will be using the dataset that contains Points representing all public-use airports in the United States. The details of each airport will be held in separate columns of a table, which you can create by executing the following query:

```
CREATE TABLE US_Airports (
  Code char(4),
  Name varchar(255),
  City varchar(255),
  County varchar(255),
  State char(2),
  Location geography,
  Elevation int
)
```

Each row of data in this table will hold details of an individual airport. For example, the following code listing inserts rows of sample data relating to four major US airports: John F. Kennedy International Airport in New York, Los Angeles International Airport, Miami International Airport, and Chicago O'Hare International Airport:

```
INSERT INTO US_Airports VALUES (
  'JFK',
  'JOHN F KENNEDY INTL',
  'NEW YORK',
  'QUEENS',
  'NY',
  0xE6100000010CA94A5B5CE3514440430070ECD97152C0,
  13)

INSERT INTO US_Airports VALUES (
  'LAX',
  'LOS ANGELES INTL',
  'LOS ANGELES',
  'LOS ANGELES',
  'CA',
  0xE6100000010CCF11F92EA5F840401492CCEA1D9A5DC0,
  126)

INSERT INTO US_Airports VALUES (
  'MIA',
  'MIAMI INTL',
  'MIAMI',
  'DADE',
  'FL',
  0xE6100000010C0BB1FA230CCB39403D47E4BB941254C0,
  11)
```

```
INSERT INTO US_Airports VALUES (
  'ORD',
  'CHICAGO O''HARE INTL',
  'CHICAGO',
  'COOK',
  'IL',
  0xE6100000010C9D29745E63FD4440321AF9BCE2F955C0,
  668)
```

Because there are over 6,500 airports in the United States, I won't list details of all of them here. Instead, you can find the SQL script necessary to create and populate the US_Airports table in the Source Code/Download area of the Apress web site (http://www.apress.com). Loading and executing this script in SQL Server Management Studio will create all the source data necessary for the application described in this chapter.

Creating a New Web Site Project

Before we begin to write any code, we need to create a new web application. For this example, I'll create a web project using Visual Studio 2008, but you may also use Visual Studio 2005 or Visual Studio Web Developer Express if you prefer (the latter of which you can download for free from http://www.microsoft.com/express/vwd/).

To create an empty project, follow these steps:

1. From the Visual Studio main menu bar, select File ➤ New ➤ Web Site.

2. In the New Web Site dialog box, choose the Empty Web Site template.

3. Select a location in which to save the project.

4. Click OK.

Creating the HTML Page

The HTML page will be the main web page that is served to the user's browser. It includes references to the API of the web mapping service provider, together with the additional JavaScript functions contained in the static JScript.js file, and defines the overall structure of the page. To add the HTML page to the project, follow these steps:

1. Select Website ➤ Add New Item.

2. In the Add New Item dialog box, select HTML Page.

3. Type a name for the new file if you wish, or accept the default HTMLPage.htm.

4. Click Add.

The default template for a new HTML page appears:

```
<!DOCTYPE html PUBLIC "-//W3C//DTD XHTML 1.0 Transitional//EN"
"http://www.w3.org/TR/xhtml1/DTD/xhtml1-transitional.dtd">
<html xmlns="http://www.w3.org/1999/xhtml" >
<head>
    <title></title>
</head>
<body>

</body>
</html>
```

We will insert elements into this empty template to create the main interface for the airport locator application. In the following section I'll introduce each element in turn, before showing you the full HTML page.

Referencing the API

Before we can use either map service, we must first include a reference to the relevant API. Both maps are controlled using JavaScript APIs, so we include them using the `<script>` tag, contained within the `<head>` element of the HTML page, as demonstrated in Listings 9-1 and 9-2 for the respective APIs.

Listing 9-1. *Referencing the Google AJAX API*

```
<script src="http://www.google.com/jsapi?key=INSERTYOURAPIKEYHERE"
type="text/javascript"></script>
```

Listing 9-2. *Referencing the Virtual Earth API*

```
<script src="http://dev.virtualearth.net/mapcontrol/mapcontrol.ashx?v=6.2"
type="text/javascript"></script>
```

If you're using Google Maps API on a live web site, you must edit the URL specified in the `src` attribute in Listing 9-1 to specify your Google Maps Developer API key where indicated. If you are only testing Google Maps on a web page on a local machine (accessed via `http://localhost`), you do not need to supply a key. If you need to obtain a key, you can sign up for one by following the instructions in Chapter 8 or online at `http://code.google.com/apis/maps/signup.html`. You do not need a key to use the Virtual Earth API, so if you choose to use this provider, you may enter the code in Listing 9-2 exactly as shown.

░**Note** When you include a reference to the Virtual Earth API, you must state the version of the API that you want to use by appending the value of the v parameter to the end of the URL string. At the time of writing, the latest release of the Virtual Earth API is version 6.2, which you can specify by using the URL `http://dev.virtualearth.net/mapcontrol/mapcontrol.ashx?v=6.2`.

Including the JavaScript

Although it is possible to write JavaScript functions as inline statements contained within an HTML document (as used in the example web pages that consume GeoRSS feeds in Chapter 8), it is normally considered best practice to keep the JavaScript and HTML elements of a web page in separate files. In this example, we will store the JavaScript functions in a file called JScript.js, so we must include a reference to that file from within the <head> section of the HTML document, as shown in Listing 9-3.

Listing 9-3. *Including the JavaScript File*

```
<script src="JScript.js" type="text/javascript"></script>
```

Specifying the Map Container

The body of the HTML document defines the structure of the resulting web page, including the position and size of the map control itself. This is determined by the placement and styling of a special HTML <div> element in the body of the page, as shown in Listing 9-4.

Listing 9-4. *Specifying the Map Container*

```
<div id="divMap" style="position:absolute; width:640px; height:480px;"></div>
```

Notice that the <div> element is empty (i.e., there is nothing contained between the <div> and </div> tags)—the map is dynamically inserted into this container by the constructor function of the map API. Even though there is no content within the <div>, there are two important properties assigned to the opening <div> tag:

id: This is the unique identifier used to refer to this element through the Document Object Model (DOM). The id property of the <div> element must be supplied to the constructor function so that the API knows where to insert the map. In Listing 9-4, the map is assigned the id *divMap*.

style: This property defines the Cascading Style Sheets (CSS) declarations that are applied to style the map. CSS is used to set the position, color, font, and spacing of elements on a web page. In the example in Listing 9-4, the CSS properties are applied to position the map on the page with a width of 640 pixels and a height of 480 pixels. This is a size that should appear comfortably in most browsers, although you may change these to other values if you prefer.

Because the constructor functions for both Virtual Earth and Google Maps operate in a very similar way, we can use the same code in Listing 9-4 for both cases.

Tip For more information on DOM and CSS, consult the World Wide Web Consortium pages at http://www.w3.org/DOM/ and http://www.w3.org/Style/CSS/, respectively.

Reviewing the HTML Page

Once you have made the changes listed in the previous sections, the HTMLPage.htm document should appear as shown in Listings 9-5 and 9-6.

Listing 9-5. *The HTML Page for Google Maps*

```
<!DOCTYPE html PUBLIC "-//W3C//DTD XHTML 1.0 Transitional//EN"
"http://www.w3.org/TR/xhtml1/DTD/xhtml1-transitional.dtd">
<html xmlns="http://www.w3.org/1999/xhtml">
<head>
  <title>Google Maps & SQL Server 2008</title>
  <script src="http://www.google.com/jsapi?key=INSERTYOURAPIKEYHERE"
type="text/javascript" ></script>
  <script src="JScript.js" type="text/javascript"></script>
</head>
<body>
  <div id="divMap" style="position:absolute; width:640px; height:480px;"></div>
</body>
</html>
```

Listing 9-6. *The HTML Page for Virtual Earth*

```
<!DOCTYPE html PUBLIC "-//W3C//DTD XHTML 1.0 Transitional//EN"
"http://www.w3.org/TR/xhtml1/DTD/xhtml1-transitional.dtd">
<html xmlns="http://www.w3.org/1999/xhtml">
<head>
    <title>Virtual Earth & SQL Server 2008</title>
    <script src="http://dev.virtualearth.net/mapcontrol/mapcontrol.ashx?v=6.2"
type="text/javascript"></script>
    <script src="JScript.js" type="text/javascript"></script>
</head>
<body>
  <div id="divMap" style="position:absolute; width:640px; height:480px;"></div>
</body>
</html>
```

Creating the Static JavaScript

To create the JavaScript functions, we must first add the new JScript.js file to the project, as follows:

1. Click Website ➤ Add New Item.

2. In the Add New Item dialog box, select JScript File.

3. Ensure that the file is named the same as the reference you included in the HTML file above (JScript.js, by default).

4. Click Add.

A blank window appears, into which we will add the JavaScript functions required to control the map. Remember that the JScript.js file will contain only methods that configure the general map appearance and functionality—the JavaScript methods that add data to the map must be created dynamically based on the results of the SQL Server query, and so must be generated by the .NET handler. As with the HTML file created earlier in this chapter, I'll show you snippets of each element of the JScript.js file, before listing the full code required.

■**Caution** JavaScript may be implemented in slightly different ways depending on the browser in which it is executed. This can make it hard to debug errors in JavaScript code, since different browsers may react to errors with different behavior. Be careful to avoid any mistakes when entering the code listings in this section, and remember that JavaScript is a case-sensitive language.

Declaring the Map Object

Firstly, we will declare a variable that will hold the object representing the map embedded in the web page. We will call this variable map. Since the methods and values associated with the map object will be accessed by many different functions within the script, it must be declared as a global variable—one that exists outside the scope of any particular function.

To declare the global map variable, you should include the line of code shown in Listing 9-7 at the top of the JScript.js file.

Listing 9-7. *Declaring the Global Map Variable*

```
var map = null;
```

■**Caution** You must declare the map variable in the main body of the JavaScript file, outside of the code declaring any particular function. If you declare the map variable within a function, the map object is only available locally within the scope of that function, and cannot be accessed by other code.

Loading the Google Maps API

The Google AJAX API loader, http://www.google.com/jsapi, which you included in the <script> tag of the HTML page (if you chose to base your application on the Google Maps interface), allows you to embed modules from many different Google services on the same page. Once you have included a reference to the generic loader (as shown in the "Referencing the API" section earlier in the chapter), you must then specify which modules you want to use with the google.load() method. You can load the latest release of version 2 of the Google Maps API by including the line of code provided in Listing 9-8.

Listing 9-8. *Loading the Google Maps API*

```
google.load("maps", "2");
```

If you choose to use Virtual Earth as your mapping provider, the URL used to reference the Virtual Earth API, `http://dev.virtualearth.net/mapcontrol/mapcontrol.ashx?v=6.2`, automatically loads the library of methods specifically related to Virtual Earth, so no equivalent to the `google.load()` method is required.

Setting Callbacks to Load and Unload the Map

Once the API has finished loading, we want to call the JavaScript function that will create and initialize the map object. The Google Maps API includes a dedicated method, `setOnLoadCallback()`, that can be used to specify a function to run immediately after `google.load()` has finished loading the API. For Virtual Earth, no such method exists, so we will instead attach an event to the load event of the browser window, which requires slightly different syntax depending on the user's browser.

For both cases, we will set the callback to run the `getMap()` function. The code shown in Listings 9-9 and 9-10 illustrates how to call the `getMap()` function for each API.

Listing 9-9. *Setting the Load Callback in Google Maps*

```
google.setOnLoadCallback(getMap);
```

Listing 9-10. *Setting the Load Callback in Virtual Earth*

```
if (window.addEventListener) { window.addEventListener("load", getMap, false); }
else if (window.attachEvent) { window.attachEvent("onload", getMap); }
```

When the web page is unloaded again (i.e., when the user leaves the page or closes their browser), we will set another callback, which will call the `disposeMap()` function. This function will ensure that all of the objects created by the map are removed, to prevent the risk of memory leaks. To set the callback, we can use the same code for both Google Maps and Virtual Earth to attach the `disposeMap()` method to the unload event, as shown in Listing 9-11.

Listing 9-11. *Setting the Unload Callback*

```
if(window.addEventListener) {window.addEventListener("unload", disposeMap, false);}
else if (window.attachEvent) {window.attachEvent("onunload", disposeMap);}
```

Creating the Map

To create a new map and assign it to the `map` variable, you must use the *constructor method* of the relevant API to create a `Map2` (Google Maps) or `VEMap` (Virtual Earth) object. We will place a call to these methods within the `getMap()` function, which is the function called immediately after the page is loaded. In both Google Maps and Virtual Earth, the constructor method for the map must be passed an argument that specifies the `id` of the `<div>` element on the page in which the map should be created. In the HTML page you just created, the `id` assigned to the map container is called `divMap`, so you create a map in that container using the code shown in Listings 9-12 and 9-13.

Listing 9-12. *Initializing a New Map in Google Maps*

```
function getMap() {
  map = new google.maps.Map2(document.getElementById('divMap'));
}
```

Listing 9-13. *Initializing a New Map in Virtual Earth*

```
function getMap() {
  map = new VEMap('divMap');
}
```

Once the map object is created, it is assigned to the variable map. That means that from now on, all of the methods used to control the map can be applied directly to the map object, using the map.method() syntax.

Configuring the Map

Having constructed a new map object on the page, we next need to configure the basic properties of that map. This must be done to set the initial state of the newly constructed map object, before calling any other operations on it. Configuring the map involves setting the following parameters:

- The point on which the map should be centered

- How far the map should be zoomed in

- What style of map display (e.g., road map, aerial imagery) should be shown

To set the initial view of the map object, you can use the setCenter() (Google Maps) or LoadMap() (Virtual Earth) method, as shown in Listings 9-14 and 9-15, respectively.

Listing 9-14. *Configuring the Initial Map View in Google Maps*

```
map.setCenter(new google.maps.LatLng(34, -118), 8, G_NORMAL_MAP);
```

Listing 9-15. *Configuring the Initial Map View in Virtual Earth*

```
map.LoadMap(new VELatLong(34, -118), 8, VEMapStyle.Road);
```

You should add the relevant method to the getMap() function, immediately after the map is constructed. The parameters provided to each method are as follows:

The first parameter states the latitude and longitude of a point at which the map should be centered. google.maps.LatLng(34, -118) and VELatLong(34, -118) demonstrate the syntax required to specify a point at latitude 34°N and longitude 118°W, which is the approximate location of Los Angeles.

The second parameter represents the zoom level, which is an integer value between 0 and 19. Zoom level 0 applies the least zoom, where the entire world can be seen on the map at once, whereas zoom level 19 is the most zoomed in, where individual buildings are visible. At each successive zoom level, the precision of the map is doubled in both horizontal and vertical dimensions. The zoom level chosen for this example, 8, will display a map covering a range of approximately 350 km wide by 250 km high.

The final parameter is used to set the style of map display. Table 9-1 shows the styles of map display that can be specified for each API.

Table 9-1. *Comparison of Google Maps and Virtual Earth Map Styles*

Virtual Earth	Google Maps	Description
VEMapStyle.Road	G_NORMAL_MAP	Road map style.
VEMapStyle.Shaded	G_PHYSICAL_MAP	Shaded map style. Appears as road map, but with shaded contour lines depicting the relief of the underlying terrain.
VEMapStyle.Aerial	G_SATELLITE_MAP	Aerial map style.
VEMapStyle.Hybrid	G_HYBRID_MAP	Hybrid style, combining aerial map style with label overlays.
VEMapStyle.Birdseye		An oblique, bird's-eye view taken from an overhead angle.
VEMapStyle.Oblique		Same as VEMapStyle.Birdseye.
VEMapStyle.BirdseyeHybrid		Oblique imagery as in the VEMapStyle.Birdseye view, combined with a label overlay.

Adding Interactivity

Now that we have added a map to the page and configured some initial settings, we need to make the map *do* something. To achieve this, both APIs provide *event handlers* that are triggered when certain events happen, which can be used to call particular functions. Examples of events that can trigger functions are when the user clicks or double-clicks a mouse button, presses a key on the keyboard, pans the map, or zooms in or out.

Listings 9-16 and 9-17 illustrate how to attach an event handler that calls the loadAirportData() function whenever a user clicks on the map. Add the appropriate code to the end of the getMap() function.

Listing 9-16. *Adding an Event Handler in Google Maps*

```
google.maps.Event.addListener(map, "click", loadAirportData);
```

Listing 9-17. *Adding an Event Handler in Virtual Earth*

```
map.AttachEvent("onclick", loadAirportData);
```

When we attach an event handler based on a mouseclick event, the handler automatically passes a parameter to the function attached to the event with the coordinates representing the position of the mouse at the time the event was fired. We can then pass these coordinates as parameters to the stored procedure in order to find nearby airports.

▓**Caution** In JavaScript, as in SQL Server, methods are invoked by stating the method name followed by round brackets, such as `loadAirportData()`. However, the last parameter of each method shown in Listings 9-16 and 9-17 is a *pointer* to the function that should be triggered on a given event, not a call to the function itself. You should therefore not include brackets following the name of the `loadAirport` function.

Creating an AJAX Object

AJAX (Asynchronous JavaScript and XML) is a popular technique used on the World Wide Web to enable web applications to interact with a server without needing to refresh the entire web page. AJAX uses JavaScript to make background requests to a server over HTTP and dynamically evaluates the response in the browser. In this application, we will use AJAX techniques to request new data from SQL Server when the user clicks on the map, and then execute the response using the `eval()` function to add features to the map.

All modern web browsers can support AJAX. Unfortunately, since AJAX is not based on an agreed standard, different web browsers implement it in different ways. To use AJAX in our application, we will first add a function, `GetXmlHttp()`, which will create the appropriate AJAX XMLHTTP object based on the user's browser. This function is shown in Listing 9-18.

Listing 9-18. *Creating a Cross-Browser AJAX Object*

```
function GetXmlHttp() {
  var xmlHttp;
  try { xmlHttp = new XMLHttpRequest(); } // Firefox, Opera 8.0+, Safari
  catch (e) {
    try { xmlHttp = new ActiveXObject("Msxml2.XMLHTTP"); } // IE 6.0+
    catch (e) {
      try { xmlHttp = new ActiveXObject("Microsoft.XMLHTTP"); } // Older IE
      catch (e) {
        alert("Your browser does not support AJAX!");
        return false;
      }
    }
  }
  return xmlHttp;
}
```

Note Although originally designed for transport of XML content, AJAX methods can actually be used for transferring any kind of text data from a server to a client's browser.

Requesting Data

The loadAirportData() function will be called every time the user clicks the left mouse button on the map. It will be used to request and retrieve information from SQL Server, via a .NET handler, relating to airports within a 50 km radius of that point using AJAX, and execute the resulting output using the JavaScript eval() statement. The loadAirportData() function itself receives an argument from the triggering event handler representing the coordinates of the point on which the mouse was clicked—e for Virtual Earth, and latlng in Google Maps (or overlaylatlng if the cursor was positioned over an object when the mouse button was clicked). This information is passed on using parameters in the headers of a GET request to the .NET handler. The maptype parameter is also supplied to the handler, which is used to specify the map provider for which the output should be created.

Listings 9-19 and 9-20 demonstrate the code listing required to create the loadAirportData() function for Google Maps and Virtual Earth, respectively.

Listing 9-19. *Requesting Data via AJAX with Google Maps*

```
function loadAirportData(overlay, latlng, overlaylatlng) {
  if (overlay) { latlng = overlaylatlng; }
  var xmlhttp = GetXmlHttp();
  if (xmlhttp) {
    var url = "./Handler.ashx";
    var params = "lat=" + latlng.lat() + "&long=" + latlng.lng() + "&maptype=GM";
    xmlhttp.open("GET", url + '?' + params, true);
        xmlhttp.onreadystatechange = function() {
          if (xmlhttp.readyState == 4)
          {
            var result = xmlhttp.responseText;
            eval(result);
          }
        }
        xmlhttp.send(null);
    }
  }
```

Listing 9-20. *Requesting Data via AJAX with Virtual Earth*

```
function loadAirportData(e) {
  var pos = map.PixelToLatLong(new VEPixel(e.mapX, e.mapY));
  var xmlhttp = GetXmlHttp();
```

```
 if (xmlhttp) {
   var url = "./Handler.ashx";
   var params = "lat=" + pos.Latitude + "&long=" + pos.Longitude + "&maptype=VE";
   xmlhttp.open("GET", url + '?' + params, true);
   xmlhttp.onreadystatechange = function() {
     if (xmlhttp.readyState == 4)
     {
       var result = xmlhttp.responseText;
       eval(result);
     }
   }
   xmlhttp.send(null);
 }
}
```

Note Apart from a few syntactic differences, the loadAirportData() function contains very little function-ality specific to either API—it simply requests and executes the appropriate response from the .NET handler. It is the role of the .NET handler to return the appropriate code required to generate map elements for the API in question.

Clearing the Map

Every time the user clicks on the map, the loadAirportData() function is triggered, which calls the stored procedure and returns new data to the map. Before adding the locations of any airports lying within the new search zone, we need to remove any existing shapes on the map. We can do this by adding the code shown in Listings 9-21 and 9-22 at the beginning of the loadAirportData() function, before making the AJAX request to the .NET handler.

Listing 9-21. *Removing Existing Map Elements Using Google Maps*

```
map.clearOverlays();
```

Listing 9-22. *Removing Existing Map Elements Using Virtual Earth*

```
map.DeleteAllShapes();
```

Updating the Status Window

The code required to create the geometries on the map will be returned from SQL Server to the .NET handler and converted into JavaScript, but it will only actually be executed by the client's browser. Depending on the complexity of the geometry in question, and the power of the client's machine, there may be a short delay while this information is processed before the user sees the relevant information displayed on the map. In order to inform the user that their request is being processed, we will issue a simple status bar message immediately before executing the dynamic output created by the .NET handler. To do so, insert the code shown in Listing 9-23 immediately before the eval(result) statement in the loadAirportData() function.

Listing 9-23. *Updating the Status Bar Prior to Loading Shapes*

```
window.status = 'Loading Data... Please Wait.';
```

Once the eval() function has executed the functions returned by the .NET handler to add features to the map, we should update the status bar message again. However, because we are evaluating dynamically created code, there is an increased chance of error, and we should not simply assume that the eval() function successfully loaded the data. Instead, we will wrap the eval() statement in a try… catch block. This means that the browser will attempt to execute the eval() statement contained in the try block. If it is successful, then we can update the user with a success message. If the eval() statement fails, execution will jump to the catch block, and we can change the message accordingly. The necessary code is shown in Listing 9-24.

Listing 9-24. *Updating the Status Bar After Loading Is Complete*

```
try {
  eval(result);
  window.status = 'Data Loaded!';
}
catch (e) {
  window.status = 'Data could not be loaded.';
}
```

Tip You can use try… catch to implement error-checking routines in your application.

Disposing of the Map

One very important (and often overlooked) aspect to consider when designing any application is to ensure that it exits cleanly and releases any resources that had been allocated to it. In the example in this chapter, the JavaScript code that provides the map interface runs within a browser, which uses memory and processing power of the client's computer. When the user stops using the application, we should ensure that these resources are no longer assigned to the application, and are made available to any other processes that may require them. Failure to properly release resources is known as a *memory leak*, which, in a very severe case, could consume all of the system memory on the client's machine, forcing the system to be restarted.

In theory, all modern browsers perform *garbage collection* of JavaScript processes—automatically reclaiming memory when it is no longer required. However, not all browsers succeed in this task quite as well as they should. Thus, due to the complex nature of the JavaScript code used by Virtual Earth and Google Maps, if a memory leak were to occur, it could noticeably degrade the performance of the client's system. Fortunately, rather than having to rely on automatic garbage collection by the browser, both APIs include a simple method that can be called to delete the map object and perform the necessary cleanup operations. We will call these methods from the disposeMap() function, which is triggered by the page unload event (i.e., the user leaves the page or closes their browser). The relevant code to create this function is shown in Listings 9-25 and 9-26.

Listing 9-25. *Disposing of a Google Maps Map Object*

```
function disposeMap()
{
  google.maps.Unload();
  map = null;
}
```

Listing 9-26. *Disposing of a Virtual Earth Map Object*

```
function disposeMap()
{
  map.Dispose();
  map = null;
}
```

Reviewing the JavaScript

Once you have added all of the elements listed above to the JavaScript page, the JScript.js file
should be as shown in Listings 9-27 and 9-28, as appropriate (comments have been added to
remind you of the function performed by each line of code).

Listing 9-27. *The JScript.js Page for Google Maps*

```
// Declare the global map object
var map = null;

// Load the Google Maps API
google.load("maps", "2");

// Set the onLoad callback
google.setOnLoadCallback(getMap);

// Set the unLoad callback
if (window.addEventListener) {window.addEventListener("unload", disposeMap, false);}
else if (window.attachEvent) {window.attachEvent("onunload", disposeMap);}

// This function is called when the page has been loaded
function getMap() {

  // Create a new map object in the divMap container
  map = new google.maps.Map2(document.getElementById("divMap"));

  // Configure the initial map view
  map.setCenter(new google.maps.LatLng(34, -118), 8, G_NORMAL_MAP);
```

```
  // Call the loadAirportData function when the user clicks the map
  google.maps.Event.addListener(map, "click", loadAirportData);
}

// This function creates a cross-browser AJAX object
function GetXmlHttp() {
  var xmlHttp;
  try { xmlHttp = new XMLHttpRequest(); } // Firefox, Opera 8.0+, Safari
  catch (e) {
    try { xmlHttp = new ActiveXObject("Msxml2.XMLHTTP"); } // IE 6.0+
    catch (e) {
      try { xmlHttp = new ActiveXObject("Microsoft.XMLHTTP"); } // Older IE
      catch (e) {
        alert("Your browser does not support AJAX!");
        return false;
      }
    }
  }
  return xmlHttp;
}


// This function is called when the mouse is clicked
function loadAirportData(overlay, latlng, overlaylatlng) {

  // Retrieve the latitude / longitude of the cursor location
  if (overlay) { latlng = overlaylatlng; }

  // Clear the map
  map.clearOverlays();

  //Get the appropriate XMLHTTP object for the browser
  var xmlhttp = GetXmlHttp();

  // If we have a valid XMLHTTP object
  if (xmlhttp) {
    // Define the URL of the handler
    var url = "./Handler.ashx";
    // Build the parameters that must be passed
    var params = "lat=" + latlng.lat() + "&long=" + latlng.lng() + "&maptype=GM";
    // Open the XmlHTTP request
    xmlhttp.open("GET", url + '?' + params, true);
    // Fire this when the readyState of the request changes
    xmlhttp.onreadystatechange = function() {
      // readystate 4 indicates that the request is complete
      if (xmlhttp.readyState == 4) {
        // Read in the JavaScript response from the handler
        var result = xmlhttp.responseText;
        // Update the status message
        window.status = 'Loading Data... Please Wait.';
```

```
        try {
          // Execute the dynamically created JavaScript
          eval(result);
          // Update the status message
          window.status = 'Data Loaded!';
        }
        catch (e) {
          // If the response cannot be evaluated
          window.status = 'Data could not be loaded.';
        }
      }
    }
    // Send the XMLHTTP request
    xmlhttp.send(null);
  }
}


// This function is called when the page is unloaded
function disposeMap() {

  // Dispose of the map object
  google.maps.Unload();

  // Unset the map variable
  map = null;
}
```

Listing 9-28. *The JScript.js Page for Virtual Earth*

```
// Declare the global map object
var map = null;

// Set the Load callback
if (window.addEventListener) {window.addEventListener("load", getMap, false);}
else if (window.attachEvent) {window.attachEvent("onload", getMap);}

// Set the unLoad callback
if (window.addEventListener) {window.addEventListener("unload", disposeMap, false);}
else if (window.attachEvent) {window.attachEvent("onunload", disposeMap);}

// This function is called when the page has been loaded
function getMap() {

  // Create a new map object in the divMap container
  map = new VEMap('divMap');
```

```
  // Configure the initial map view
  map.LoadMap(new VELatLong(34, -118), 8, VEMapStyle.Road);

  // Call the loadAirportData function when the user clicks the map
  map.AttachEvent("onclick", loadAirportData);
}

Insert 3
// This function creates a cross-browser AJAX object
function GetXmlHttp() {
  var xmlHttp;
  try { xmlHttp = new XMLHttpRequest(); } // Firefox, Opera 8.0+, Safari
  catch (e) {
    try { xmlHttp = new ActiveXObject("Msxml2.XMLHTTP"); } // IE 6.0+
    catch (e) {
      try { xmlHttp = new ActiveXObject("Microsoft.XMLHTTP"); } // Older IE
      catch (e) {
        alert("Your browser does not support AJAX!");
        return false;
      }
    }
  }
  return xmlHttp;
}


function loadAirportData(e) {

  // Clear the map
  map.DeleteAllShapes();

  // Convert the cursor location to latitude / longitude
  var pos = map.PixelToLatLong(new VEPixel(e.mapX, e.mapY));

  //Get the appropriate XMLHTTP object for the browser
  var xmlhttp = GetXmlHttp();

  // If we have a valid XMLHTTP object
  if (xmlhttp) {
    // Define the URL of the handler
    var url = "./Handler.ashx";
    // Build the parameters that must be passed
    var params = "lat=" + pos.Latitude + "&long=" + pos.Longitude + "&maptype=VE";
    // Open the XmlHTTP request
    xmlhttp.open("GET", url + '?' + params, true);
    // Fire this when the readyState of the request changes
    xmlhttp.onreadystatechange = function() {
      // readystate 4 indicates that the request is complete
      if (xmlhttp.readyState == 4) {
```

```
          // Read in the JavaScript response from the handler
          var result = xmlhttp.responseText;
          // Update the status message
          window.status = 'Loading Data...';
          try {
            // Execute the dynamically created JavaScript
            eval(result);
            // Update the status message
            window.status = 'Data Loaded!';
          }
          catch (e) {
            // If the response cannot be evaluated
            window.status = 'Data could not be loaded.';
          }
        }
      }
    // Send the XMLHttp request
    xmlhttp.send(null);
  }
}

// This function is called when the page is unloaded
function disposeMap() {

  // Release all resources assigned to the map
  map.Dispose();

  // Unset the map variable
  map = null;
}
```

Creating the Stored Procedure

Let's take a little break from Visual Studio and move back to SQL Server. In the application we are building, the elements added to the map will be retrieved from SQL Server by using a stored procedure. Recall that in the last chapter we integrated data from SQL Server onto a map by creating a stored procedure that used SELECT ... FOR XML and XQuery to style the results into a GeoRSS feed, which could then be consumed directly by Virtual Earth and Google Maps. In this chapter, instead of creating output in GeoRSS format, we need to convert the results into the native JavaScript function calls of the Virtual Earth and Google Maps APIs. We therefore can't use either of the previous two methods, which apply only to XML-based data. Instead, we will create a simple SELECT statement that returns each of the columns of data that we want to display, including a spatial column of the geography datatype returned in WKT format (using the STAsText() method). These results will then be passed to the .NET handler, which will perform the hard work of transforming from the WKT output to the relevant Google Maps or Virtual Earth API functions. Although this approach is a little harder to set up, it allows us a far greater degree of flexibility to determine exactly how the data should appear on the map.

■**Note** The approach described in this chapter uses the same underlying technology stack as in Chapter 8—
the results of a SQL Server stored procedure are returned to a .NET handler and then parsed by the methods
of a web mapping service provider. However, whereas in the previous example SQL Server performed the
majority of the work to convert spatial information into the GeoRSS format that could be directly imported by
the API, in this example the .NET handler performs the conversion to programmatically create the JavaScript
functions from the WKT representation of each map element.

Listing 9-29 illustrates the T-SQL code required to create a new stored procedure,
uspAirportLocator, which returns the necessary columns of data required by the .NET handler.

Listing 9-29. *Creating the uspAirportLocator Stored Procedure*

```
CREATE PROCEDURE [dbo].[uspAirportLocator]
    @latitude float,
    @longitude float
AS
BEGIN

-- Create a point representing the location where the user clicked
DECLARE @Point geography
SET @Point = geography::Point(@latitude, @longitude, 4326)

-- Use BufferWithTolerance() to create an approx. 50km search radius about the point
DECLARE @SearchArea geography
SET @SearchArea = @Point.BufferWithTolerance(50000,1000,0)

-- Select any airports that intersect the search area
SELECT
  Location.STGeometryType() AS GeometryType,
  Location.STAsText() AS WKT,
  Name AS Title,
  'City: ' + CITY + '<br/>' +
  'County:' + COUNTY + '<br/>' +
  'State: ' + STATE + '<br/>' AS Description
FROM
 US_Airports
WHERE
  Location.STIntersects(@SearchArea) = 1
```

```
-- Also select the search area itself
UNION ALL SELECT
  @SearchArea.STGeometryType() AS GeometryType,
  @SearchArea.STAsText() AS WKT,
  'Search Area' AS Title,
  'The 50km radius search area in which to identify features.' AS Description

END
GO
```

This stored procedure executes a SELECT statement, with a few important features:

The procedure accepts two floating-point parameters, @latitude and @longitude. These are supplied by the .NET handler when the procedure is called, based on the latitude and longitude of the point clicked on the map.

The two parameters are used in conjunction with the geography Point() method to create a local variable, @Point, which is a Point object of the geography datatype positioned at the coordinates given. The coordinates returned from both the Google Maps and Virtual Earth APIs are expressed using the EPSG:4326 spatial reference system, so we state the value 4326 in the third parameter passed to the Point() method.

The BufferWithTolerance() method is used to create an approximately circular Polygon of radius 50 km around the point that the user clicked. For more information on the BufferWithTolerance() method, see Chapter 12.

Both the type (geog.STGeometryType()) and the WKT representation (geog.STAsText()) of each geometry are returned in separate columns, called GeometryType and WKT, respectively. These will be used by the .NET handler to create the features on the map. The SELECT statement also returns Title and Description columns, which will be used to display descriptive information relating to each airport in a pop-up window.

Within the WHERE clause, a condition is placed to select only those airports from the table that *intersect* the 50 km search area, using Location.STIntersects(@SearchArea) = 1. The STIntersects() method will be discussed in more detail in Chapter 13, but for this example you can think of it as filtering the result set to show only those rows in which the Point geometry representing an airport lies within the Polygon specified by the 50 km circular search area.

Caution It is important that you use the same column name aliases listed here (GeometryType, WKT, Title, and Description), since the .NET handler will retrieve the columns of data from the stored procedure by these names. If you want to retrieve different columns of data, you must also modify the .NET handler as appropriate.

Creating the Web Handler

The role of the .NET handler will be to create the JavaScript code that adds to the map features representing each item of data. To do so, it must first connect to SQL Server, execute the stored procedure, and retrieve the results. It will then read through each row of data in the response, determining the appropriate geometry type and the transformations that must be made to structure it into the appropriate Virtual Earth or Google Maps shape. The output of the .NET handler will be a stream of JavaScript functions that will be read back via AJAX and executed using the eval() function.

You can add a new handler to the Visual Studio project as follows:

1. Select Website ➤ Add New Item (or press Ctrl+Shift+A).

2. In the Add New Item dialog box, select Generic Handler in the list of available templates.

3. Type a name for the handler. For this example, I'll use the default name—Handler.ashx. If you choose a different name, you must change the value of the associated url variable specified in the loadAirportData() function of the JScript.js file.

4. Select Visual Basic as the language in the Language drop-down list box.

5. Click Add.

Note In this section, I'll only show the code required to create the handler using the VB .NET language. The C# equivalent is included in the Source Code/Download area of the Apress web site, http://www.apress.com.

A new handler will be created and added to the project. The main window will open to show the contents of the new handler file, which currently contains the default "Hello World!" template. We will modify this file to insert VB .NET code that performs the following actions:

1. Creates a connection to SQL Server.

2. Executes the stored procedure, passing the latitude and longitude parameters of the point that was clicked on the map.

3. Retrieves the results.

4. Processes each row of the returned dataset to create a string of JavaScript functions that, when executed by the client's browser, performs the following actions:

   a. Defines a new geometry from the WKT representing that feature.

   b. Adds that geometry to the map.

   c. Attaches descriptive information relating to the item.

   d. Customizes the appearance of each geometry.

In the following section, I'll show you snippets of code that demonstrate each of these steps in turn, before listing the full handler code.

Note The .NET handler cannot directly control any element of the map. Instead, it creates output containing the JavaScript functions that, when executed within the client's browser, add or style items on the map.

Creating a Connection to SQL Server

Creating a connection from .NET to SQL Server is very straightforward, only requiring two lines of code. First, you declare a new `SqlConnection` object that defines the parameters of the connection—the name of the server, the credentials used to connect, and the initial database. Then you simply call the `Open()` method on the new `SqlConnection` object. This is illustrated in Listing 9-30.

Listing 9-30. *Creating a Connection to SQL Server*

```
Dim myConn = New SqlConnection("server=SERVERNAME;" & _
                               "Trusted_Connection=yes;" & _
                               "database=Spatial")
myConn.Open()
```

Listing 9-30 assumes that your SQL Server is configured to use a trusted connection—that is, you connect to SQL Server using your existing Windows logon credentials. If you connect to SQL Server using a specified username and password, you should use the syntax shown in Listing 9-31 instead.

Listing 9-31. *Creating a Connection to SQL Server with a Specified Username/Password*

```
Dim myConn = New SqlConnection("server=SERVERNAME;" & _
                               "Uid=USERNAME; Pwd=PASSWORD" & _
                               "database=Spatial")
myConn.Open()
```

Calling the Stored Procedure

Having established a connection to the server, you then need to specify the stored procedure that you want to execute, and pass in the appropriate parameters representing the longitude and latitude of the point that was clicked on the map. These are set as properties of a .NET `SqlCommand` object. The command is executed by calling the `ExecuteReader()` method, and the results can then be accessed by using a data reader. This process is shown in Listing 9-32.

Listing 9-32. *Executing the Stored Procedure*

```
'Define the stored procedure to execute
Dim myQuery As String = "dbo.uspAirportLocator"
Dim myCMD As New SqlCommand(myQuery, myConn)
myCMD.CommandType = Data.CommandType.StoredProcedure

'Send the point the user clicked on to the stored proc
myCMD.Parameters.Add("@Latitude", Data.SqlDbType.Float)
myCMD.Parameters("@Latitude").Value = context.Request.Params("lat")
myCMD.Parameters.Add("@Longitude", Data.SqlDbType.Float)
myCMD.Parameters("@Longitude").Value = context.Request.Params("long")

'Create a reader to read the results
Dim myReader As SqlDataReader = myCMD.ExecuteReader()
```

Constructing the Geometry Elements

Both mapping APIs have the ability to define shapes equivalent to the Point, LineString, and Polygon geometries supported by SQL Server 2008. In order to construct each geometry on the map, we need to convert the WKT syntax returned by the STAsText() method in the stored procedure into the appropriate calls to the map API constructor methods. Since the method of construction differs for each type of geometry, our handler will first check what type of geometry is being described, by using the value of the GeometryType column, and then execute different sections of code depending on the geometry type in question.

Points

The Point is the simplest geometry to construct, and is the building block on which more complex geometries are based. In this example, we will use Points to represent the location of each airport. In Virtual Earth, a feature representing a single point is called a *Pushpin*, whereas in Google Maps it is called a *Marker*. Table 9-2 illustrates the syntax used to define a Point geometry in the WKT format (as returned by the uspAirportLocator stored procedure), and the equivalent constructor method in Virtual Earth and Google Maps.

Table 9-2. *Comparison of Constructors for Point Geometries*

Format	Representation
WKT	POINT('-122.4 37.7')
Virtual Earth	VEShape(VEShapeType.Pushpin,new VELatLong(37.7, -122.4))
Google Maps	GMarker(new GLatLng(37.7, -122.4))

Caution The coordinates in the WKT format are written in longitude–latitude order, whereas the geometry constructor methods for the Virtual Earth and Google Maps APIs specify coordinates in latitude–longitude order.

The necessary code to convert from the WKT representation, WKT, to the equivalent Virtual Earth Pushpin or Google Maps Marker constructor, using VB .NET, is shown in Listing 9-33.

Listing 9-33. *Constructing a Google Maps Marker or Virtual Earth Pushpin from a WKT Point*

```
'Declare the source WKT string representation
Dim WKT = "POINT('-122.4 37.7')"

'Remove the POINT declaration and the opening bracket
WKT = Replace(WKT, "POINT (", "")

'Remove the closing bracket
WKT = Replace(WKT, ")", "")

'Create an array of the coordinates
Dim Coords() As String = Split(WKT, " ")

'Build the Google Maps Marker constructor representing this point
Dim Marker = "new GMarker(new GLatLng(" +(Coords(1))+ "," +(Coords(0))+ "));"

'Build the Virtual Earth Pushpin constructor representing this point
Dim Pushpin = "new VELatLong(" +(Coords(1))+ "," +(Coords(0))+ ");"
```

Note All Google Maps objects belong to the google.maps namespace. For the sake of clarity, I won't restate the namespace for every example in this section, but prefix each object name with the letter G instead. So, in the preceding example, GMarker refers to google.maps.Marker.

LineStrings

LineString geometries (called *Polylines* in Virtual Earth and Google Maps) are formed from an array of pairs of coordinates for each point in the LineString. For WKT, this is simply a comma-delimited list of the coordinate tuples of each point. For Virtual Earth and Google Maps, the Polyline is constructed from a comma-separated list of the equivalent individual coordinate elements, VELatLong or LatLng, respectively. Table 9-3 illustrates the comparison between the syntax of a LineString geometry in WKT, and the equivalent Polyline constructor in Virtual Earth and Google Maps.

Although not required in this particular example (since none of the features plotted on the map will be represented by a LineString), the code in Listing 9-34 illustrates how to construct a Virtual Earth or Google Maps Polyline from the WKT syntax of a LineString, WKT, using VB .NET.

Table 9-3. *Comparison of Constructors for LineString Geometries*

Format	Representation
WKT	`LINESTRING('-122.5 44.5, -112.3 48.5')`
Virtual Earth	`VEShape(VEShapeType.Polyline,` `[new VELatLong(44.5, -122.5), new VELatLong(48.5, -112.3)])`
Google Maps	`GPolyline([new GLatLng(44.5, -122.5), new GLatLng(48.5, -112.3)])`

Listing 9-34. *Converting a WKT LineString to a Google Maps Polyline Constructor*

```
'Declare a new WKT LineString
Dim WKT = LINESTRING('-122.5 44.5, -112.3 48.5')

'Remove the LINESTRING definition and opening bracket
WKT = Replace(WKT, "LINESTRING (", "")

'Remove the last bracket
WKT = Replace(WKT, ")", "")

'Create an array of each point in the LineString
Dim PointArray() As String = Split(WKT, ",")

'Loop through each point in the array
Dim i As Integer = 0
While i <= PointArray.Length - 1
  'Split the point into individual coordinates
  Dim Coords() = Split(Trim(PointArray(i)), " ")
  'Build an array of the equivalent point definitions
  If (MapType = "GM") Then
    VEGM = VEGM + "new GLatLng(" + Coords(1) + "," + Coords(0) + "),"
  ElseIf (MapType = "VE") Then
    VEGM = VEGM + "new VELatLong(" + Coords(1) + "," + Coords(0) + "),"
  End If
'Go on to the next point
  i = i + 1
End While

'Remove the trailing comma after the last point added by the loop above
VEGM GM = Left(VEGM GM, GM.Length - 1)

'Build the Google Maps Polyline constructor for this LineString
Dim GMPolyline = "new GPolyline( [" + GM + "]);"

'Build the Virtual Earth Polyline constructor for this LineString
Dim VEPolyline = "new VEShape(VEShapeType.Polyline, [" + VE + "]);"
```

Polygons

Polygon geometries are formed in a very similar way to LineString geometries, using a comma-separated array of each point contained in the Polygon ring. Remember that Polygons must be closed, so the list of points must start and end with the same point. Table 9-4 demonstrates the syntax required to create a Polygon from a series of points representing the exterior ring using WKT, Virtual Earth, and Google Maps.

Table 9-4. *Comparison of Constructors for Polygon Geometries*

Format	Representation
WKT	POLYGON(('-90.8 44.5, -75.3 48.5, -80.4 35.5, -95.8 30.5'))
Virtual Earth	VEShape(VEShapeType.Polygon, [new VELatLong(44.5, -90.8), new VELatLong(48.5, -75.3), new VELatLong(35.5, -80.4), new VELatLong(30.5, -95.8), new VELatLong(44.5, -90.8)])
Google Maps	GPolygon([new GLatLng(44.5, -90.8), new GLatLng(48.5, -75.3), new GLatLng(35.5, -80.4), new GLatLng(30.5, -95.8), new GLatLng(44.5, -90.8)])

In the example airport locator application described in this chapter, the stored procedure will return the WKT representation of a Polygon geometry representing the circular search area within which the airports displayed on the map were located. The code required to convert from a WKT Polygon representation to the equivalent Google Maps or Virtual Earth constructor is shown in Listing 9-35.

Listing 9-35. *Converting a WKT Polygon Representation to Virtual Earth and Google Maps*

```
'Declare a WKT Polygon
Dim WKT = POLYGON(('-90.8 44.5, -75.3 48.5, -80.4 35.5, -95.8 30.5'))

'Replace the double brackets which surround the coordinate point pairs
WKT = Replace(WKT, "POLYGON ((", "")

'Remove the last double brackets
WKT = Replace(WKT, "))", "")

'Create an array of each point in the Polygon
Dim PointArray() As String = Split(WKT, ",")

Dim i As Integer = 0
While i <= PointArray.Length - 1
  'Split the point into individual coordinates
  Dim Coords() = Split(Trim(PointArray(i)), " ")
  'Build an array of the equivalent point definitions
  If (MapType = "GM") Then
    VEGM = VEGM + "new GLatLng(" + Coords(1) + "," + Coords(0) + "),"
```

```
   ElseIf (MapType = "VE") Then
      VEGM = VEGM + "new VELatLong(" + Coords(1) + "," + Coords(0) + "),"
   End If
'Go on to the next point
   i = i + 1
End While

'Remove the last trailing comma
VEGM = Left(VEGM, VEGM.Length - 1)

'Build the Google Maps constructor for this Polygon
Dim GMPolygon = "new GPolygon( [" + GM + "]);"

'Build the Virtual Earth constructor for this Polygon
Dim VEPolygon = "new VEShape(VEShapeType.Polygon, [" + VE + "]);"
```

▪**Caution** Neither Virtual Earth nor Google Maps currently supports Polygons containing interior rings. In Listing 9-35, the array of points created from WKT therefore only represents those points defining the exterior ring of the Polygon.

Multielement Geometries

Neither Virtual Earth nor Google Maps directly supports multielement geometries. Although not required for this example, you could extend the handler to account for Geometry Collections by breaking them into an array of individual elements before plotting them on the map as separate geometries using the previous techniques.

Adding a Shape to the Map

Once a new geometry, shape, has been constructed using the methods described previously, it must be added to the map using the appropriate method, as shown in Listings 9-36 and 9-37.

Listing 9-36. *Adding the New Geometry to the Map Using Google Maps*

```
map.addOverlay(shape);
```

Listing 9-37. *Adding the New Geometry to the Map Using Virtual Earth*

```
map.AddShape(shape);
```

Attaching Descriptive Information to the Map

In addition to adding to the map markers that highlight the locations of all airports within a 50 km range of a given location, our application will also display an information box containing descriptive information about each airport. When using Virtual Earth, each shape may have an optional title and description attached to it, which are automatically displayed in a tool tip-style box when you hover the mouse cursor on that element. In Google Maps, we will use the `openInfoWindowHtml()` method instead, which allows us to manually specify a point at which an information box should appear, and the HTML content that it should contain. We will attach an event handler to each marker that will populate the information box with the columns of data returned from SQL Server that contain details of that airport. The event handler will cause the information box to be displayed whenever the mouse cursor hovers over the marker, to replicate the same functionality as in Virtual Earth. Listings 9-38 and 9-39 demonstrate the syntax required for each of these methods.

Listing 9-38. *Attaching Airport Information Using Google Maps*

```
google.maps.Event.addListener(Marker, "mouseover", function() {
  Marker.openInfoWindowHtml('ShapeTitle' + 'ShapeDescription');
});
```

Listing 9-39. *Attaching Airport Information Using Virtual Earth*

```
shape.SetTitle('ShapeTitle');
shape.SetDescription('ShapeDescription');
```

Customizing the Appearance of Map Elements

Each shape has a number of properties that can be altered to affect how it is displayed on the map; for example, you can change the icon used to represent Points, or adjust the coloring and shading of LineStrings and Polygons. Listings 9-40 and 9-41 illustrate how to use the methods provided by Google Maps and Virtual Earth to display geometries with a 50 percent transparent red fill and fully opaque blue borders.

Listing 9-40. *Customizing Shape Appearance Using Google Maps*

```
shape.setStrokeStyle( {color:"#0000FF", weight:10, opacity:1} )
shape.setFillStyle( {color:"#FF0000", opacity:0.5} );
```

Listing 9-41. *Customizing Shape Appearance Using Virtual Earth*

```
shape.SetLineColor(new VEColor(0, 0, 255, 1.0));
shape.SetFillColor(new VEColor(255, 0, 0, 0.5));
```

By adjusting the appearance of each geometry based on the value of an associated item of data, you can visually represent properties of the underlying feature represented by that geometry. These may be based on a discrete set of values (e.g., using different symbols to represent Points on a weather forecast map indicating rain, sun, or snow) or a continuous range of values (such as shading Polygons to represent the average rainfall received by an area of land).

Tip The methods shown in Listings 9-40 and 9-41 illustrate how to change the appearance of an existing element on the map, shape. You can also customize the style of a shape as it is created, by passing the previous relevant style properties as parameters directly to the constructor method.

Reviewing the .NET Handler

Now it's time to bring together all the individual components outlined in the preceding sections to make up the complete .NET handler. Since a lot of the code is generic to both map providers, we will create a single handler that can be used to provide dynamic data to either Google Maps or Virtual Earth. For those sections of code that are specific to one API, we will implement a Case (or If) statement to only execute certain blocks of code depending on the value of the maptype parameter provided when the handler is called. Edit the contents of your Handler.ashx file to appear as shown in Listing 9-42, changing the line highlighted in bold to supply the relevant details required to connect to your SQL Server instance.

Note For this application, because the only types of features represented on the map will be instances of Points (representing airports) or Polygons (representing the search area), the code in Listing 9-42 only handles these types of geometry. To extend this handler to deal with LineStrings or multielement geometries, you can add more cases to the Select Case myReader("GeometryType").ToString expression.

Listing 9-42. *The Complete VB .NET Handler (Handler.ashx)*

```vb
<%@ WebHandler Language="VB" Class="Handler" %>
Imports System
Imports System.Web
Imports System.Data.SqlClient
Imports System.Configuration
Imports System.Text

Public Class Handler : Implements IHttpHandler

  Public Sub ProcessRequest(ByVal context As HttpContext) _
    Implements IHttpHandler.ProcessRequest

    'Declare the global script variables
    Dim Output As String = "" 'The JavaScript response sent back to the Map API
    Dim MapType As String = context.Request.Params("maptype") 'API to use (VE/GM)
```

```vb
'Declare the variables used to create each feature
Dim WKT As String = "" 'The WKT representation provided by the stored proc
Dim VEGM As String = "" 'The VE/GMaps equivalent representation
Dim Shape As String = "" 'The unique name of each shape
Dim ShapeTitle As String = "" 'The title to display for each shape
Dim ShapeDescription As String = "" 'The description attached to the shape
Dim id As Integer = 0 'Shape counter
Dim LineStyle As String = "" 'The line style
Dim FillStyle As String = "" 'The fill style

'Set up a connection to SQL server
Dim myConn = New SqlConnection("server=ENTERSERVERNAMEHERE;" & _
                               "Trusted_Connection=yes;" & _
                               "database=Spatial")

'Open the connection
myConn.Open()

'Define the stored procedure to execute
Dim myQuery As String = "dbo.uspAirportLocator"
Dim myCMD As New SqlCommand(myQuery, myConn)
myCMD.CommandType = Data.CommandType.StoredProcedure

'Send the point the user clicked on to the stored proc
myCMD.Parameters.Add("@Latitude", Data.SqlDbType.Float)
myCMD.Parameters("@Latitude").Value = context.Request.Params("lat")
myCMD.Parameters.Add("@Longitude", Data.SqlDbType.Float)
myCMD.Parameters("@Longitude").Value = context.Request.Params("long")

'Create a reader for the result set
Dim myReader As SqlDataReader = myCMD.ExecuteReader()

'Go through the results
While myReader.Read()

  'Set a unique variable name for this shape
  Shape = "shape" + id.ToString

  'Set the title for the shape
  ShapeTitle = myReader("Title").ToString
```

```
'Set the description for the shape
ShapeDescription = myReader("Description").ToString

'Set the appropriate styling options for each shape
Select Case MapType
  Case "GM"
    'Set the color and opacity for fills
    FillStyle = """"#0000ff""", 0.5"
    'Set the color, weight, and opacity for lines
    LineStyle = """"#ffffff""", 2, 0.7"
  Case "VE"
    'Set the color and opacity for fills
    FillStyle = "new VEColor(0, 0, 255, 0.5)"
    'Set the color and opacity for lines
    LineStyle = "new VEColor(255, 255, 255, 0.7)"
End Select

'Convert from WKT to the relevant API constructor for the type of geometry
Select Case myReader("GeometryType").ToString
  Case "Point"
    'Get the WKT representation of the object
    WKT = myReader("WKT").ToString
    'Replace the double brackets that surround the coordinate point pair
    WKT = Replace(WKT, "POINT (", "")
    'Remove the closing double brackets
    WKT = Replace(WKT, ")", "")
    'Build the appropriate Pushpin/GMarker object from the coordinates
    VEGM = ""
    Dim Coords() = Split(Trim(WKT), " ")
    If (MapType = "GM") Then
      VEGM = VEGM + "new google.maps.LatLng(" +Coords(1) +"," +Coords(0) +")"
      Output += "var " + Shape + "=new google.maps.Marker(" + VEGM + ");"
    ElseIf (MapType = "VE") Then
      VEGM = VEGM + "new VELatLong(" + Coords(1) + "," + Coords(0) + ")"
      Output += "var " + Shape + _
        "=new VEShape(VEShapeType.Pushpin, " + VEGM + ");"
    End If
    'Display descriptive airport information when mouse hovers over point
    Select Case MapType
      Case "GM"
        'Display the shape title and description in an InfoWindow
        Output += "google.maps.Event.addListener(" + Shape + _
        ", ""mouseover"", function() {" + Shape + ".openInfoWindowHtml(""" + _
        ShapeTitle + "<br/>" + ShapeDescription + """);" + "});"
```

```
            Case "VE"
              'Set the shape title
              Output += Shape + ".SetTitle('" + ShapeTitle + "');"
              'Set the shape description
              Output += Shape + ".SetDescription('" + ShapeDescription + "');"
          End Select

        Case "Polygon"
          'Get the WKT representation of the object
          WKT = myReader("WKT").ToString
          'Replace the double brackets that surround the coordinate point pairs
          WKT = Replace(WKT, "POLYGON ((", "")
          'Remove the closing double brackets
          WKT = Replace(WKT, "))", "")
          'Create an array of each point in the Polygon
          Dim PointArray() As String = Split(WKT, ",")
          'Build the appropriate VE/GMaps Polygon object from the coordinates
          VEGM = ""
          Dim i As Integer = 0
          While i <= PointArray.Length - 1
            Dim Coords() = Split(Trim(PointArray(i)), " ")
            If (MapType = "GM") Then
              VEGM = VEGM + "new google.maps.LatLng(" +Coords(1) +"," +Coords(0) +"),"
            ElseIf (MapType = "VE") Then
              VEGM = VEGM + "new VELatLong(" + Coords(1) + "," + Coords(0) + "),"
            End If
            i = i + 1
          End While
          'Remove the last trailing comma
          VEGM = Left(VEGM, VEGM.Length - 1)
          'Add the constructor for the Polygon, and apply styling options
          If (MapType = "GM") Then
            Output += "var " + Shape + _
            "=new google.maps.Polygon([" + VEGM + "], " + _
            LineStyle + ", " + FillStyle + ");"
          ElseIf (MapType = "VE") Then
            Output += "var " + Shape + _
            "=new VEShape(VEShapeType.Polygon, [" + VEGM + "]);"
            Output += Shape + ".SetLineColor(" + LineStyle + ");"
            Output += Shape + ".SetFillColor(" + FillStyle + ");"
            Output += Shape + ".HideIcon();"
          End If
      End Select
```

```
    'Add the shape to the map
    Select Case MapType
      Case "GM"
        Output += "map.addOverlay(" + Shape + ");"
      Case "VE"
        Output += "map.AddShape(" + Shape + ");"
    End Select

    'Increment the shape counter
    id = id + 1
  End While

  'Close the reader
  myReader.Close()
  'Close the connection
  myConn.Close()

  'Tell the browser to handle the response as JavaScript
  context.Response.ContentType = "text/JavaScript"
  'Do not cache the results, so always load new data
  context.Response.CacheControl = "no-cache"
  'Make the response expire immediately
  context.Response.Expires = -1
  'Return the constructed JavaScript
  context.Response.Write(Output)

End Sub

ReadOnly Property IsReusable() As Boolean Implements IHttpHandler.IsReusable
  Get
    Return False
  End Get
End Property

End Class
```

■**Caution** Remember to change the line highlighted in bold in Listing 9-42 to reflect the connection string required to connect to your SQL Server instance.

Viewing the Page

Once you have completed the four components—the HTML document, the JavaScript file, the stored procedure, and the .NET handler—ensure that the changes to the files have been saved, by selecting File ➤ Save All (or by pressing Ctrl+Shift+S). You can then test the finished application by right-clicking the HTMLPage.htm file in the Solution Explorer pane of Visual Studio and selecting the View in Browser option from the menu.

Your default web browser will open, containing the relevant map from your chosen provider centered on the city of Los Angeles. You can scroll around and zoom the map using the control icons in the top left corner or the keyboard cursor keys. Clicking on any point in the United States triggers the .NET handler to retrieve data from SQL Server and display a series of markers representing the location of all airports within a 50 km distance. Moving the mouse cursor over any marker displays an information box containing descriptive information about that airport. Figures 9-2 and 9-3 show the finished web page in Google Maps and Virtual Earth, respectively, with John Wayne Airport in Orange County selected.

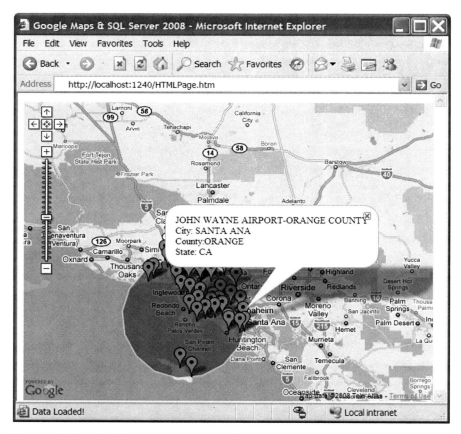

Figure 9-2. *Viewing the airport locator application in Google Maps*

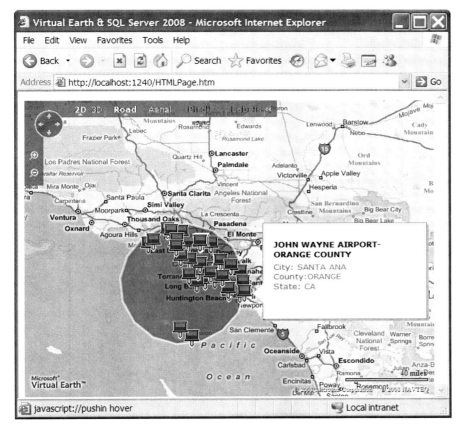

Figure 9-3. *Viewing the airport locator application in Virtual Earth*

Further Applications

The example application shown in this chapter could easily be extended and modified to cover a wide range of scenarios. For instance, by including conditional code statements in the .NET handler, you can customize the appearance of items on the map to use different symbols, colors, or sizes, based on the value of associated columns of data. To demonstrate this, Figure 9-4 illustrates a Virtual Earth map that displays election data from the 2008 US presidential election retrieved from SQL Server. This application uses a modified version of the example shown in this chapter, combining state outline data from the US Census Bureau (http://www.census.gov) with voting data as reported by *USA Today* (http://www.usatoday.com/news/politics/election2008/president.htm). Each state is shaded according to the value of the party that received the greatest number of electoral votes in that state.

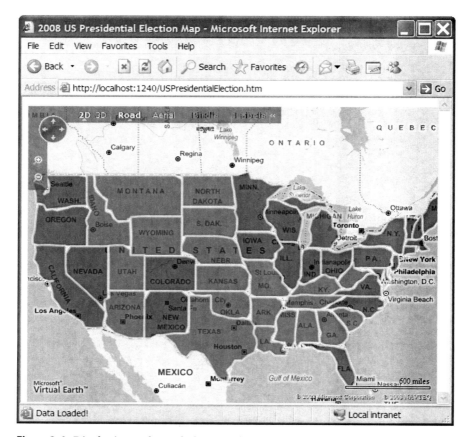

Figure 9-4. *Displaying a choropleth map of the 2008 US presidential election results*

Listing 9-43 demonstrates the code required to achieve this effect, using a VB .NET handler to generate JavaScript for the Virtual Earth API. This code sets the color of an element on the map, shape, based on the value of a column called ELEC_VOTE. If the value of ELEC_VOTE is "Democrat" then the shape (representing the state outline) is colored blue, or if it is "Republican" then the corresponding shape is colored red.

Listing 9-43. *Setting the Color of a Shape Based on a Condition Using VB .NET*

```
Select Case myReader("ELEC_VOTE").ToString
  Case "Democrat"
    JS = "shape.SetFillColor(new VEColor(0,0,255,0.7);"
  Case "Republican"
    JS = "shape.SetFillColor(new VEColor(255,0,0,0.7);"
End Select
```

Note The type of map illustrated in Figure 9-4, which shades elements according to the value of a particular property of that element, is known as a *choropleth* map.

Here are some other ideas of ways in which you can extend the example application in this chapter:

- Allow the user to manually specify the radius of the search area, and pass this as a parameter to the stored procedure.

- Apply different colors or icons to the markers on the map to reflect how many runways an airport has, or whether it has a helipad.

- Add more types of features to the dataset created—the National Transportation Atlas Database, used to source the airport data in this example, also contains details of railway tracks, hazardous material routes, ferry terminals, and many more elements that could be added to the map.

- Combine the application with the geocoding function demonstrated in Chapter 7 so that, instead of clicking a point on the map, the user types in an address as the center point of the search.

- Rather than click on a specific point location, let the user trace a line or draw a shape on the map, and have the application return those airports that lie within that area.

- Add further event handlers to make the map respond to a wider range of user actions—such as right-clicks, double-clicks, or key presses.

Summary

In this chapter, I taught you how to use two of the most popular web mapping services—Google Maps and Virtual Earth—to provide a powerful interface for displaying and interacting with spatial data from SQL Server 2008. Specifically, you learned the following:

- Microsoft Virtual Earth and Google Maps provide JavaScript APIs—libraries of methods that you can import into your own web application to create and configure a dynamic map display.

- Event handlers that enable functions to be run in response to user's actions can be added to each map.

- When used in conjunction with AJAX, an event handler can call a .NET handler that connects to SQL Server, executes a stored procedure, and dynamically adds new data to the map based on the result set returned.

- String functions in .NET can be used to manipulate the WKT output from SQL Server into the appropriate JavaScript functions required by the Google Maps and Virtual Earth APIs, These functions can then be executed within the user's browser using the eval() method.

- Each API offers a number of different methods to customize individual elements of the map, including setting different colors, styles, and icons.

- Additional descriptive information can be attached to each feature on the map and displayed in a tool tip-style window.

CHAPTER 10

Visualizing Query Results in Management Studio

In the last chapter, I showed you how to display spatial information from SQL Server 2008 using rich web mapping services such as Virtual Earth and Google Maps. While these are powerful tools for presenting spatial information for end users of web applications or reports, there are many times when, as a developer, you want to be able to quickly visualize spatial data in a simple, graphical way without the complexity of using external APIs. In this chapter, I'll show you how you can use the new Spatial Results tab in SQL Server Management Studio to visualize the results of a SELECT query containing spatial data.

SQL Server Management Studio

SQL Server Management Studio is the tool supplied with SQL Server to help you run day-to-day tasks against your database, such as performing administrative actions and writing and executing ad hoc queries. You've already used Management Studio several times in this book, to execute T-SQL code that creates, inserts, or selects spatial data. Normally, the results of any T-SQL SELECT statements executed in SQL Server Management Studio are displayed in the Results tab, represented in tabular format. This means that spatial information must be shown in WKT, WKB, or GML format, or in SQL Server's native binary format. However, SQL Server Management Studio 2008 includes a new feature, the Spatial Results tab, that provides an alternative method of viewing the results of a SELECT query in a graphical fashion.

Visualizing Spatial Results

To access the Spatial Results tab, you simply need to execute any query that selects at least one item of geometry or geography data. To illustrate this, try executing the following T-SQL statement:

```
SELECT
  'The Pentagon' AS Label,
  geography::STPolyFromText(
    'POLYGON(
      (
        -77.0532238483429 38.870863029297695 ,
        -77.05468297004701 38.87304314667469 ,
        -77.05788016319276 38.872800914712734 ,
        -77.05849170684814 38.870219840133124 ,
        -77.05556273460388 38.8690670969385 ,
        -77.0532238483429 38.870863029297695
      ),
      (
        -77.05582022666931 38.8702866652523 ,
        -77.0569360256195 38.870737733163644 ,
        -77.05673217773439 38.87170668418343 ,
        -77.0554769039154 38.871848684516294 ,
        -77.05491900444031 38.87097997215688 ,
        -77.05582022666931 38.8702866652523
      )
    )',
    4326
  ) AS Geometry
```

■**Note** The Spatial Results tab is displayed only after you execute a SELECT query that returns at least one item of raw geometry or geography data in the result set. It does not visualize the text representation of a geometry expressed by the STAsText() method, for instance.

In the Results tab at the bottom of the screen, you will see the standard tabular results showing the name of the feature and SQL Server's binary representation of the geometry. However, notice that two others tabs are available at the top of the Results pane: Spatial Results and Messages, as illustrated in Figure 10-1.

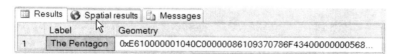

Figure 10-1. *The three tabs of the Results pane*

To visualize the results of a spatial query, simply click the Spatial Results tab. You will see the results illustrated in Figure 10-2.

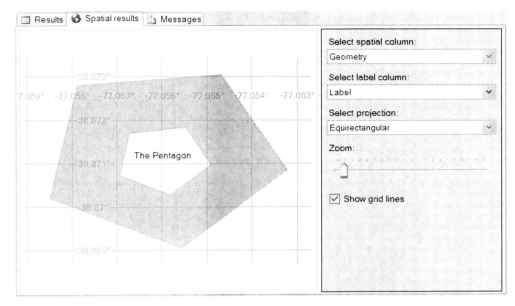

Figure 10-2. *Viewing results using the Spatial Results tab*

The map view on the left side of the Spatial Results tab displays a visual representation of the results of the SELECT query—in this case, a five-sided polygon containing an interior ring, representing the US Department of Defense Pentagon building. Although in this case we have only plotted a single Polygon geometry, the Spatial Results tab will display up to 5,000 distinct items from any one spatial column (geography or geometry) contained in the result set of a SELECT query. Each individual geometry will be automatically colored and, optionally, labeled to enable you to identify it.

Tip You can pan around the features rendered by the Spatial Results tab by clicking and dragging within the map view. Hovering the mouse cursor over a feature will bring up a tool tip window that displays additional columns of data relating to that feature.

Choosing Visualization Options

The right pane of the Spatial Results tab contains five controls that affect how the results are displayed, as described in the following list:

Select spatial column: Allows you to select which column of geography or geometry from the result set should be plotted on the map display. The Spatial Results tab can plot data from only one geometry or geography column at a time—it is not possible to overlay data from multiple columns. If you execute a SELECT query that returns more than one column of the geometry or geography datatype, then you must choose which column should be displayed.

Select label column: Each feature on the map may be labeled according to the value of another column included in the result set. For instance, if there is a column that records the name or unique identifier associated with each geometry, you can choose to label each geometry in the display with the associated value contained in that column.

Select projection (geography only): The geography datatype is used to store unprojected angular coordinates from a geographic coordinate system. However, in order to display them on any flat map, such as that used in the Spatial Results tab, they must be projected in some way. The Select Projection drop-down list offers four common projection methods: Equirectangular, Mercator, Robinson, and Bonne. (See the following section, "Supported Projections," for details on how each projection works.) Note that if you are plotting results from the geometry datatype, no projection is required because the coordinates of each geometry already lie in a flat plane.

Zoom: This slider allows you to zoom in and out of the rendered map.

Show grid lines: Checking this check box allows you to plot the graticule of latitude and longitude (or x and y) values over the map window.

Note The Spatial Results tab is designed to be used only for development purposes within SQL Server Management Studio—you cannot export the map or embed it in any other applications.

Supported Projections

As previously mentioned, when using the Spatial Results tab to display data from the geography datatype, you may choose to display it using one of the following projections:

- Equirectangular
- Mercator
- Robinson
- Bonne

Note Data stored in the geometry datatype has already been projected onto a flat plane and will be presented according to the projection with which it was created. It is not possible to use the Spatial Results tab to reproject geometry data into another projection.

Equirectangular

The equirectangular projection is one of the simplest map projections, in both its definition and construction. In an equirectangular projection, the values of the angular coordinates of latitude and longitude are plotted directly onto the y and x axes of the projected map. This creates a rectangular map image, with a width:height ratio of 2:1, as illustrated in Figure 10-3.

Originally developed nearly 2,000 years ago, the equirectangular projection is still used today, especially in the production of thematic or overview world maps. The projection is not widely used for in-depth spatial analysis, however, because the features of the projected map feature a large degree of distortion (particularly near the poles).

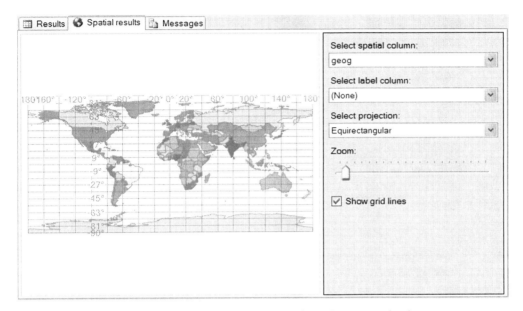

Figure 10-3. *The equirectangular projection of results from the geography datatype*

Note Since the equirectangular projection assigns latitude and longitude values directly to the y and x axes, maps created using the equirectangular projection are sometimes known as *unprojected* maps.

Mercator

The Mercator projection is one of the most commonly seen and well-recognized map projections in the Western world, owing to its widespread use in newspapers, television, and other popular media. For many years it was also the standard projection used in elementary school geography textbooks.

The Mercator projection uses a cylindrical projection method that, like the equirectangular projection, accurately depicts elements close to the equator, but introduces an increasing level of distortion toward the poles. Despite this, it is still widely used in world maps and, particularly, in nautical navigation. Figure 10-4 illustrates the Mercator projection.

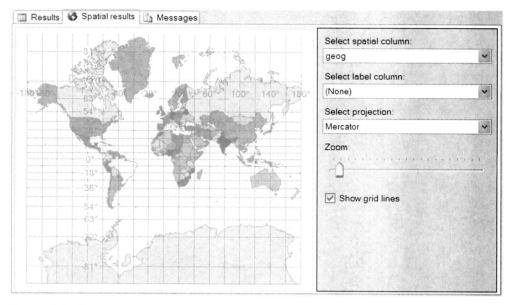

Figure 10-4. *The Mercator projection of results from the geography datatype*

■**Note** For more information on the equirectangular or Mercator projections, refer to the discussion of projection methods in Chapter 1.

Robinson

The Robinson projection was developed by Dr. Arthur Robinson, in 1963. In contrast to most other projection methods, which are defined by mathematical formulae, the Robinson projection is created by referring to a table of specified values. These values define the relative length of each parallel of latitude and the distance at which that parallel should be depicted from the equator. The exact values on which the Robinson projection is based were manually determined by Dr. Robinson to provide the most "correct-looking" representation of the whole world.

Unlike some projections that retain one particular property of the features of the map (e.g., area, distance, shape) while sacrificing others, the Robinson projection does not completely preserve any one aspect, but rather aims to balance distortion across all aspects of the projected image. As a result, the Robinson projection is sometimes referred to as a "compromise" projection. The Robinson projection of the world is illustrated in Figure 10-5.

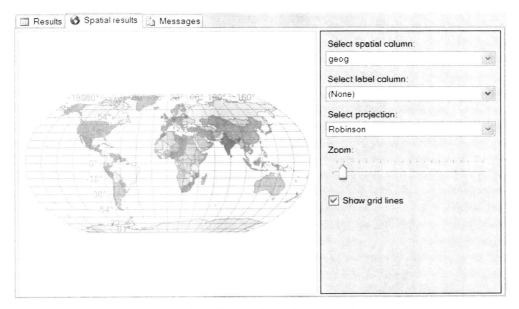

Figure 10-5. *The Robinson projection of results from the geography datatype*

Bonne

The Bonne projection is named after Rigobert Bonne (ca. 1729–1795), the Royal Hydrographer to the French court. Although named after Bonne, the key features of this projection are based on earlier models, including those proposed by Ptolemy and Sylvanus. The Bonne projection is a *pseudoconic* projection, in which all the parallels of latitude are constructed from the concentric arcs of a circle. To construct the map, a parallel is chosen at which a conic tangent is made, and, depending on the parallel chosen, there may be significant differences in the appearance of the constructed map image. When chosen in the Spatial Results tab, the Bonne projection is displayed based on a central parallel at the center of the region being mapped. When plotting a dataset covering the whole of the earth, this leads to a projected map image, as shown in Figure 10-6.

The map image created by the Bonne projection is twice as wide as it is tall, retains the relative areas of elements of the map, and accurately depicts the horizontal distance between any two points. However, it distorts other elements of the image—particularly the shape of features.

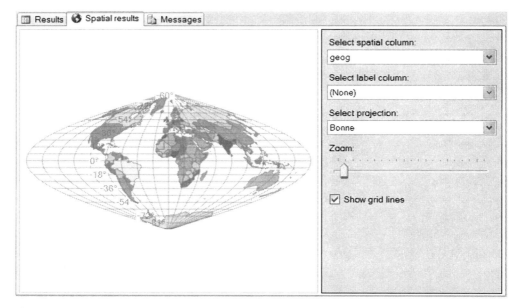

Figure 10-6. *The Bonne projection of results from the geography datatype*

For comparative purposes, the Bonne projection illustrated in Figure 10-6 plots the same world country data used in the previous figures in this chapter. However, in practice, the Bonne projection is rarely used to project the whole world in this manner. Instead, it is normally used to project features lying in a single hemisphere, and is most commonly seen in atlas maps of continents. Figure 10-7 illustrates how the Bonne projection can be used to create a map representing South America. The projection in this case is centered about the line of latitude that lies at the center of the data plotted on the map—which is approximately 70°W. Note how, when based on different parameters (as in this case), the overall appearance of the image created by the Bonne projection is very different from that shown in Figure 10-6.

■**Tip** The country features shown in Figures 10-3 to 10-6 represent administrative boundary shapefile data available from the United Nations GEO Data Portal, which you can download from `http://geodata.grid.unep.ch/mod_download/download_geospatial.php?selectedID=290&temptxt=download/admin98_li_shp.zip`. For more information on how to import shapefile data in SQL Server 2008, refer to Chapter 6.

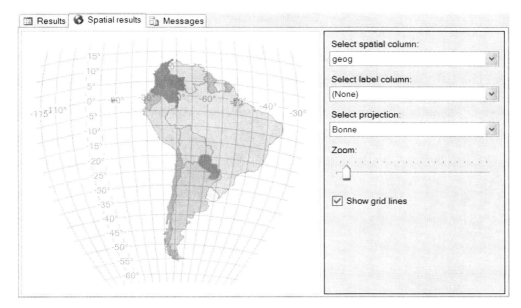

Figure 10-7. *The Bonne projection of South America*

Note At first glance, it may appear that the features represented in Figure 10-7 are not centered, and the projection should instead be based on the line of longitude at about 60°W, which divides the central landmass of South America into two approximately equal parts. However, the South American dataset illustrated includes Easter Island (a territory of Chile), which lies at approximately 109°W, -28°S. Although too small to be visible in the Spatial Results tab, this outlying island must be considered when determining the central point of the Bonne projection.

Summary

In this chapter, I showed you how to use the Spatial Results tab in SQL Server Management Studio 2008 to display graphical results of ad hoc spatial queries. Important features of the Spatial Results tab are as follows:

- To access the Spatial Results tab, you must execute a SELECT query that returns at least one item of geometry or geography data.

- The Spatial Results tab can display up to 5,000 items of geography or geometry data in a zoomable, pannable window.

- Each item displayed may be optionally labeled with a value from an associated column of data in the results.

- You cannot save or export the display shown in the Spatial Results tab—it is intended for ad hoc use within SQL Server Management Studio only.

- To display results of the geography datatype, features may be projected using any one of four projection methods: equirectangular, Mercator, Robinson, or Bonne.

PART 4

Analyzing Spatial Data

This part of the book discusses the different methods available to filter and analyze data in the geography and geometry datatypes, which provide the key functionality you need to use to exploit the power of spatial data in your applications.

The methods have been divided into three categories, one per chapter: Chapter 11 deals with those methods that analyze and return information concerning the properties of individual spatial objects, Chapter 12 covers methods that define new objects through the combination or modification of existing objects, and Chapter 13 introduces methods that test the relationship between objects, such as their equality, proximity, or intersection.

Examining Properties of Spatial Objects

There are many different questions that we might ask about any individual item of spatial data: Where is it? How big is it? What sort of object is it? Where does it start and end? Where does its center lie? In this chapter, we examine the methods that SQL Server provides to answer all of these questions and more. All of the techniques discussed in this chapter apply only to a single item of geography or geometry data, and generally do not require any parameters. The standard syntax required to access any property of an instance is therefore

```
Instance.Property
```

where `Instance` is the name of the column or variable containing the information that you want to query, and `Property` is the name of the particular property in question.

In some cases, certain aspects of information about an item are not strictly *properties* of that item, but rather the result of a method, in which case the appropriate syntax is

```
Instance.Method()
```

Note To help you understand the results obtained from each of the methods in this chapter, each method will be illustrated with an example code listing. That code listing will `SELECT` a geometry to which the method is applied, the result returned by the method when used on that geometry, and, if the result returned by the method is a `geometry` or `geography` instance, the WKT representation of that instance. I'll show the WKT representation of the output, but you can also use the Spatial Results tab in SQL Server Management Studio to visualize the results of each method.

As with the static methods discussed in Chapter 4, many of the instance methods discussed in this chapter are based on standards defined by the Open Geospatial Consortium. For such methods, the name of the method is prefixed by the letters "ST," such as `STX`, `STCentroid()`, or `STLength()`. A number of these standard methods are used to describe whether a geometry exhibits a specific behavior—is it simple, is it closed, is it empty? The answer to any of these questions is a Boolean response, meaning that it can only be true or false. In such cases, the name of the method follows the convention `STIsXXX()`, where XXX is the property or behavior being tested. The value returned by the method is either 0 or 1. If the instance does exhibit the behavior in question, the response is 1 (true), whereas if the instance does not exhibit the behavior specified by the method, the result is 0 (false).

Caution Although many of the methods and properties discussed in this chapter apply to both the geometry and geography datatypes, some methods are implemented by only one type. For each method I introduce, I'll tell you which datatype (or datatypes) implements it and how it can be used on instances of that type.

Returning the Name of a Geometry Type

Remember that a single item of geography or geometry data may be represented by one of several different types of geometry—Point, LineString, Polygon, or a collection containing combinations of these three. One of the most fundamental properties of any item of spatial data is therefore the type of geometry it uses to represent the feature it describes. While this might appear obvious from a visual examination of the WKT or GML representation of the item, it is certainly not obvious from the WKB representation or SQL Server's own binary representation formats. You can use the STGeometryType() method to return a string description of the type of geometry described by an item of geography or geometry data.

Supported Datatypes

The STGeometryType() method can be used on instances of the following datatypes:

- geometry

- geography

Usage

STGeometryType() requires no parameters and can be used against an item of geography or geometry data as follows:

Instance.STGeometryType()

The value returned by the STGeometryType() method is an nvarchar(4000) string that contains the name of the type of geometry represented by a particular instance. The possible values returned by STGeometryType() are as follows:

- Point

- LineString

- Polygon

- MultiPoint

- MultiLineString

- MultiPolygon
- GeometryCollection

Example

To demonstrate the STGeometryType() method, consider the following example, which creates an item of geometry data representing Mexico City using SRID 32614 (UTM Zone 14N), and then selects the type of geometry represented by that item using the STGeometryType() method:

```
DECLARE @MexicoCity geometry
SET @MexicoCity = geometry::STGeomFromText('POINT(486321 2146238)', 32614)
SELECT
  @MexicoCity AS Shape,
  @MexicoCity.STGeometryType() AS GeometryType
```

The result of the STGeometryType() method is as follows:

```
Point
```

This result confirms that the @MexicoCity variable uses a Point geometry to represent the location of Mexico City.

Returning the Number of Dimensions Occupied by a Geometry

Like STGeometryType(), the STDimension() method can be used to provide basic information describing the type of geometry used to represent a feature. However, rather than return a string description of the type of geometry represented by a geography or geometry instance, STDimension() returns an integer value representing *the number of dimensions* which that geometry type occupies. As discussed in Chapter 1, Points are zero-dimensional geometries, LineStrings occupy one dimension, and Polygons occupy two dimensions.

Supported Datatypes

The STDimension() method can be used on instances of the following datatypes:

- geometry
- geography

Usage

The STDimension() method does not require any parameters and can be used against an item of geography or geometry data as follows:

```
Instance.STDimension()
```

The result of the STDimension() method is an integer value representing the number of dimensions occupied by Instance, as follows:

- For a Point or MultiPoint, STDimension() returns 0.

- For a LineString or MultiLineString, STDimension() returns 1.

- For a Polygon or MultiPolygon, STDimension() returns 2.

- For empty geometries of any type (that is, a geometry that contains no points in its definition), STDimension() returns −1.

- For a Geometry Collection containing several different types of geometry, STDimension() returns the maximum number of dimensions of any element contained within that collection.

Note Single-element and multielement instances of the same type of geometry occupy the same number of dimensions.

Example

The usage of the STDimension() method is demonstrated in the following code, which creates an instance of a LineString geometry and then tests the number of dimensions it occupies:

```
DECLARE @LineString geometry
SET @LineString = geometry::STGeomFromText('LINESTRING(-120 48, -122 47)', 0)
SELECT
  @LineString AS Shape,
  @LineString.STDimension() AS Dimension
```

The result of the STDimension() method is as follows:

1

Testing Whether a Geometry Is of a Particular Type

The InstanceOf() method complements the functionality of the STGeometryType() method in describing the type of geometry specified by an instance. Whereas STGeometryType() is used as a descriptive method to return the name of the type of geometry represented by an instance, the InstanceOf() method is used to test whether an instance represents a given type of geometry. The type of geometry against which to test the instance must be provided as a parameter to the InstanceOf() method, as a text string.

Supported Datatypes

The InstanceOf() method can be used on instances of the following datatypes:

- geometry

- geography

Usage

The syntax for using the InstanceOf() method is as follows:

Instance.InstanceOf(geometry_type)

The parameter geometry_type, which is an nvarchar(4000) string, enables you to determine whether an instance is an example of the specified geometry type. For example, Instance. InstanceOf('LineString') returns 1 if Instance is a LineString geometry, or 0 otherwise.

The InstanceOf() method returns 1 when Instance is of the exact type of geometry specified by geometry_type, or if Instance is of any of the types of geometry that are *descended from* geometry_type. The value of geometry_type against which an instance is tested can be not only one of the seven instantiable object types, as returned by STGeometryType(), but also the additional Curve, Surface, MultiCurve, and MultiSurface geometry types from which LineStrings, Polygons, MultiLineStrings, and MultiPolygons are descended, respectively. You can also specify the generic *geometry* type from which all specific geometry types are descended. The full list of possible values for geometry_type is as follows:

- Geometry
- Point
- MultiPoint
- Curve
- LineString
- MultiCurve
- MultiLineString
- Surface
- Polygon
- MultiSurface
- MultiPolygon
- GeometryCollection

Note For an illustration of the inheritance tree that shows the relationship between these geometry types, refer to Chapter 3.

Example

The following example defines a geometry MultiPoint instance, and then uses InstanceOf() to test whether it is an example of the Geometry Collection geometry type:

```
DECLARE @MultiPoint geometry
SET @MultiPoint = geometry::STGeomFromText('MULTIPOINT(0 0, 51 2, -3 10.5)', 0)
SELECT
  @MultiPoint AS Shape,
  @MultiPoint.InstanceOf('GEOMETRYCOLLECTION') AS InstanceOfGeomColl
```

The result of the InstanceOf() method is as follows:

1

This confirms that @MultiPoint is an instance of a type descended from the Geometry Collection geometry type.

Note Remember that MultiPoint, MultiLineString, and MultiPolygon geometries are all homogenous multielement geometries descended from the generic multielement Geometry Collection object. If using the InstanceOf() method to test whether any of these types are examples of a Geometry Collection, the result will be 1 (true).

Comparing the Results of STDimension(), STGeometryType(), and InstanceOf()

Table 11-1 illustrates a comparison of the results obtained from the STGeometryType(), STDimension(), and InstanceOf() methods for various types of geometry. The STGeometryType() and STDimension() columns show the result obtained from calling each method on instances of the type of geometry shown in the Geometry column. The InstanceOf() column shows the values of geometry_type that will return a value of 1 (true) when called on that type of geometry.

Table 11-1. *Comparing Results of STGeometryType(), STDimension(), and InstanceOf() Methods*

Geometry	STGeometryType()	STDimension()	InstanceOf()
Point	Point	0	Geometry, Point
LineString	LineString	1	Geometry, Curve, LineString
Polygon	Polygon	2	Geometry, Surface, Polygon
MultiPoint	MultiPoint	0	Geometry, GeometryCollection, MultiPoint
MultiLineString	MultiLineString	1	Geometry, GeometryCollection, MultiCurve, MultiLineString
MultiPolygon	MultiPolygon	2	Geometry, GeometryCollection, MultiSurface, MultiPolygon
GeometryCollection	GeometryCollection	–1, 0, 1, 2[a]	Geometry, GeometryCollection
Empty Point	Point	–1	Geometry, Point
Empty LineString	LineString	–1	Geometry, Curve, LineString
Empty Polygon	Polygon	–1	Geometry, Surface, Polygon

[a] *When you use the* STDimension() *method on a GeometryCollection instance, it returns the greatest number of dimensions of any element in that particular collection. For instance, for a Geometry Collection containing only Point and LineString elements, the result of* STDimension() *would be 1.*

Note An empty geometry is one that contains no points.

An illustration of the results obtained from these methods is shown in Figure 11-1.

STGeometryType() = Point
STDimension() = 0
InstanceOf('Geometry') = 1
InstanceOf('Point') = 1

STGeometryType() = MultiPoint
STDimension() = 0
InstanceOf('Geometry') = 1
InstanceOf('GeometryCollection') = 1
InstanceOf('MultiPoint') = 1

STGeometryType() = LineString
STDimension() = 1
InstanceOf('Geometry') = 1
InstanceOf('Curve') = 1
InstanceOf('LineString') = 1

STGeometryType() = MultiLineString
STDimension() = 1
InstanceOf('Geometry') = 1
InstanceOf('GeometryCollection') = 1
InstanceOf('MultiCurve') = 1
InstanceOf('MultiLineString') = 1

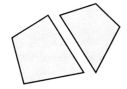

STGeometryType() = Polygon
STDimension() = 2
InstanceOf('Geometry') = 1
InstanceOf('Surface') = 1
InstanceOf('Polygon') = 1

STGeometryType() = MultiPolygon
STDimension() = 2
InstanceOf('Geometry') = 1
InstanceOf('GeometryCollection') = 1
InstanceOf('MultiSurface') = 1
InstanceOf('MultiPolygon') = 1

Figure 11-1. *Comparing results of the STGeometryType(), STDimension(), and InstanceOf() methods*

Testing Whether a Geometry Is Simple

Back in Chapter 1, I introduced the criteria that different geometries must meet in order to be classified as *simple* geometries. Essentially, the requirement for simplicity is that the geometry cannot self-intersect—that is, it cannot contain the same point more than once (except in the case of the start/end point of a LineString). Some specific examples are as follows:

- Point objects are always simple. MultiPoint objects are simple, so long as they do not contain the same Point twice.

- LineStrings and MultiLineStrings are simple so long as the path drawn between the points does not cross itself.

- Polygons and MultiPolygons are always simple.

The STIsSimple() method can be used against any instance of geometry data to test whether it meets the relevant criteria for the type of geometry in question.

Supported Datatypes

The STIsSimple() method can be used on instances of the following datatype:

- `geometry`

Usage

The STIsSimple() method requires no parameters and can be invoked on an instance of geometry data as follows:

```
Instance.STIsSimple()
```

If the geometry represented by the instance is simple, the STIsSimple() method returns the value 1. If the geometry fails to meet the criteria required for simplicity, then the method returns 0. Figure 11-2 illustrates the results obtained from the STIsSimple() method when used on different types of geometries.

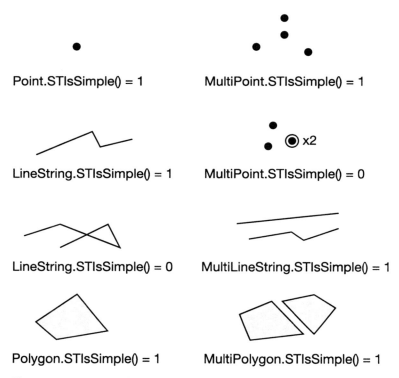

Figure 11-2. *Testing whether a geometry is simple by using STIsSimple()*

Example

The following example creates a LineString geometry representing the route taken by a delivery van through the center of New York City. The LineString contains six points, representing the individual locations at which the van stops to make a delivery (for the purposes of illustration, we'll assume that the van takes the shortest straight-line journey between each location). The STIsSimple() method is then used to test whether the LineString geometry is simple.

```
DECLARE @DeliveryRoute geometry
SET @DeliveryRoute = geometry::STLineFromText(
  'LINESTRING(586960 4512940, 586530 4512160, 585990 4512460,
  586325 4513096, 587402 4512517, 587480 4512661)', 32618)
SELECT
  @DeliveryRoute AS Shape,
  @DeliveryRoute.STIsSimple() AS IsSimple
```

The result of the `STIsSimple()` method is as follows:

0

This result tells us that the LineString geometry `@DeliveryRoute` is not simple—during its journey, the van must cross back over part of the route it has previously traveled. This might be an indication that the route represented by `@DeliveryRoute` is not the optimal route between the destinations. To demonstrate, suppose that the van had started at the same point but then had taken a different route between the remaining points, as follows:

```
DECLARE @DeliveryRoute geometry
SET @DeliveryRoute = geometry::STLineFromText(
  'LINESTRING(586960 4512940, 587480 4512661, 587402 4512517,
  586325 4513096, 585990 4512460, 586530 4512160)', 32618)
SELECT
  @DeliveryRoute AS Shape,
  @DeliveryRoute.STIsSimple() AS IsSimple
```

In this case, the LineString connects the same six points, but does not cross back on itself. The result of the `STIsSimple()` method used on this geometry is therefore 1.

By making the route simple, we have eliminated the need for the van to recross its path, reducing the total distance traveled from 3.6 km to 3.3 km. You can confirm this by using the `STLength()` method, which is introduced later in this chapter.

Testing Whether a Geometry Is Closed

A geometry can be defined as *closed* based on the following rules:

- A Point geometry is not closed.

- A LineString is closed only if the start and end points are the same.

- All Polygon instances are closed.

- A Geometry Collection containing any unclosed geometry (a Point, or an unclosed LineString) is, itself, not closed.

The `STIsClosed()` method can be used to test whether a geometry instance meets these criteria. Figure 11-3 illustrates the results of the `STIsClosed()` method when used on examples of different types of geometry.

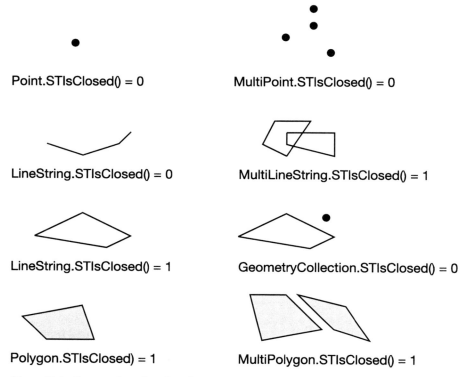

Figure 11-3. *Comparing closed and not closed geometries using STIsClosed()*

Supported Datatypes

The STIsClosed() method can be used on instances of the following datatype:

- geometry

Usage

The STIsClosed() method requires no parameters and can be used as follows:

```
Instance.STIsClosed()
```

The STIsClosed() method returns a value of 1 if the instance is closed, or 0 if the instance is not closed.

Example

The summit of Mount Snowdon (*Yr Wyddfa*, in Welsh) is the highest mountain peak in Wales, at an altitude of just over 3,500 ft above sea level. The following example creates a MultiLineString geometry containing a number of LineStrings representing contour lines around the summit of the mountain. A contour line is a line that connects points of equal elevation, so, in addition to stating longitude and latitude coordinates, the points of each LineString all specify a z coordinate value equal to the height of the contour which that LineString represents, measured in feet above sea level. The STIsClosed() method is then used to test whether the MultiLineString instance is closed.

```
DECLARE @Snowdon geometry
SET @Snowdon = geometry::STMLineFromText(
'MULTILINESTRING(
 (-4.07668 53.06804 3445,  -4.07694 53.06832 3445,  -4.07681 53.06860 3445,
  -4.07668 53.06869 3445,  -4.07651 53.06860 3445,  -4.07625 53.06832 3445,
  -4.07661 53.06804 3445,  -4.07668 53.06804 3445),
 (-4.07668 53.06776 3412,  -4.07709 53.06795 3412,  -4.07717 53.06804 3412,
  -4.07730 53.06832 3412,  -4.07730 53.06860 3412,  -4.07709 53.06890 3412,
  -4.07668 53.06898 3412,  -4.07642 53.06890 3412,  -4.07597 53.06860 3412,
  -4.07582 53.06832 3412,  -4.07603 53.06804 3412,  -4.07625 53.06791 3412,
  -4.07668 53.06776 3412),
 (-4.07709 53.06768 3379,  -4.07728 53.06778 3379,  -4.07752 53.06804 3379,
  -4.07767 53.06832 3379,  -4.07773 53.06860 3379,  -4.07771 53.06890 3379,
  -4.07728 53.06918 3379,  -4.07657 53.06918 3379,  -4.07597 53.06890 3379,
  -4.07582 53.06879 3379,  -4.07541 53.06864 3379,  -4.07537 53.06860 3379,
  -4.07526 53.06832 3379,  -4.07556 53.06804 3379,  -4.07582 53.06795 3379,
  -4.07625 53.06772 3379,  -4.07668 53.06757 3379,  -4.07709 53.06768 3379))',
 4326)
SELECT
  @Snowdon AS Shape,
  @Snowdon.STIsClosed() AS IsClosed
```

Since each LineString element contained within the MultiLineString ends at the same point from which it started, @Snowdon is a closed geometry, as shown by the following result returned by the STIsClosed() method:

1

■**Tip** Remember that you can use the optional z coordinate to state the elevation, or height, of each point in a geometry.

Testing Whether a LineString Is a Ring

Rings are LineString geometries that are both simple and closed. The STIsRing() method is the OGC-compliant method for testing whether a geometry instance is an example of a ring. Figure 11-4 illustrates the results of the STIsRing() method when used on a sample of different LineString geometries.

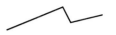  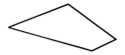

LineString.STIsRing() = 0   LineString.STIsRing() = 0   LineString.STIsRing() = 1

Figure 11-4. *Testing whether a LineString geometry is a ring*

Supported Datatypes

The STIsRing() method can be used on instances of the following datatype:

- geometry

Usage

The syntax for using the STIsRing() method is as follows:

```
Instance.STIsRing()
```

When used on a LineString instance, STIsRing() returns a value of 1 if the instance meets the criteria for a ring, or 0 if the instance is not a ring. When used against any sort of geometry other than a LineString, the method returns NULL.

Tip Since a ring is a closed, simple LineString, STIsRing() = 1 is logically equivalent to InstanceOf('LineString') = 1 AND STIsClosed() = 1 AND STIsSimple() = 1.

Example

The following example creates a LineString geometry representing the track of the Indianapolis Motor Speedway, home of the Indy 500 race. It then uses the STIsRing() method to test whether the geometry created is a ring.

```
DECLARE @Speedway geometry
SET @Speedway = geometry::STLineFromText(
    'LINESTRING(565900 4404737, 565875 4405861, 565800 4405987, 565670 4406055,
        565361 4406050, 565222 4405975, 565150 4405825, 565170 4404760, 565222 4404617,
        565361 4404521, 565700 4404524, 565834 4404603, 565900 4404737)', 32616)
SELECT
  @Speedway AS Shape,
  @Speedway.STIsRing() AS IsRing
```

The result of the STIsRing() method is as follows:

This result confirms that the geometry representing the oval-shaped track is a ring—it starts and ends at the same point, and does not cross itself.

Counting the Number of Points in a Geometry

The STNumPoints() method returns the number of points contained in the definition of a geography or geometry instance. Every point listed in the geometry definition is counted; if the geometry definition includes the same point several times, it will be counted multiple times. For instance, remember that in a Polygon definition, the start and end points of each linear ring are the same. Since this point is duplicated, when used against a Polygon geometry, the result of STNumPoints() will always be greater than the number of sides of the shape.

Figure 11-5 illustrates the results of the STNumPoints() method when used against a variety of geometry types.

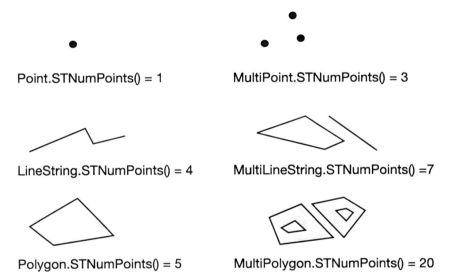

Point.STNumPoints() = 1   MultiPoint.STNumPoints() = 3

LineString.STNumPoints() = 4   MultiLineString.STNumPoints() =7

Polygon.STNumPoints() = 5   MultiPolygon.STNumPoints() = 20

Figure 11-5. *Counting the number of points in a geometry using STNumPoints()*

Supported Datatypes

The STNumPoints() method can be used on instances of the following datatypes:

- geometry
- geography

Usage

The STNumPoints() method does not take any parameters and can be used on an item of geography or geometry data as follows:

```
Instance.STNumPoints()
```

The result of the method is an integer value representing the number of points contained in the instance.

Example

The Bermuda Triangle is an area of the Atlantic Ocean famed for causing the unexplained disappearance of ships and aircraft as they pass through. Although the exact location of the triangle varies between different sources, it is popularly defined as being the area contained within the points of Miami, Florida; San Juan, Puerto Rico; and the island of Bermuda. The following example creates a geography Polygon representing the Bermuda Triangle based on these points, and then calls the STNumPoints() method on the resulting instance:

```
DECLARE @BermudaTriangle geography
SET @BermudaTriangle = geography::STPolyFromText(
  'POLYGON((-66.07 18.45, -64.78 32.3, -80.21 25.78, -66.07 18.45))',
  4326)

SELECT
  @BermudaTriangle AS Shape,
  @BermudaTriangle.STNumPoints() AS NumPoints
```

The result of the STNumPoints() method is as follows:

4

Caution The definition of a three-sided polygon, such as the Bermuda Triangle, contains four points, *not* three, as you might think! This is because the STNumPoints() method counts the point at the start and end of the polygon ring twice.

Testing Whether a Geometry Is Empty

An *empty* geometry is one that does not contain any points. Even though it contains no points, an empty geometry is still nominally assigned a particular type—so you may have an empty Point or an empty LineString geometry, for instance.

You might be wondering why you would ever create an empty geometry—why create a shape to represent a feature with no points defined? One way of thinking about this is as follows: if geometries represent the position (and therefore, by implication, the presence) of features on the earth's surface, then empty geometries denote the *absence* of any such features. Empty geometries can be used as a response to the question "Where is x?" when the answer is "nowhere on Earth."

For instance, the STIntersection() method can be used to return a geometry that represents the set of points shared between two geometry instances. If those instances have no points in common (they are *disjoint*), then the result of the STIntersection() method will be an empty geometry.

Note An empty geometry is not the same as NULL. A NULL value in a geometry or geography column suggests that a result has not been evaluated. An empty geometry suggests that a result has been evaluated, but that it does not represent a feature on the earth.

Supported Datatypes

The STIsEmpty() method can be used on instances of the following datatypes:

- geometry
- geography

Usage

The STIsEmpty() method does not require any parameters and can be used against an instance of either the geometry or geography datatype as follows:

```
Instance.STIsEmpty()
```

If the geometry represented by Instance does not contain any points, then the result of STIsEmpty() will be 1. Otherwise, the result will be 0.

Tip STIsEmpty() = 1 is logically equivalent to STNumPoints() = 0.

Example

The following example creates two parallel LineString geometries, @LineString1 and @LineString2. It then uses the STIntersection() method to create the geometry formed from the intersection of the two LineStrings, before calling the STIsEmpty() method on the resulting geometry.

```
DECLARE @LineString1 geometry
DECLARE @LineString2 geometry
SET @LineString1 = geometry::STLineFromText('LINESTRING(2 4, 10 6)', 0)
SET @LineString2 = geometry::STLineFromText('LINESTRING(0 2, 8 4)', 0)
SELECT
  @LineString1.STUnion(@LineString2) AS Shape,
  @LineString1.STIntersection(@LineString2).STIsEmpty() AS IsEmpty
```

The result of the STIsEmpty() method is as follows:

1

This shows that the geometry created by the STIntersection() method is an empty geometry, containing no points. In other words, @LineString1 and @LineString2 do not contain any points in common with each other (because they are parallel).

Note This example uses the STUnion() method to combine both LineString geometries (for display on the Spatial Results tab), and the STIntersection() method to calculate the intersection between the two geometries. Both of these methods are explained in more detail in Chapter 12.

Returning Cartesian Coordinate Values

Remember that points in the geometry datatype are defined by Cartesian coordinates of x and y, or easting and northing coordinates from a projected spatial reference system. In order to retrieve the individual coordinate values from which any point has been created, you can use the STX and STY properties of the geometry type.

The STX and STY properties apply to Point geometries instantiated directly from static methods, such as STPointFromText(), as well as to points that are returned as the result of other methods, such as STPointN() or STCentroid(), discussed later in this chapter.

Note The coordinate values of a Point geometry are accessed as properties of that object rather than via a method. When retrieving a coordinate value using STX or STY, you should not include round brackets after the property name.

Supported Datatypes

The STX and STY properties can be used with Point instances of the following datatype:

- geometry

Usage

The STX and STY properties apply to Point instances of the geometry datatype as follows:

```
Instance.STX
Instance.STY
```

The result obtained from either of these properties is a floating-point number representing the relevant coordinate value.

Example

The following example creates a Point geometry from projected coordinates representing the location of Johannesburg, South Africa, using the UTM projection (Zone 35, Southern Hemisphere). It then uses the STX and STY properties to retrieve the x (easting) and y (northing) coordinates of that Point.

```
DECLARE @Johannesburg geometry
SET @Johannesburg = geometry::STGeomFromText('POINT(604931 7107923)', 32735)
SELECT
  @Johannesburg.STX AS X,
  @Johannesburg.STY AS Y
```

The results are as follows:

```
X        Y
604931   7107923
```

Returning Geographic Coordinate Values

When using the geography datatype, the equivalent functionality of the STX and STY properties is provided by the Lat and Long properties instead. These properties work in exactly the same way as STX and STY, except that, instead of relating to x and y planar coordinate values of a Point, they retrieve the geographic coordinates of latitude and longitude.

Supported Datatypes

The Lat and Long properties can be used on Point instances of the following datatype:

- geography

Usage

The Lat and Long properties can be applied to Point instances of the geography datatype as follows:

```
Instance.Lat
Instance.Long
```

The result of each method is a floating-point number representing the relevant coordinate value.

Example

The following example creates a Point using the geography datatype from a Well-Known Binary representation, corresponding to the location of Colombo, Sri Lanka. It then uses the Long and Lat properties to retrieve the longitude and latitude coordinates of that Point.

```
DECLARE @Colombo geography
SET @Colombo =
  geography::STGeomFromWKB(0x0101000000666666666666F65340B81E85EB51B81B40, 4326)
SELECT
  @Colombo.Long AS Longitude,
  @Colombo.Lat AS Latitude
```

The results are as follows:

```
Longitude   Latitude
79.85         6.93
```

Returning Extended Coordinate Values

Remember that, in addition to the required x and y (or longitude and latitude) coordinates, each point in a geometry definition may also contain optional z and m coordinates. The z coordinate is used to store the height, or elevation, of the point. The m coordinate is any measure that can be associated with the point, expressed as a floating-point number. In order to retrieve these additional coordinate values, you can use Z and M properties on a Point instance.

Supported Datatypes

The Z and M properties can be used on Point instances of the following datatypes:

- geometry
- geography

Usage

The m and z coordinate properties can be accessed on a geometry or geography Point instance as follows:

```
Instance.M
Instance.Z
```

The result is a floating-point number representing the appropriate z or m coordinate value. If no coordinate value is defined, the result is NULL.

Example

The Federal Communications Commission maintains a database of antenna structures registered for wireless telecommunications communication within the United States. The database contains a variety of fields, including the latitude and longitude of each antenna and the overall height above ground level. You can search the database online at http://wireless2.fcc.gov/UlsApp/AsrSearch/asrRegistrationSearch.jsp.

The following example creates a Point geometry representing one such antenna, located at a latitude of 39°49'54"N and a longitude of 89°38'52"W, using the EPSG:4269 spatial reference system. The antenna extends to 34.7 m above ground level, which is represented by the z coordinate of the point. The m coordinate is assigned a value of 1000131, which represents a reference number assigned to this antenna. The example then demonstrates how the z and m properties can be used to retrieve the corresponding coordinate values of the Point.

```
DECLARE @Antenna geography
SET @Antenna =
  geography::STPointFromText('POINT(-89.64778 39.83167 34.7 1000131)', 4269)
SELECT
  @Antenna.M AS M,
  @Antenna.Z AS Z
```

The results are as follows:

M	Z
1000131	34.7

Returning a Specific Point from a Geometry

The STPointN() method can be used to isolate and return any individual point from the definition of a geometry. Based on the value of a parameter supplied to the method, n, the STPointN() method returns the *n*th point contained in the definition of any type of geometry.

Figure 11-6 illustrates the results of the STPointN() method when used on a variety of different types of geometry.

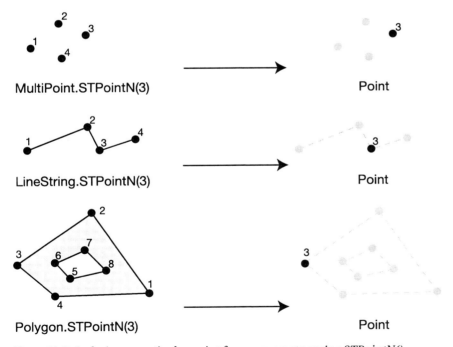

Figure 11-6. *Isolating a particular point from a geometry using STPointN()*

Note In order to use the STPointN() method effectively, you must understand the order in which the points of a geometry are defined. If the geometry was created by a user, the points are listed in the order of the representation originally passed to the static method that created the instance. If the geometry was created by SQL Server 2008 (from the result of another method), points are ordered first by instance, then by ring within the instance, and then by point within each ring.

Supported Datatypes

The STPointN() method can be used on instances of the following datatypes:

- geometry

- geography

Usage

STPointN() must be supplied with an integer parameter, n, which specifies the ordinal number of the point that should be returned from the geometry. The syntax is as follows:

```
Instance.STPointN(n)
```

Instance.STPointN(n) returns the *n*th point of the geometry represented by Instance. Valid values for the parameter n range from 1 (the first point of the geometry) to the value of STNumPoints() (the final point in the geometry). The return value of the method is a geography or geometry Point object—matching the datatype of the instance on which the method was called.

Example

The following example creates a LineString geography instance, representing the route taken by runners competing in the London Marathon. The individual points from which the LineString is constructed are evenly spaced at every mile along the course. The STPointN() method is used to select the 14th point in the LineString, which represents the approximate halfway point of the race.

```
DECLARE @LondonMarathon geography
SET @LondonMarathon = geography::STLineFromText(
  'LINESTRING(0.0112 51.4731, 0.0335 51.4749, 0.0527 51.4803, 0.0621 51.4906,
   0.0448 51.4923, 0.0238 51.4870, 0.0021 51.4843, -0.0151 51.4814,
   -0.0351 51.4861, -0.0460 51.4962, -0.0355 51.5011, -0.0509 51.5013,
   -0.0704 51.4989, -0.0719 51.5084, -0.0493 51.5098, -0.0275 51.5093,
   -0.0257 51.4963, -0.0134 51.4884, -0.0178 51.5003, -0.0195 51.5046,
   -0.0087 51.5072, -0.0278 51.5112, -0.0472 51.5099, -0.0699 51.5084,
   -0.0911 51.5105, -0.1138 51.5108, -0.1263 51.5010, -0.1376 51.5031)',
   4326)
```

```
SELECT
  @LondonMarathon AS Shape,
  @LondonMarathon.STPointN(14) AS Point14,
  @LondonMarathon.STPointN(14).STAsText() AS WKT
```

The WKT representation of the result returned by the STPointN() method is as follows:

```
POINT (-0.0719 51.5084)
```

Finding the Start and End Points of a Geometry

STStartPoint() and STEndPoint() are "shortcut" methods that provide the same result as STPointN() when used in the specific cases of returning the first point and the last point of a geometry, respectively:

- STStartPoint() is equivalent to STPointN(1)

- STEndPoint() is equivalent to STPointN(STNumPoints())

Supported Datatypes

The STStartPoint() and STEndPoint() methods can be used on instances of the following datatypes:

- geometry

- geography

Usage

The STStartPoint() and STEndPoint() methods can be used on instances of either the geography or geometry datatype as follows:

```
Instance.STStartPoint()
Instance.STEndPoint()
```

The result of each method is a Point object of either geography or geometry datatype—matching the type of the instance on which the method was called.

Example

In May 1919, a crew of five aviators completed the first successful transatlantic flight, under the command of Lieutenant Commander Albert Read. Starting in Rockaway Naval Air Station, Read first piloted his NC-4 aircraft to Halifax, Nova Scotia, then on to Trepassey, Newfoundland, before crossing to the island of Horta in the Azores. From the Azores, the crew then set off to Lisbon, Portugal, and made a short stop in Spain before finally completing the journey at Plymouth, England. The following example creates a LineString geometry representing the

approximate route taken, and then uses the STStartPoint() and STEndPoint() methods to
return the points at the start and end of the journey.

```
DECLARE @TransatlanticCrossing geography
SET @TransatlanticCrossing = geography::STLineFromText('
LINESTRING(
  -73.88 40.57, -63.57 44.65, -53.36 46.74, -28.63 38.54,
  -28.24 38.42, -9.14 38.71,  -8.22 43.49,  -4.14 50.37)',
  4326
)
SELECT
  @TransatlanticCrossing AS Shape,
  @TransatlanticCrossing.STStartPoint().STAsText() AS StartPoint,
  @TransatlanticCrossing.STEndPoint().STAsText() AS EndPoint
```

The results, representing Rockway Naval Air Station and Plymouth, are as follows:

```
StartPoint            EndPoint
POINT (-73.88 40.57)  POINT (-4.14 50.37)
```

Finding the Centroid of a geometry Polygon

The centroid of a Polygon can be thought of as its "center of gravity"—the point around which
the area contained by the Polygon is evenly distributed. The position of the centroid is derived
mathematically from a calculation based on the overall shape of the geometry. In SQL Server 2008,
you can use the STCentroid() method to return a Point instance representing the centroid of
any geometry Polygon.

Figure 11-7 illustrates the centroid of several Polygons of the geometry datatype.

Caution The centroid of a Polygon is not necessarily contained inside the Polygon itself—it is the point
around which the Polygon is evenly distributed.

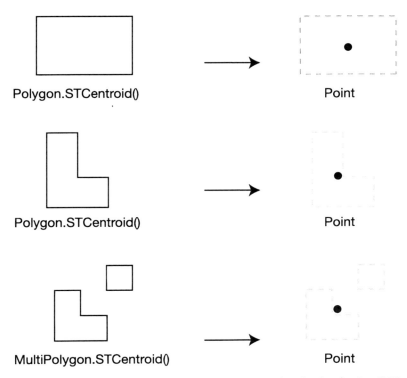

Figure 11-7. *Centroids of different Polygon geometries obtained using STCentroid()*

Supported Datatypes

The STCentroid() method can be used on instances of the following datatype:

- geometry

Usage

The STCentroid() method requires no parameters and can be used on a Polygon or MultiPolygon instance of the geometry datatype as follows:

Instance.STCentroid()

The result of the STCentroid() method is a Point geometry, declared using the same SRID as the instance on which it was called. Note that STCentroid() can be used only on Polygon or MultiPolygon geometries—if used on a Point or LineString object, the method will return NULL.

Example

The following example creates a Polygon geometry representing the state of Colorado, and then uses the STCentroid() method to determine the centroid of that Polygon:

```
DECLARE @Colorado geometry
SET @Colorado = geometry::STGeomFromText('POLYGON((-102.0423 36.9931, -102.0518
41.0025, -109.0501 41.0006, -109.0452 36.9990, -102.0423 36.9931))', 4326)
SELECT
  @Colorado AS Shape,
  @Colorado.STCentroid() AS Centroid,
  @Colorado.STCentroid().STAsText() AS WKT
```

The WKT representation of the result of the STCentroid() method is as follows:

```
POINT (-105.54621375420314 38.998581021101813)
```

This represents a location a few miles north of Elevenmile Canyon Reservoir, in the center of the state.

Finding the Center of a geography Instance

The STCentroid() method cannot be applied to the geography datatype. However, similar functionality is provided by the EnvelopeCenter() method. EnvelopeCenter() averages the position vectors that describe the location of each point in the geometry from the center of the earth, and returns the Point geometry plotted at the resulting position. This is a very simple approximation of the center point of any type of geometry. In contrast to the STCentroid() method of the geometry datatype, which can be used only on Polygon instances, the EnvelopeCenter() method can be used against any type of geometry of the geography datatype.

> **Note** The result of the EnvelopeCenter() method is based on the average of each unique point in a geometry. If the same point is defined twice in a geography instance, such as the start point and end point of a closed ring, this point is included only once in the calculation.

Figure 11-8 illustrates the method by which the result of EnvelopeCenter() is calculated for a geography Polygon defined by the points P1, P2, P3, and P4. The position vectors describing the location of each point from the center of the earth are averaged (with the point, P1, which represents both the start and end points of the Polygon ring, included only once), and the method returns the point located at the resulting position.

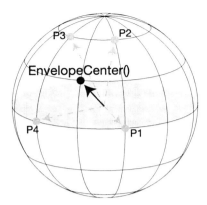

Figure 11-8. *Using the EnvelopeCenter() method to calculate the center point of a geography instance*

Supported Datatypes

The EnvelopeCenter() method can be used on instances of the following datatype:

- geography

Usage

The EnvelopeCenter() method can be used against any type of geography instance as follows:

```
Instance.EnvelopeCenter()
```

The result of the method is a Point object of the geography datatype, defined using the same spatial reference system as that in which the original instance was supplied.

Example

The following example creates a geography Polygon representing the state of Utah, and then uses the EnvelopeCenter() method to return the point at the center of the Polygon:

```
DECLARE @Utah geography
SET @Utah = geography::STPolyFromText(
  'POLYGON((-109 37, -109 41, -111 41, -111 42, -114 42, -114 37, -109 37))', 4326)
SELECT
  @Utah AS Shape,
  @Utah.EnvelopeCenter() AS EnvelopeCenter,
  @Utah.EnvelopeCenter().STAsText() AS WKT
```

The WKT representation of the Point returned by the EnvelopeCenter() method is as follows:

```
POINT (-111.33053985766453 40.018634026864916)
```

Figure 11-9 illustrates the location of this point relative to the Polygon representing the overall shape of the state, projected using the Mercator projection. The point obtained from EnvelopeCenter() lies slightly to the northeast of the simple geometric center obtained from averaging the minimum and maximum coordinate values. This is due to the fact that the Polygon definition contains a greater density of points defining the concave corner at the northeast of the state than the single point at each of the other three corners. This causes the average vector calculated by EnvelopeCenter() to be weighted to the northeast.

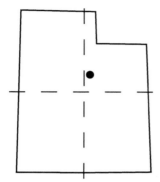

Figure 11-9. *Using the EnvelopeCenter() method on a Polygon representing the state of Utah*

■**Caution** EnvelopeCenter() only calculates an approximation of the center point of a geography instance.

Returning an Arbitrary Point from a Geometry

The STPointOnSurface() method can be used against any type of geometry to return an arbitrary point that lies within the interior of that geometry. The result of the STPointOnSurface() method depends on the type of geometry on which the method is called:

- For LineStrings or MultiLineStrings, the result is a Point that lies on the LineString(s).

- For Polygons, the result is a Point within the exterior ring (and not contained within an interior ring).

- For Points, the result is the Point itself, or in the case of MultiPoints, any one Point contained within the MultiPoint collection.

- For empty geometries, the method returns NULL.

"Why would you want to return a single, *arbitrary* point from a geometry?" you might be wondering. "Surely, if you wanted to obtain a Point geometry representing the overall shape of a Polygon geometry, for instance, wouldn't you be better off using the STCentroid() method instead, since that will return a point in the middle of the shape?" The answer to this question is that, while the STCentroid() method can be used to determine the point representing the centroid of a geometry, the resulting Point is not necessarily contained *within* the geometry itself. In contrast, the result of the STPointOnSurface() method will always be a single Point that is guaranteed to lie in the interior of the geometry on which the method is called.

Supported Datatypes

The STPointOnSurface() method can be used on instances of the following datatype:

- geometry

Usage

The STPointOnSurface() method can be used only with instances of the geometry datatype as follows:

```
Instance.STPointOnSurface()
```

Example

The following example creates a Polygon geometry, and then uses the STPointOnSurface() method to return an arbitrary point contained within that Polygon:

```
DECLARE @Polygon geometry
SET @Polygon = geometry::STGeomFromText('POLYGON((10 2,10 4,5 4,5 2,10 2))',0)
SELECT
  @Polygon AS Shape,
  @Polygon.STPointOnSurface() AS PointOnSurface,
  @Polygon.STPointOnSurface().STAsText() AS WKT
```

The WKT representation of the result of the STPointOnSurface() method is as follows:

```
POINT (8.3333333333333339 3.3333333333333335)
```

Note The STPointOnSurface() method returns an *arbitrary* point, not a *random* point. If you were to execute the preceding example several times, you would receive the same result on each occasion.

Measuring the Length of a Geometry

STLength() returns the length of a geometry. When used on a LineString, this gives the total length of the line segments joining the points. When used on a Polygon, this represents the total length of all defined rings of the Polygon. For a Polygon containing only one ring, STLength() therefore returns the length of the perimeter of the Polygon. If used on a multielement type such as MultiLineString, the STLength() method returns the total length of all elements in an instance, or of all the instances within a Geometry Collection. Figure 11-10 illustrates the result of the STLength() method when used on different types of geometry.

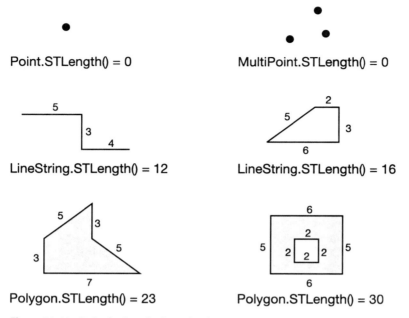

Figure 11-10. *Calculating the length of various types of geometry*

Supported Datatypes

The STLength() method can be used in instances of the following datatypes:

- geometry
- geography

Usage

The STLength() method can be used on any type of geometry or geography instance as follows:

Instance.STLength()

The result is a floating-point numeric value representing the length of the geometry in question. For geography instances, the result will be stated in the units in the unit_of_measure column of the sys.spatial_reference_systems table corresponding to the SRID in which the coordinates were stated. For geometry instances, the result will be stated in the same units of measure as the coordinates of the geometry themselves.

Example

The Royal Mile is the straight route connecting Edinburgh Castle with the Palace of Holyrood House, which runs along some of the oldest streets in Edinburgh. The following example creates a LineString geometry representing the Royal Mile, and then uses the STLength() method to determine its length:

```
DECLARE @RoyalMile geography;
SET @RoyalMile = geography::STLineFromText(
  'LINESTRING(-3.20001 55.94821, -3.17227 55.9528)', 4326)
SELECT
  @RoyalMile AS Shape,
  @RoyalMile.STLength() AS Length
```

The result of the STLength() method is as follows:

```
1806.77067641223
```

Since the coordinates of the @RoyalMile LineString were defined using the EPSG:4326 spatial reference system, the result is stated in the unit of measurement for that system, which is the meter.

Prior to 1824, the result of 1,807 meters, as measured along the length of the Royal Mile, was the definition of a *Scottish mile*. This is longer than the mile in common usage today, which is equal to approximately 1,609 meters.

Calculating the Area Contained by a Geometry

The STArea() method is used to calculate and return the total area occupied by an object. If used on a zero- or one-dimensional object (i.e., a Point or a LineString), then the method will return 0.

When used with the geography datatype, the results of the STArea() method will be returned in the square of the unit of measure defined by the spatial reference system of the geography instance. For example, when used against a geography object specified with SRID 4326, the result returned by STArea() will be expressed in square meters, whereas when using SRID 4157, the unit of measure will be square feet. When used against a geometry object, the unit of measure will be the square of the unit in which the coordinates were supplied. Figure 11-11 illustrates the result of the STArea() method when used against different types of geometry.

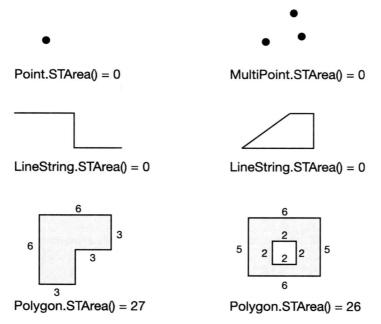

Figure 11-11. *Calculating the area of various types of geometry using STArea()*

Supported Datatypes

The STArea() method can be used on instances of the following datatypes:

- geometry
- geography

Usage

The STArea() method can be called on any item of geography or geometry as follows:

```
Instance.STArea()
```

The result of STArea() will be 0 for any geometry unless it contains at least one Polygon geometry.

Example

The following example creates a geometry Polygon representing a plot of land in the south of France, using the UTM Zone 31N projection (EPSG:32631). The plot has an associated cost, represented by the variable @Cost. By dividing the total cost of the plot by the result of the STArea() method applied to the geometry, we can work out the cost per square meter of land.

```
DECLARE @Cost money = 80000
DECLARE @Plot geometry
SET @Plot = geometry::STPolyFromText(
  'POLYGON((633000 4913260, 633000 4913447, 632628 4913447, 632642 4913260,
    633000 4913260))',
  32631)
SELECT
  @Plot AS Shape,
  @Cost / @Plot.STArea() AS PerUnitAreaCost
```

Since the coordinate values supplied in EPSG:32631 are measured in meters, the result represents the cost per square meter of land, as follows:

1.1720753058384

Setting or Retrieving the SRID of a Geometry

Every instance of any type of geometry from either the geography or geometry datatype has an associated spatial reference identifier. This determines the system in which the coordinates were obtained, and enables them to uniquely identify a position on the earth. The STSrid property can be used to return, or set, the SRID of any object.

Supported Datatypes

The STSrid property can be used on instances of the following datatypes:

- geometry
- geography

Usage

The STSrid property can be used against both the geometry and geography datatypes as follows:

Instance.STSrid

STSrid is unusual in that, whereas most spatial properties are read-only, STSrid can also be used to *set* the SRID of an object.

Example

Suppose you wanted to import some spatial data stated in projected coordinates from an unknown source that did not state the spatial reference system in which the coordinates had been defined. Since projected coordinates operate on a flat plane, you can initially import the data into a field of geometry datatype using SRID 0, as follows:

```
CREATE TABLE #Imported_Data (
  Location geometry
)
```

```
INSERT INTO #Imported_Data VALUES
  (geometry::STGeomFromText('LINESTRING(122 74, 123 72)', 0)),
  (geometry::STGeomFromText('LINESTRING(140 65, 132 63)', 0))
```

You can check the SRID of the items contained in the Location column of the #Imported_Data table by selecting the STSrid property, as follows:

```
SELECT
  Location.STAsText(),
  Location.STSrid
FROM #Imported_Data
```

The following are the results:

```
LINESTRING (122 74, 123 72)    0
LINESTRING (140 65, 132 63)    0
```

Now suppose that, having inserted the data, you discover that the data relates to projected coordinates based on the EPSG:32731 reference system. You therefore want to update the SRID of all your records to reflect this. To do this, you can set the value of the STSrid property using an UPDATE statement, as follows:

```
UPDATE #Imported_Data
  SET Location.STSrid = 32731
```

If you now select the value of the STSrid property once more, you find that the SRID of each geometry in the table has been updated to the correct value:

```
SELECT
  Location.STAsText(),
  Location.STSrid
FROM #Imported_Data
```

The results are as follows:

```
LINESTRING (122 74, 123 72)    32731
LINESTRING (140 65, 132 63)    32731
```

Note Specifying a different SRID does not cause the coordinate values of a geometry to be reprojected into that system. It only provides the description of the system in which those coordinates have been defined.

Isolating the Exterior Ring of a Geometry Polygon

As previously discussed, Polygons may contain a number of internal rings that define "holes"—areas of space cut out of the main Polygon geometry. Sometimes, however, it can be useful to return just the exterior ring of a Polygon, ignoring any interior rings that might be defined within it. The STExteriorRing() method can be used in this case. It returns a LineString object representing the exterior perimeter boundary of a Polygon shape. Figure 11-12 illustrates the results of the STExteriorRing() method when used against a geometry Polygon instance.

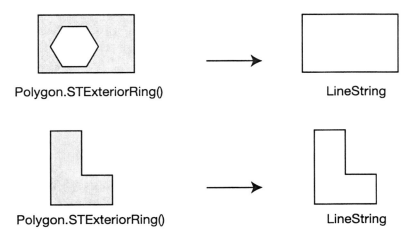

Polygon.STExteriorRing()   LineString

Polygon.STExteriorRing()   LineString

Figure 11-12. *Isolating the exterior ring of a Polygon geometry*

Supported Datatypes

The STExteriorRing() method can be used in instances of the following datatype:

- geometry

Usage

The STExteriorRing() method can be used against Polygons of the geometry datatype as follows:

```
Instance.STExteriorRing()
```

The result of the STExteriorRing() method is a LineString geometry, defined using the same SRID as the instance on which the method is called.

Example

The following example creates a geometry Polygon in the shape of a capital letter *A*. The STExteriorRing() method is then used to return the exterior ring of the Polygon.

```
DECLARE @A geometry
SET @A = geometry::STPolyFromText(
  'POLYGON((0 0, 4 0, 6 5, 14 5, 16 0, 20 0, 13 20, 7 20, 0 0),
          (7 8,13 8,10 16,7 8))',
  0)
SELECT
  @A AS Shape,
  @A.STExteriorRing() AS ExteriorRing,
  @A.STExteriorRing().STAsText() AS WKT
```

The following WKT representation of the STExteriorRing() method is illustrated in Figure 11-13:

```
LINESTRING (0 0, 4 0, 6 5, 14 5, 16 0, 20 0, 13 20, 7 20, 0 0)
```

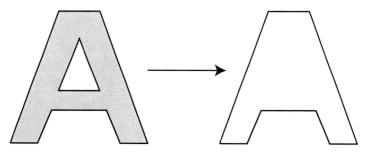

Figure 11-13. *Using STExteriorRing() to isolate the exterior ring of the capital letter A*

▪**Note** The STExteriorRing() method returns a LineString geometry of the exterior ring—*not* the Polygon that encloses the area within that ring. It does not simply "fill in the holes" created by the internal rings within a Polygon.

Counting the Interior Rings of a Geometry

The STNumInteriorRing() method is used to return an integer value that represents the total number of internal rings defined within a Polygon geometry. If a Polygon contains no interior rings, then the result of the method is 0. Figure 11-14 illustrates the results of the STNumInteriorRing() method when used on two different MultiPolygon instances.

Polygon.STNumInteriorRing() = 1   MultiPolygon.STNumInteriorRing() = 3

Figure 11-14. *Counting the number of interior rings in a Polygon geometry*

Supported Datatypes

The STNumInteriorRing() method can be used on instances of the following datatype:

- geometry

Usage

STNumInteriorRing() requires no parameters and can be used against a geometry Polygon instance as follows:

Instance.STNumInteriorRing()

The result is an integer value, equal to or greater than zero, representing the number of interior rings in the Polygon geometry.

Example

The following example creates a Polygon and then uses the STNumInteriorRing() method to confirm the number of interior rings contained within the Polygon:

```
DECLARE @Polygon geometry
SET @Polygon = geometry::STPolyFromText('
  POLYGON(
    (0 0, 20 0, 20 10, 0 10, 0 0),
    (3 1,3 8,2 8,3 1),
    (14 2,18 6, 12 4, 14 2))',
    0)
SELECT
  @Polygon AS Shape,
  @Polygon.STNumInteriorRing() AS NumInteriorRing
```

The result of the STNumInteriorRing() method is as follows:

Isolating an Interior Ring from a Polygon

The STInteriorRingN() method isolates the *n*th interior ring from a Polygon. Since the rings of a Polygon are made up of closed LineStrings, the result of the method will always be a simple, closed LineString. Figure 11-15 illustrates the resulting LineString created by the STInteriorRingN() method when used on a variety of different Polygon geometries.

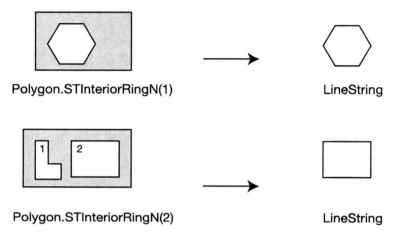

Polygon.STInteriorRingN(1)   **LineString**

Polygon.STInteriorRingN(2)   **LineString**

Figure 11-15. *Isolating an interior ring from a Polygon geometry*

Supported Datatypes

The STInteriorRingN() method can be used on instances of the following datatype:

- geometry

Usage

The syntax for the STInteriorRingN() method, used on an instance of a geometry Polygon, is as follows:

```
Instance.STInteriorRingN(n)
```

This will return the LineString geometry that represents the *n*th ring of Instance. Valid values for n range from 1 (the first interior ring) to the result of STNumInteriorRing() (the final interior ring).

Example

The following example creates a geometry Polygon in the shape of a capital letter *A*. It then uses the STInteriorRingN() method to isolate the first (and only) interior ring from the geometry.

```
DECLARE @A geometry
SET @A = geometry::STPolyFromText(
  'POLYGON((0 0, 4 0, 6 5, 14 5, 16 0, 20 0, 13 20, 7 20, 0 0),
    (7 8,13 8,10 16,7 8))', 0)
SELECT
  @A AS Shape,
  @A.STInteriorRingN(1) AS InteriorRing1,
  @A.STInteriorRingN(1).STAsText() AS WKT
```

The result of the STInteriorRingN() method, expressed in WKT here, is illustrated in Figure 11-16:

```
LINESTRING (7 8, 13 8, 10 16, 7 8)
```

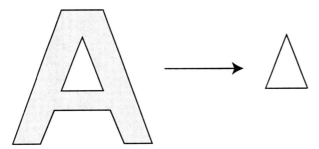

Figure 11-16. *Isolating the interior ring from a geometry Polygon of the capital letter* A

Counting the Rings in a geography Polygon

When using the geography datatype, which defines positions on a round model of the earth, you cannot sensibly assign the rings of a Polygon into the categories of "interior" and "exterior"— every ring divides space into those areas contained within the Polygon and those excluded from it. For this reason, the geography datatype does not implement the STNumInteriorRing() method, but rather has a separate method, NumRings(). The NumRings() method counts the total number of rings in a geography Polygon instance, without drawing any distinction between whether they are "interior" or "exterior." Figure 11-17 illustrates the results of the NumRings() method on two different geography Polygons.

Polygon.NumRings() = 1   Polygon.NumRings() = 2

Figure 11-17. *Counting the total number of rings in a geography Polygon using NumRings()*

Supported Datatypes

The NumRings() method can be used on instances of the following datatype:

- geography

Usage

The NumRings() method can be used on a geography Polygon instance as follows:

```
Instance.NumRings()
```

The result is an integer value representing the total number of defined rings for the Polygon.

Example

The following code creates a geography Polygon containing two rings, representing the US Department of Defense Pentagon building. It then uses the NumRings() method to count the number of rings in the instance.

```
DECLARE @Pentagon geography
SET @Pentagon = geography::STPolyFromText(
  'POLYGON(
    (
      -77.0532238483429 38.870863029297695 ,
      -77.05468297004701 38.87304314667469 ,
      -77.05788016319276 38.872800914712734 ,
      -77.05849170684814 38.870219840133124 ,
      -77.05556273460388 38.8690670969385 ,
      -77.0532238483429 38.870863029297695
    ),
    (
      -77.05582022666931 38.8702866652523 ,
      -77.0569360256195 38.870737733163644 ,
      -77.05673217773439 38.87170668418343 ,
      -77.0554769039154 38.871848684516294 ,
      -77.05491900444031 38.87097997215688 ,
      -77.05582022666931 38.8702866652523
    )
  )',
  4326
)
SELECT
  @Pentagon AS Shape,
  @Pentagon.NumRings() AS NumRings
```

The result of the NumRings() method confirms that the Polygon, @Pentagon, contains two rings:

2

Isolating a Ring from a geography Polygon

Just as the geography datatype implements the NumRings() method rather than
STNumInteriorRing(), it also defines its own method for isolating any given ring from a Polygon,
without classification of "interior" or "exterior." The method used to isolate any ring from a
geography Polygon is RingN(). Figure 11-18 illustrates the result of the RingN() method when
used on a Polygon of the geography datatype.

Polygon.RingN(2)   LineString

Figure 11-18. *Isolating a particular ring from a geography Polygon using RingN()*

Supported Datatypes

The RingN() method can be used on Polygon instances of the following datatype:

- geography

Usage

The RingN() method must be supplied with a parameter, n, which specifies the ring to return
from the geography instance, as follows:

Instance.RingN(n)

The value of n must be an integer number, between 1 and the total number of rings contained
by the instance (which can be determined using the NumRings() method). The result of the
method will be a LineString of the geography datatype, defined using the SRID of Instance.

Example

The following code creates a geography Polygon containing two rings, representing the US
Department of Defense Pentagon building. It then uses the RingN() method to isolate just the
first ring from the definition, before returning the WKT representation of the result.

```
DECLARE @Pentagon geography
SET @Pentagon = geography::STPolyFromText(
  'POLYGON(
    (
      -77.0532238483429 38.870863029297695 ,
```

```
          -77.05468297004701 38.87304314667469 ,
          -77.05788016319276 38.872800914712734 ,
          -77.05849170684814 38.870219840133124 ,
          -77.05556273460388 38.8690670969385 ,
          -77.0532238483429 38.870863029297695
      ),
      (
          -77.05582022666931 38.8702866652523 ,
          -77.0569360256195 38.870737733163644 ,
          -77.05673217773439 38.87170668418343 ,
          -77.0554769039154 38.871848684516294 ,
          -77.05491900444031 38.87097997215688 ,
          -77.05582022666931 38.8702866652523
      )
   )',
   4326
)
SELECT
   @Pentagon AS Shape,
   @Pentagon.RingN(1) AS Ring1,
   @Pentagon.RingN(1).STAsText() AS WKT
```

The WKT representation of the result of the RingN() method is as follows:

```
LINESTRING (
   -77.0532238483429 38.870863029297695,
   -77.054682970047011 38.873043146674689,
   -77.057880163192763 38.872800914712734,
   -77.058491706848145 38.870219840133124,
   -77.055562734603882 38.8690670969385,
   -77.0532238483429 38.870863029297695
)
```

Identifying the Boundary of a Geometry

The STBoundary() method returns the geometry representing the boundary of a geometry instance. In spatial data, the word *boundary* does not mean the outer perimeter of the geometry, as you might expect, but has a specific definition depending on the type of geometry in question:

- Point and MultiPoint instances do not have a boundary.

- LineString and MultiLineString have a boundary formed from the start points and end points of the geometry, removing any points that occur an even number of times.

- The boundary of a Polygon is formed from the LineStrings that represent each of its rings.

Figure 11-19 illustrates the result of the STBoundary() method when used on a variety of different types of geometry.

Tip If used against a Polygon containing no interior rings, the geometry created by STBoundary() is the same as the ExteriorRing() method. However, whereas STExteriorRing() returns the points of the ring in the order they were defined, STBoundary() returns points starting with the smallest coordinate value.

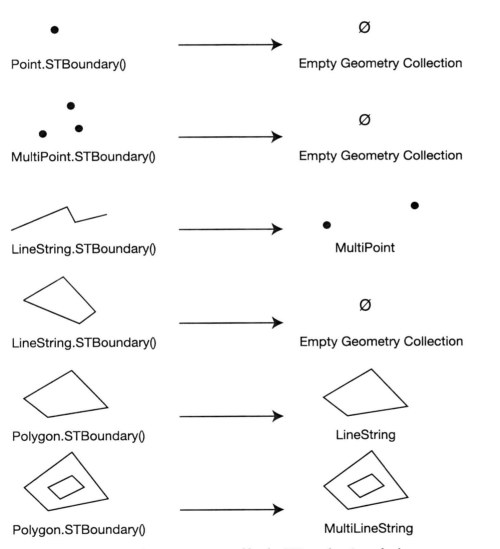

Figure 11-19. *Examples of geometries created by the STBoundary() method*

Supported Datatypes

The STBoundary() method can be used on instances of the following datatype:

- geometry

Usage

The STBoundary() method does not require any parameters, so it can be invoked on an instance of the geometry datatype as follows:

Instance.STBoundary()

The type of geometry represented by the result of the STBoundary() method will depend on the type of geometry of the instance on which it was called.

Example

The following example creates a Polygon geometry in the shape of a capital letter *A*, and then uses the STBoundary() method to identify the boundary of the Polygon:

```
DECLARE @A geometry
SET @A = geometry::STPolyFromText(
  'POLYGON((0 0, 4 0, 6 5, 14 5, 16 0, 20 0, 13 20, 7 20, 0 0),
    (7 8,13 8,10 16,7 8))', 0)
SELECT
  @A AS Shape,
  @A.STBoundary() AS Boundary,
  @A.STBoundary().STAsText() AS WKT
```

The following is the WKT representation of the result of the STBoundary() method, which is illustrated in Figure 11-20.

```
MULTILINESTRING ((7 8, 10 16, 13 8, 7 8), (0 0, 4 0, 6 5, 14 5, 16 0, 20 0, 13 20,
7 20, 0 0))
```

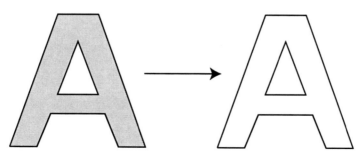

Figure 11-20. *Identifying the boundary of a Polygon geometry of the capital letter* A *using STBoundary()*

Calculating the Bounding Box of a Geometry

The envelope of a geometry represents the smallest axis-aligned rectangle that completely encompasses every part of the geometry. It is also referred to as a *bounding box*. If MinX, MaxX, MinY, and MaxY are the minimum and maximum x and y coordinates of any point contained in the geometry, then the bounding box is the Polygon defined by the following WKT representation:

```
POLYGON( (MinX, MinY), (MaxX, MinY), (MaxX, MaxY), (MinX, MaxY), (MinX, MinY) )
```

The STEnvelope() method can be used to return the bounding box of any type of instance of the geometry datatype. Figure 11-21 illustrates the bounding boxes created by the STEnvelope() method when used on a variety of geometry instances.

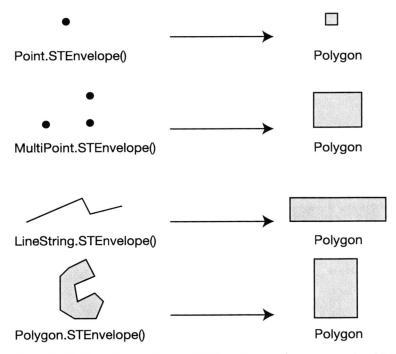

Point.STEnvelope()   Polygon

MultiPoint.STEnvelope()   Polygon

LineString.STEnvelope()   Polygon

Polygon.STEnvelope()   Polygon

Figure 11-21. *Creating envelopes of different types of geometry using STEnvelope()*

■**Note** If you use the STEnvelope() method on a Point instance, SQL Server will create the Polygon of smallest area around that point. For example, the result of STEnvelope() around a Point at POINT(30 20) is POLYGON ((29.999999 19.999999, 30.000001 19.999999, 30.000001 20.000001, 29.999999 20.000001, 29.999999 19.999999)).

Supported Datatypes

The STEnvelope() method can be used on instances of the following datatype:

- geometry

Usage

The STEnvelope() method requires no parameters and can be used on any instance of the geometry datatype as follows:

```
Instance.STEnvelope()
```

The result of the STEnvelope() method will always be a Polygon, defined using the same SRID as the geometry instance on which it was invoked.

Example

The following code creates a Polygon geometry in the shape of a capital letter *A*, and then creates the bounding box around the Polygon using the STEnvelope() method:

```
DECLARE @A geometry
SET @A = geometry::STPolyFromText(
  'POLYGON((0 0, 4 0, 6 5, 14 5, 16 0, 20 0, 13 20, 7 20, 0 0),
  (7 8,13 8,10 16,7 8))', 0)
SELECT
  @A AS Shape,
  @A.STEnvelope() AS Envelope,
  @A.STEnvelope().STAsText() AS WKT
```

The following is the result of the STEnvelope() method in WKT format, which is illustrated in Figure 11-22:

```
POLYGON ((0 0, 20 0, 20 20, 0 20, 0 0))
```

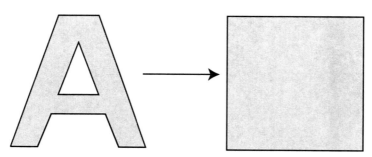

Figure 11-22. *Creating a bounding box of a geometry Polygon of the capital letter* A *using STEnvelope()*

Calculating the Envelope of a geography Object

The STEnvelope() method is not supported by objects of the geography datatype, since the straight axis-aligned lines of a simple rectangular bounding box cannot be applied to an elliptical model of the earth. However, SQL Server provides an alternative method to describe the extent of a geography instance, through the use of the EnvelopeCenter() and EnvelopeAngle() methods.

I have already introduced the EnvelopeCenter() method earlier in this chapter. It is used to return the Point object calculated from the vector average of all the points in the geography instance, which can be used to approximate the center of any type of geometry. The EnvelopeAngle() method returns the angle between the point obtained from the EnvelopeCenter() method and the point in the geometry that lies furthest from the EnvelopeCenter() point. The resulting value is a measure of the extent to which the points of a geography instance are spread out around a central point. Figure 11-23 illustrates the method by which EnvelopeCenter() is calculated.

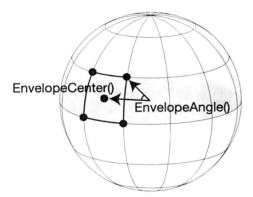

Figure 11-23. *Calculating the extent of a geography instance using EnvelopeAngle()*

Supported Datatypes

The EnvelopeAngle() method can be used on instances of the following datatype:

- geography

Usage

The EnvelopeAngle() method does not require any parameters and can be called on an instance of the geography datatype as follows:

Instance.EnvelopeAngle()

The result of the EnvelopeAngle() method is a floating-point number, representing the value of the angle measured in degrees. The maximum value that can be returned by this method is 90, since any angle greater than this would imply that the points of the Polygon occupy more than a single hemisphere—one of the restrictions of the geography datatype.

Example

The following example creates a geography Polygon, @NorthernHemisphere, representing the Northern Hemisphere. @NorthernHemisphere is defined as the area contained within an exterior ring of points lying just above the equator, at a latitude of 0.1 degrees north. The example then uses the EnvelopeAngle() method to calculate the greatest angle between any point in the @NorthernHemisphere Polygon and the center of the envelope.

```
DECLARE @NorthernHemisphere geography
SET @NorthernHemisphere =
  geography::STGeomFromText('POLYGON((0 0.1,90 0.1,180 0.1, -90 0.1, 0 0.1))',4326)
SELECT
  @NorthernHemisphere AS Shape,
  @NorthernHemisphere.EnvelopeAngle() AS EnvelopeAngle
```

Since the center of the envelope is the North Pole, and the Polygon extends to just above the equator, the result shows an angle close to 90°:

89.90000000000014

Tip The result of the EnvelopeAngle() method cannot exceed 90, since the associated geography instance would exceed a single hemisphere.

Counting the Elements in a Geometry Collection

The STNumGeometries() method operates in much the same way as the STNumPoints() method introduced earlier in this chapter, except that instead of counting the number of *points* within a geometry, STNumGeometries() returns an integer value that states the number of *geometries* contained within a geometry or geography instance. When used against a Geometry Collection, the result will be the number of elements contained within the collection. If used against a single-element instance—Point, LineString, or Polygon—the result of STNumGeometries() will be 1. If used on an empty instance of any type, the result will be 0. Figure 11-24 illustrates the results of the STNumGeometries() method when used on a variety of geometry instances.

Point.STNumGeometries() = 1   MultiPoint.STNumGeometries() = 4

MultiLineString.STNumGeometries() = 2   MultiPolygon.STNumGeometries() = 2

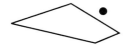

GeometryCollection.STNumGeometries() = 2

Figure 11-24. *Counting the number of geometries in an instance using STNumGeometries()*

Supported Datatypes

The STNumGeometries() method can be used on instances of the following datatypes:

- geometry
- geography

Usage

The STNumGeometries() method does not require any parameters and can be called on a geography or geometry instance as follows:

Instance.STNumGeometries()

Example

In the following example, a Geometry Collection is created that contains a MultiPoint element (consisting of two Points), a LineString, and a Polygon. The STNumGeometries() method is then used to count the total number of elements in the collection.

```
DECLARE @Collection geometry
SET @Collection = geometry::STGeomFromText('
  GEOMETRYCOLLECTION(
    MULTIPOINT((32 2), (23 12)),
    LINESTRING(30 2, 31 5),
    POLYGON((20 2, 23 2.5, 21 3, 20 2))
  )',
  0)
```

```
SELECT
  @Collection AS Shape,
  @Collection.STNumGeometries() AS NumGeometries
```

The result of the STNumGeometries() method is as follows:

3

Note that, even though the MultiPoint element contains two Point geometries, the result of STNumGeometries() for the entire collection is 3, since the MultiPoint geometry is counted as only one element in the Geometry Collection. STNumGeometries() counts single-element geometries (Point, LineString, and Polygon), multielement geometries (MultiPoint, MultiLineString, and MultiPolygon), and empty geometries of any type contained within a collection as single items.

Retrieving an Individual Geometry from a Geometry Collection

The STGeometryN() method returns the *n*th geometry from a Geometry Collection. It can be used on either the generic Geometry Collection object or one of the specific subtypes of collection—MultiPoint, MultiLineString, or MultiPolygon. Figure 11-25 illustrates the use of the STGeometryN() method to isolate individual elements from a range of geometries.

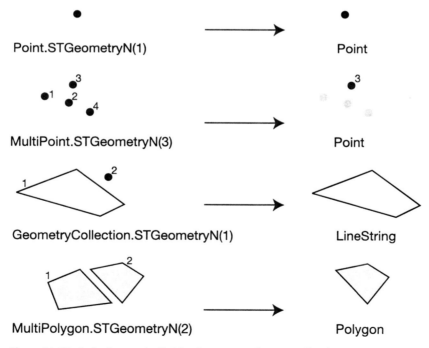

Figure 11-25. *Isolating an individual geometry from a collection*

Note You can use STGeometryN(1) on a single-element geometry, in which case the result of the method is the geometry itself.

Supported Datatypes

The STGeometryN() method can be used on instances of the following datatypes:

- geometry
- geography

Usage

The STGeometryN() method must be supplied with a single parameter, n, using the following syntax,

```
Instance.STGeometryN(n)
```

where n is the ordinal number of the geometry from the collection that you want to retrieve. The value of n must be between 1 and the total number of elements in the Geometry Collection (which you can obtain from the value of STNumGeometries()).

Example

The following example creates a MultiLineString geometry representing the seven runways at Dallas/Fort Worth International Airport, which has the greatest number of runways of any airport in the world. It then uses the STGeometryN() method to isolate and return a LineString geometry representing a single runway.

```
DECLARE @DFWRunways geography
SET @DFWRunways = geography::STMLineFromText(
  'MULTILINESTRING(
    (-97.0214781 32.9125542, -97.0008442 32.8949814),
    (-97.0831328 32.9095756, -97.0632761 32.8902694),
    (-97.0259706 32.9157078, -97.0261717 32.8788783),
    (-97.0097789 32.8983206, -97.0099086 32.8749594),
    (-97.0298833 32.9157222, -97.0300811 32.8788939),
    (-97.0507357 32.9157992, -97.0509261 32.8789717),
    (-97.0546419 32.9158147, -97.0548336 32.8789861)
  )', 4326)
SELECT
  @DFWRunways AS Shape,
  @DFWRunways.STGeometryN(3) AS Geometry3,
  @DFWRunways.STGeometryN(3).STAsText() AS WKT
```

The WKT representation of the result of the STGeometryN() method is as follows:

```
LINESTRING (-97.0259706 32.9157078, -97.0261717 32.8788783)
```

Summary

Table 11-2 gives a summary of all the methods and properties introduced in this chapter, together with the datatypes that support those methods.

Table 11-2. *Methods to Return Properties of Spatial Objects Supported by Each Datatype*

Method	Description	geometry	geography
Describing a Geometry			
STGeometryType()	Returns the name of the type of geometry (e.g., Point, LineString)	•	•
InstanceOf()	Tests whether an instance is of a particular geometry type	•	•
STDimension()	Returns the number of dimensions an instance occupies	•	•
STIsSimple()	Determines whether an instance is simple	•	
STIsClosed()	Determines whether an instance is closed	•	
STIsRing()	Determines whether an instance is a ring	•	
STNumPoints()	Returns the number of points in an instance	•	•
STIsEmpty()	Determines whether a geometry is empty (i.e., contains no points)	•	•
Returning the Coordinates Values of a Point			
STX	Returns the x coordinate of a geometry Point instance	•	
STY	Returns the y coordinate of a geometry Point instance	•	
Lat	Returns the latitude of a geography Point instance		•
Long	Returns the longitude of a geography Point instance		•
M	Returns the m (measure) coordinate of a Point instance	•	•
Z	Returns the z (elevation) coordinate of a Point instance	•	•

Table 11-2. *Methods to Return Properties of Spatial Objects Supported by Each Datatype*

Method	Description	geometry	geography
Returning Individual Points from a Geometry			
STPointN()	Returns the *n*th point in the definition of an instance	•	•
STStartPoint()	Returns the first point in the definition of an instance	•	•
STEndPoint()	Returns the last point in the definition of an instance	•	•
STCentroid()	Returns the geometric center point of a Polygon instance	•	
EnvelopeCenter()	Returns the envelope center point of a geography instance		•
STPointOnSurface()	Returns an arbitrary point from the interior of an instance	•	
STLength()	Returns the length of all lines in an instance	•	•
STArea()	Returns the area contained by an instance	•	•
STSrid	Sets or retrieves the SRID of the spatial reference system in which an instance is defined	•	•
Handling Rings of a Polygon			
STExteriorRing()	Returns the exterior ring of a Polygon instance	•	
STNumInteriorRing()	Returns the number of interior rings in a Polygon instance	•	
STInteriorRingN()	Returns the specified interior ring of a Polygon instance	•	
NumRings()	Returns the number of rings in a geography Polygon instance		•
RingN()	Returns a specific ring from a geography Polygon instance		•
Describing the Extent of a Geometry			
STBoundary()	Returns the boundary of an instance	•	
STEnvelope()	Returns the envelope (bounding box) of a geometry instance	•	
EnvelopeAngle()	Returns the angle between the center of a geography instance and the most outlying point		•
Working with Multielement Geometries			
STNumGeometries()	Returns the number of geometries in an instance	•	•
STGeometryN()	Returns a specific geometry from a multielement geometry	•	•

The Microsoft reference for all of these methods is available at Microsoft SQL Server 2008 Books Online:

geography Data Type Method Reference can be found at `http://msdn2.microsoft.com/en-us/library/bb933802(SQL.100).aspx`.

geometry Data Type Method Reference can be found at `http://msdn2.microsoft.com/en-us/library/bb933973(SQL.100).aspx`.

CHAPTER 12

Modifying Spatial Objects

In this chapter, I will introduce you to a number of methods that can be used to create new items of geography or geometry data by modifying, combining, or selecting parts of other existing instances. Note that these methods are different from static methods, which instead create new items of data from a representation of the coordinate points of a geometry.

Some of the methods discussed in this chapter act upon a single instance, such as STBuffer(), STConvexHull(), and Reduce(). Other methods act upon more than one geometry, such as STIntersection() and STDifference(). As is the case with all the spatial functionality in SQL Server 2008, the geometry and geography datatypes do not necessarily have the same set of methods available, or implement them in the same way. For each method introduced in this chapter, I'll show you how and when it can be used.

Note The methods discussed in this chapter do not make any changes to the original instance on which they are invoked. Rather, they create a new instance based on a modification of the geometry represented by that instance.

Ensuring That an Object Is Valid

When you create a geometry or geography instance from a static method, SQL Server checks the representation supplied to ensure that it describes a well-formed geometry. For example, you cannot use a static method to create any of the following poorly formed geometries:

- A LineString containing less than two points

- A Polygon defined by less than four points

- A Polygon containing an unclosed ring (in which the first and last points of the ring are not the same)

- A Polygon that defines an interior ring that intersects the exterior ring

Attempting to create any of these geometries will lead to an exception in the SQLCLR (which SQL Server uses to provide the functionality of the geometry and geography datatypes), and will generate an error message containing details of the rule breached by the geometry in question.

Even though it is not possible to create poorly formed geometries such as those previously listed, it is still possible to use static methods to create examples of geometries that are not valid according to the OGC specifications (for example, a Polygon geometry that overlaps itself). Although you can store and retrieve these invalid geometries using the geometry datatype, you cannot perform any methods that operate on them. As such, it is important to be able to identify and deal with geometries that do not validate to the OGC standards.

Note Although it is possible to store and retrieve invalid geometries using the geometry datatype, you cannot use any methods on them.

STIsValid() tests whether an instance of the geometry datatype meets all the criteria required to be valid according to the OGC specifications for the type of geometry in question. In order to use any methods on a geometry instance, the result of the STIsValid() method must confirm that the geometry is valid.

Supported Datatypes

The STIsValid() method can be used on instances of the following datatype:

- geometry

Note The STIsValid() method applies only to instances of the geometry datatype. All instances of the geography datatype are considered "valid."

Usage

The STIsValid() method requires no parameters and can be used to test the validity of any instance of the geometry datatype as follows:

```
Instance.STIsValid()
```

If the instance meets all the criteria to be a valid instance for the type of geometry in question, the STIsValid() method returns the value 1. Otherwise, it returns the value 0.

Example

The following example creates a Polygon instance, and then tests whether that Polygon is valid by using the STIsValid() method:

```
DECLARE @Polygon geometry
SET @Polygon =
  geometry::STPolyFromText('POLYGON((0 0, 4 0, 4 3, 2 3, 2 4, 2 3, 0 3, 0 0))', 0)

SELECT
  @Polygon AS Shape,
  @Polygon.STIsValid() AS IsValid
```

SQL Server will execute this query successfully, without generating any warning or error message. However, the result of the STIsValid() method indicates that the geometry represented by @Polygon is not valid:

0

Why does the result of STIsValid() show that the geometry is invalid in this case? To answer this question, let's examine the @Polygon geometry, which is shown in Figure 12-1. Remember that you can also visualize this geometry in SQL Server Management Studio by clicking the Spatial Results tab after executing the previous query.

Figure 12-1. *An invalid Polygon geometry containing a spike*

Notice that the path drawn between the points of the exterior ring must retrace itself at the top of the Polygon, creating what is known as a "spike." Since the path drawn between the points of the exterior ring intersects itself, it is not simple, and the Polygon defined by this ring is invalid.

Although SQL Server lets us create the @Polygon geometry, it is not valid according to the OGC specifications. If you were to try to use any methods that operated on @Polygon, such as, for instance, obtaining the start point using STStartPoint(),

```
SELECT @Polygon.STStartPoint()
```

you would receive the following error message:

```
System.ArgumentException: 24144: This operation cannot be completed because the
instance is not valid. Use MakeValid to convert the instance to a valid instance.
Note that MakeValid may cause the points of a geometry instance to shift slightly.
```

Caution Even if STIsValid() returns 1 (meaning that a geometry is well formed according to the OGC standards), this does not necessarily mean that a geometry instance makes logical sense. The following code creates a valid geometry, despite the fact that the coordinate position specified is outside the bounds of the EPSG:4326 spatial reference system used: DECLARE @Point geometry = geometry::STPointFromText('POINT(300 200)', 4326).

Validating a Geometry

In order to make any geometry instance valid according to the OGC standards, you can apply the MakeValid() method. This creates a new instance of the geometry datatype, containing the same points as specified in the original instance, represented using one or more valid geometries. In some cases, MakeValid() might also cause the coordinate values of one or more points in the resulting geometry to shift slightly from their original location.

Supported Datatypes

The MakeValid() method can be used on instances of the following datatype:

- geometry

Usage

The MakeValid() function does not require any parameters and can be invoked on any instance of the geometry datatype as follows:

```
Instance.MakeValid()
```

The resulting value is a geometry instance, defined using the same SRID as the instance on which the MakeValid() method is invoked, which complies with the OGC standards. The geometry will contain all of the points defined in the original instance, but may use one or more different types of geometry to represent those points.

Caution To contain the same points in a valid geometry, the type of geometry returned by the MakeValid() method may not be the same as that originally supplied.

Example

The following example creates an invalid Polygon geometry, @Invalid, containing a spike in the exterior ring. It then uses the MakeValid() method to set the @Valid variable to a valid geometry representing the same point set.

```
DECLARE @Invalid geometry
SET @Invalid =
  geometry::STPolyFromText('POLYGON((0 0, 4 0, 4 3, 2 3, 2 4, 2 3, 0 3, 0 0))', 0)

DECLARE @Valid geometry
SET @Valid = @Invalid.MakeValid()

SELECT
  @Invalid AS Shape,
  @Invalid.STAsText() AS WKT
UNION ALL SELECT
  @Valid AS Shape,
  @Valid.STAsText() AS WKT
```

The WKT representation of the valid geometry obtained from the MakeValid() method is as follows:

```
GEOMETRYCOLLECTION (LINESTRING (2 4, 2 3), POLYGON ((0 0, 4 0, 4 3, 2 3, 0 3, 0 0)))
```

In order to represent the points contained in the spike, the resulting geometry created by the MakeValid() method is a Geometry Collection containing both a Polygon and a LineString. This Geometry Collection is valid, which can be confirmed using the STIsValid() method as follows:

```
SELECT @Valid.STIsValid()
```

The result, confirming that @Valid is a valid geometry, is as follows:

```
1
```

Combining Spatial Objects

The STUnion() method can be used to create a single geometry that contains the combined point set of two instances of the geography or geometry datatype. The type of the resulting geometry will be the simplest type of object that is capable of representing all the points contained within both geometries. Figure 12-2 illustrates the results of the STUnion() method when used on various combinations of geometries.

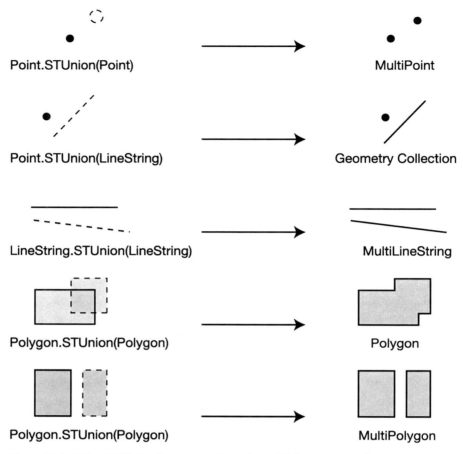

Figure 12-2. *Using STUnion() to create the union of different types of geometries*

AGGREGATING GEOMETRIES USING STUNION() AND CURSORS

The STUnion() method creates the geometry resulting from the union of points contained in two geometries. You can think of it as being roughly equivalent to using the plus (+) operator. In other words, SELECT A.STUnion(B) is similar to SELECT A + B for nonspatial data. What if you want to obtain the union of all the elements contained in an entire column of geometry or geography data? In other words, what is that spatial equivalent of SELECT SUM(A)?

SQL Server does not define any spatial aggregates such as this, but you can achieve the same result by using a cursor to fetch each row of spatial data from a table and iteratively combine them with the next item using STUnion(). To demonstrate this approach, first create a table containing three rows of sample geometry data as follows:

```
DECLARE @GeomTable TABLE (
  GeomColumn geometry
)

INSERT INTO @GeomTable VALUES
  (geometry:: STPointFromText('POINT(12 9)', 0)),
  (geometry:: STLineFromText('LINESTRING(5 2, 9 4)', 0)),
  (geometry:: STPolyFromText('POLYGON((2 0, 4 3, 5 8, 2 4, 2 0))', 0))
```

To create an aggregate from the union of each geometry in the @GeomTable table, you can use a cursor to fetch each individual geometry into a variable @Geom, one at a time, which are then unioned together into a second variable @MultiGeom, as demonstrated in the following code:

```
DECLARE @Geom geometry
DECLARE @MultiGeom geometry =
  geometry::STGeomFromText('GEOMETRYCOLLECTION EMPTY', 0)
DECLARE GeomCursor CURSOR FOR SELECT GeomColumn FROM @GeomTable
OPEN GeomCursor
FETCH NEXT FROM GeomCursor INTO @Geom
WHILE @@FETCH_STATUS = 0
BEGIN
  SET @MultiGeom = @MultiGeom.STUnion(@Geom)
  FETCH NEXT FROM GeomCursor INTO @Geom
END
CLOSE GeomCursor
DEALLOCATE GeomCursor
```

To test the result, you can select the WKT representation of the @MultiGeom instance by adding the following line to the end of the code:

```
SELECT @MultiGeom.STAsText()
```

The resulting aggregate geometry is as follows:

```
GEOMETRYCOLLECTION (
  POINT (12 9),
  LINESTRING (9 4, 5 2),
  POLYGON ((2 0, 4 3, 5 8, 2 4, 2 0))
)
```

Supported Datatypes

The STUnion() method can be used on instances of the following datatypes:

- geometry
- geography

Usage

The STUnion() method can be used to create the union of two instances of the geometry or geography datatype as follows:

```
Instance1.STUnion(Instance2)
```

The type of the resulting geometry will be the simplest geometry that is capable of representing the combined set of points from both instances. If Instance1 and Instance2 are of different types, then the resulting type will be a Geometry Collection.

Example

The following example creates two simple Polygon geometries representing the approximate shape of North Island and South Island of New Zealand. It then uses the STUnion() method to create a MultiPolygon instance representing the combined landmasses.

```
DECLARE @NorthIsland geography
SET @NorthIsland = geography::STPolyFromText(
  'POLYGON((175.3 -41.5, 178.3 -37.9, 172.8 -34.6, 175.3 -41.5))',
  4326)

DECLARE @SouthIsland geography
SET @SouthIsland = geography::STPolyFromText(
  'POLYGON((169.3 -46.6, 174.3 -41.6, 172.5 -40.7, 166.3 -45.8, 169.3 -46.6))',
  4326)
DECLARE @NewZealand geography = @NorthIsland.STUnion(@SouthIsland)

SELECT
  @NorthIsland AS Shape,
  @NorthIsland.STAsText() AS WKT
UNION ALL SELECT
  @SouthIsland AS Shape,
  @SouthIsland.STAsText() AS WKT
UNION ALL SELECT
  @NewZealand AS Shape,
  @NewZealand.STAsText() AS WKT
```

The WKT representation of the result of the STUnion() method is as follows:

```
MULTIPOLYGON (((174.30000008390141 -41.599999939451862, 172.5 -40.7,
166.30000014074497 -45.800000004311343, 169.3 -46.6, 174.30000008390141
-41.599999939451862)), ((178.29999997944242 -37.899999916892376, 172.79999991110731
-34.599999913555131, 175.3 -41.5, 178.29999997944242 -37.899999916892376)))
```

If you are surprised by the values of the coordinates contained in this result, be sure to check out the sidebar "The Imprecision of Floating-Point Spatial Data."

THE IMPRECISION OF FLOATING-POINT SPATIAL DATA

In the example of the STUnion() method (demonstrated in the "Combining Spatial Objects" section), you might have been surprised to see that the coordinate values of the first point defined in the resulting geometry were 174.30000008390141 and –41.599999939451862. The STUnion() method creates the union of the points contained in two geometries, yet this point was not contained in either of the geometries used in the method. The point (174.3, –41.6) was, however, so why did the result of STUnion() introduce an error into these coordinate values?

Remember that the spatial datatypes store coordinates as floating-point binary values. When performing floating-point calculations, there is an inherent imprecision in any results obtained, caused by the fact that certain real numbers cannot be exactly represented in a floating-point system. The exact binary floating-point value equivalent to the decimal number 0.1, for instance, would require an infinitely recurring binary sequence as follows: 11001100110011001100....

Since coordinate values must be rounded to the closest number that can be represented in floating-point binary format, errors might be introduced in the results of certain spatial methods. Note that this is not specific to spatial data, but rather applies to any calculations that use floating-point values. Consider the following example:

```
DECLARE @coordinate float = 174.3
DECLARE @a float = 1
DECLARE @b float = 1.001
SELECT @coordinate * (@b - @a) * 1000
```

Performing this calculation using rational numbers, you would expect this expression to equate to 174.3 * (0.001) * 1000, leading to a result of 174.3, the coordinate value originally supplied. However, when you execute this query, you actually obtain the answer 174.299999999981, the closest approximation obtainable using floating-point methods.

Any methods that perform operations on spatial data may lead to small changes in coordinate values such as these. The magnitude of these deviations is rarely substantial enough to have a significant effect on the accuracy of any item of data when considered in isolation, but it is important to consider when designing any applications that require precise evaluation and comparison of coordinate values.

Defining the Intersection of Two Geometries

The intersection of two geometries is created from the points that both instances have in common. If the points defined by the intersection lie in a single, congruous area, then the intersection will result in a single geometry. However, if the intersecting points are separate, then the resulting geometry may be a multielement instance. Figure 12-3 illustrates examples of intersections created between different types of geometries.

To create a geometry representing the points contained in the intersection of two geometries, as shown in Figure 12-3, you can use the STIntersection() method.

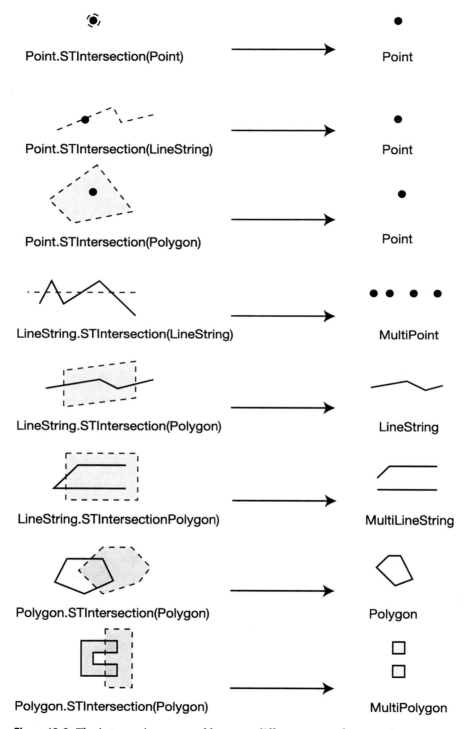

Figure 12-3. *The intersections created between different types of geometries*

Supported Datatypes

The STIntersection() method can be used on instances of the following datatypes:

- geometry
- geography

Usage

The STIntersection() method can be used to create the intersection of two instances of the geometry or geography datatype as follows:

```
Instance1.STIntersection(Instance2)
```

The result of the method will be an object of the same datatype, and using the same SRID as that defined by Instance1 and Instance2. The type of geometry represented by this object will be the simplest type of geometry that is capable of containing all of the points contained in the intersection of the two geometries.

Note STIntersection() can be used to determine the intersection between two geometry instances, or between two geography instances, but cannot be used to determine the intersection between a geometry and a geography instance.

Example

The Appian Way (or *Via Appia*) was an ancient Roman road that ran from Rome to the city of Brindisi in southeast Italy. It is arguably one of the most important roads in world history, because the Roman Army used the route to quickly deploy men and supplies throughout Italy, and it undoubtedly contributed to their consequent military success. Along the route, the Via Appia passed through the Pontine Marshes, which were infested by mosquitoes carrying the deadly malaria disease. The following example creates a LineString representing the Via Appia, and a Polygon representing the malaria-infested marsh area. It then uses the STIntersection() method to determine the treacherous section of the road that intersects the marsh area.

```
DECLARE @Marshes geography
SET @Marshes = geography::STPolyFromText(
  'POLYGON((
    12.94 41.57, 12.71 41.46, 12.91 41.39, 13.13 41.26, 13.31 41.33, 12.94 41.57))',
  4326)

DECLARE @ViaAppia geography
SET @ViaAppia = geography::STLineFromText(
  'LINESTRING(
    12.51 41.88, 13.25 41.28, 13.44 41.35, 13.61 41.25, 13.78 41.23, 13.89 41.11,
    14.22 41.10, 14.47 41.02, 14.79 41.13, 14.99 41.04, 15.48 40.98, 15.82 40.96,
    17.19 40.51, 17.65 40.50, 17.94 40.63)',
  4326)
```

```
SELECT
  @ViaAppia AS Shape,
  @ViaAppia.STAsText() AS WKT
UNION ALL SELECT
  @Marshes AS Shape,
  @Marshes.STAsText() AS WKT
UNION ALL SELECT
  @ViaAppia.STIntersection(@Marshes) AS Shape,
  @ViaAppia.STIntersection(@Marshes).STAsText() AS WKT
```

The WKT representation of the result of the STIntersection() method is as follows:

```
LINESTRING (13.2279201121493 41.298133443053509, 12.911513729116241
41.556421282182363)
```

Figure 12-4 illustrates the features used in this example on a map of Italy. The result of the STIntersection() method corresponds to the section of LineString that passes through the Polygon representing the marshes.

Figure 12-4. *Using STIntersection() to determine the section of the Appian Way that intersects the Pontine Marshes*

Identifying the Difference Between Two Geometries

The STDifference() method can be used to isolate and return the unique points contained within one geometry or geography instance that are not also contained in a second instance. Figure 12-5 illustrates the results of the STDifference() method when used to create the difference between various types of geometries.

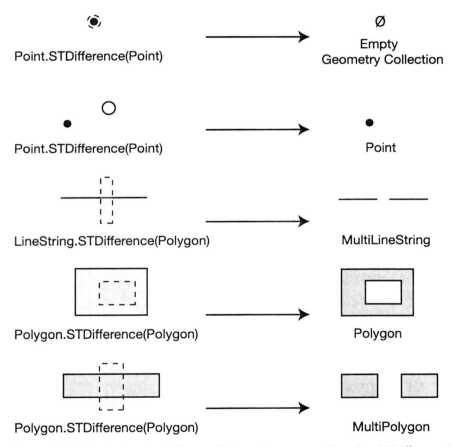

Figure 12-5. *Calculating the difference between two geometries using STDifference()*

Supported Datatypes

The STDifference() method can be used on instances of the following datatypes:

- geometry
- geography

Usage

The STDifference() method can be used to isolate the difference between one geometry or geography instance and another as follows:

Instance1.STDifference(Instance2)

The result of the STDifference() method will be a geometry or geography instance composed of the set of all the points contained within Instance1, but not in Instance2, defined using the same datatype and SRID as both instances supplied.

Example

The British Meteorological Office (commonly referred to as the "Met Office") operates a network of 16 radars that is used to continuously monitor and predict weather throughout the British Isles. Each radar is capable of providing high-quality quantitative rainfall and hydrological data within a range of approximately 75 km. The coverage provided by this radar network is illustrated in Figure 12-6.

Figure 12-6. *Weather radar coverage of the British Isles (one radar station, located on the island of Jersey, is not shown on the map)*

To demonstrate the use of the STDifference() method, the following code example creates a MultiPoint instance representing the location of all the Met Office radars. It then uses STBuffer() to create a MultiPolygon instance, containing individual Polygons centered around each point in the MultiPoint geometry, representing the area covered by that radar. Finally, it uses the STDifference() method to calculate the difference between a MultiPolygon geometry representing the British Isles and the MultiPolygon representing the area covered by radar.

```
DECLARE @Radar geography
SET @Radar = geography::STMPointFromText(
  'MULTIPOINT(
    -2.597 52.398, -2.289 53.755, -0.531 51.689, -6.340 54.500, -5.223 50.003,
    -0.559 53.335, -4.445 51.980, -4.231 55.691, -2.036 57.431, -6.183 58.211,
    -3.453 50.963, 0.604 51.295, -1.654 51.031, -2.199 49.209, -6.259 53.429,
    -8.923 52.700)', 4326)

DECLARE @RadarCoverage geography
SET @RadarCoverage = @Radar.STBuffer(75000)

DECLARE @BritishIsles geography
SET @BritishIsles = geography::STMPolyFromText(
  'MULTIPOLYGON(
    ((0.527 52.879, -3.164 56.0197, -1.626 57.631, -4.087 57.654, -2.989 58.582,
    -5.0977 58.514, -6.504 56.240, -4.746 54.670, -3.516 54.848, -3.252 53.432,
    -4.614 53.301, -4.922 51.697, -3.12 51.505, -5.625 50.032, 1.626 51.286,
    0.791 51.423, 1.890 52.291, 1.274 52.959, 0.527 52.879)),
    ((-6.548 52.123, -5.317 54.518, -7.734 55.276, -9.976 53.354, -9.888 51.369,
    -6.548 52.123)))', 4326)

SELECT
  @BritishIsles AS Shape,
  @BritishIsles.STAsText() AS WKT
UNION ALL SELECT
  @Radar AS Shape,
  @Radar.STAsText() AS WKT
UNION ALL SELECT
  @BritishIsles.STDifference(@RadarCoverage) AS Shape,
  @BritishIsles.STDifference(@RadarCoverage).STAsText() AS WKT
```

The resulting MultiPolygon of the STDifference() method represents the area of land in the British Isles *not* covered by radars operated by the Met Office, as represented by the following WKT:

```
MULTIPOLYGON (((-0.59313915744618673 50.956746798594949, -0.35419290461396075
50.99445115...
```

Calculating the Symmetric Difference Between Two Geometries

STSymDifference() calculates the *symmetric* difference between two geometries. This is the set of all the points that lie in either geometry A or geometry B, but not in both. In other words, STSymDifference() gives the union of the unique set of points from two geometries.

Note A.STSymDifference(B) is logically equivalent to A.STDifference(B).
STUnion(B.STDifference(A)).

Figure 12-7 illustrates the results obtained from the STSymDifference() method when used on several examples of different types of geometries.

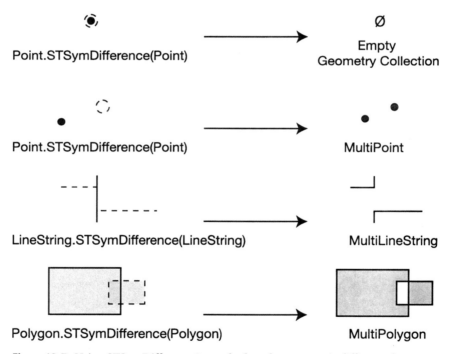

Figure 12-7. *Using STSymDifference() to calculate the symmetric difference between two geometries*

Supported Datatypes

The STSymDifference() method can be used on instances of the following datatypes:

- geometry
- geography

Usage

The STSymDifference() method can be used to calculate the symmetric difference between any two instances of the geometry or geography datatype as follows:

```
Instance1.STSymDifference(Instance2)
```

The result is a geometry containing those points representing the symmetric difference between the two geometries, using the same datatype and SRID as the instances supplied.

Note Since STSymDifference() is *symmetric*, when comparing two geometries, it does not make a difference which is supplied as Instance1 and which is supplied as Instance2.

Example

Consider a city that has two competing local radio stations, KWEST and KEAST. Each one is broadcast from its own transmitter, located in different parts of the city. The KWEST transmitter is located at a longitude of –87.88 and a latitude of 41.86. It broadcasts over a range of 10 km. The KEAST transmitter is located at a longitude of –87.79, latitude 41.89, and transmits over a range of 8 km. By representing the area of coverage of each station as a Polygon and using the STSymDifference() method to calculate the symmetric difference between them, we can work out those parts of the city that can receive one station or the other, but not both. This is demonstrated in the following code:

```
DECLARE @KWEST geography, @KEAST geography
SET @KWEST = geography::Point(41.86, -87.88, 4269).STBuffer(10000)
SET @KEAST = geography::Point(41.89, -87.79, 4269).STBuffer(8000)

SELECT
  @KEAST.STSymDifference(@KWEST) AS Shape,
  @KEAST.STSymDifference(@KWEST).STAsText() AS WKT
```

Tip This example makes use of the STBuffer() method to create circular Polygons about two Points to define the @KEAST and @KWEST areas within which each transmitter can be received. The STBuffer() method is discussed in more detail later in this chapter.

The result of the STSymDifference() method is a MultiPolygon geometry, illustrated in Figure 12-8.

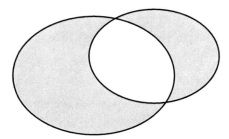

Figure 12-8. *Using STSymDifference() to calculate those areas that can receive only one radio station*

Simplifying a Geometry

The Reduce() method can be used to simplify a geometry, by reducing the number of points used to define the shape of the geometry while still attempting to maintain its overall shape. To achieve this, the Reduce() method implements the Douglas-Peucker algorithm, which is a mathematical method to remove all those points that lie within a tolerated deviation from the simpler approximation of a geometry.

Supported Datatypes

The Reduce() method can be used on any instance of the following datatypes:

- geometry

- geography

Usage

The Reduce() method can be used on an instance of either the geometry or geography datatype as follows:

```
Instance.Reduce(tolerance)
```

The single parameter, tolerance, is the tolerance factor supplied to the Douglas-Peucker algorithm. This value represents the maximum allowed deviation of any point in the resulting, simpler geometry from the original geometry. The allowed tolerance must be a positive, floating-point value, measured in linear units. When used on an instance of the geography datatype, tolerance is measured in the units defined by the unit_of_measure column of the sys.spatial_reference_systems table corresponding to the SRID in which the instance is defined. When used on an instance of the geometry datatype, tolerance is measured in the same unit of measure as the coordinate values of the instance.

The greater the tolerance, the greater the degree of simplification in the resulting geometry returned by the Reduce() method. Figure 12-9 illustrates the effect of supplying different tolerance values to the Reduce() method acting upon a Polygon instance representing Australia.

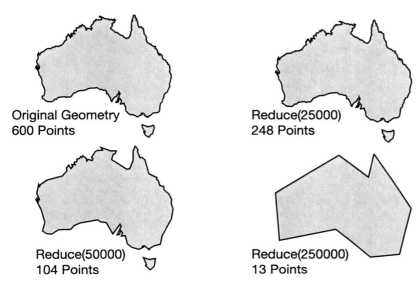

Figure 12-9. *Comparing the geometry created by the Reduce() method with different levels of tolerance*

The Douglas-Peucker algorithm is a recursive algorithm. On the first pass, it creates a first approximation of the geometry that is to be simplified by joining the start and end points directly with a straight line. On the second pass, if any of the points of the geometry lie further from this approximation than the accepted tolerance, then the point that lies furthest away is added back, creating a more refined approximation. On subsequent passes, this process is repeated over and over, further refining the approximation on each pass until all of the points contained within the original geometry lie within the accepted tolerance value from the approximation created. At this point, the Reduce() method returns the approximation created by the last iteration of the algorithm. Figure 12-10 illustrates the method by which the Douglas-Peucker algorithm used by the Reduce() method can be used to simplify a LineString geometry.

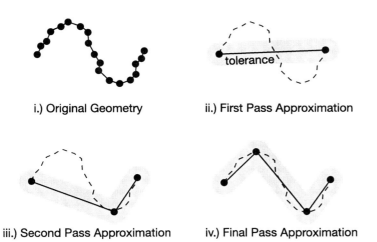

Figure 12-10. *The Douglas-Peucker reduction algorithm*

> **Note** Point and MultiPoint geometries cannot be simplified. When used on an instance of these types of geometries, the Reduce() method returns the original, unmodified geometry.

Example

The following example creates a LineString geometry and then uses the Reduce() method to obtain a simplified reduction of that geometry that deviates by no more than one unit from any point in the original geometry:

```
DECLARE @LineString geometry
SET @LineString = geometry::STLineFromText(
  'LINESTRING(130 33, 131 33.5, 131.5 32.9, 133 32.5, 135 33, 137 32, 138 31,
  140 30)',  0)

SELECT
  @LineString AS Shape,
  @LineString.STAsText() AS WKT
UNION ALL SELECT
  @LineString.Reduce(1) AS Shape,
  @LineString.Reduce(1).STAsText() AS WKT
```

The WKT representation of the simpler geometry obtained from the Reduce() method is as follows:

```
LINESTRING (130 33, 135 33, 140 30)
```

Creating a Buffer Around an Object

Every geometry defines one or more points that lie in its interior—the area of space occupied by the geometry. In addition to these points that are contained within the geometry itself, there may be many cases in which you want to consider those points that lie in the area of space immediately surrounding a particular object. To do so, you can create a "buffer." A buffer is the area formed from all of the points that lie within a given distance of a geometry. Figure 12-11 illustrates the result created by defining a buffer around different types of geometries.

The STBuffer() method is used to create the geometry resulting from buffering an object by a certain amount. When used on a single-element instance, this generally creates a Polygon whose perimeter is defined by joining the points that lie a given distance away from the object. When creating the buffer of a multielement instance, the buffer of each element in the collection is calculated separately, and then the union of all the individual buffered geometries is returned. In other words, STBuffer() does not create a buffer around the entire multielement geometry, but rather buffers each individual element.

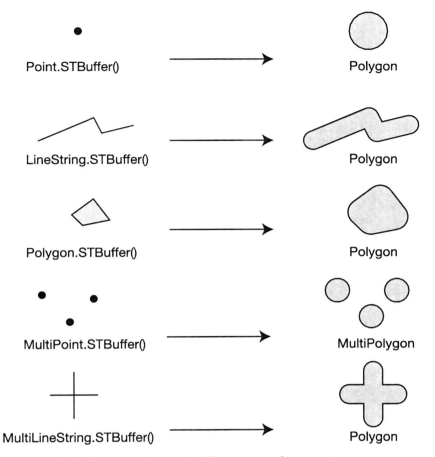

Figure 12-11. *Buffers created around different types of geometries*

Tip When you use the STBuffer() method on a Point object, you create a Polygon with a fixed radius, centered about that point. This is a useful way of creating "circular" polygon objects. (This method actually creates a many-sided polygon that approximates a circle, rather than a true circle.)

Supported Datatypes

The STBuffer() method can be used on instances of the following datatypes:

- geometry
- geography

Usage

The `STBuffer()` method can be used on any type of geometry from either the geography or geometry datatype as follows:

```
Instance.STBuffer(distance)
```

When calling the `STBuffer()` method, you must pass in a single floating-point value, `distance`, which represents the radius of the buffer zone to be created. When used on items of the geography datatype, the value of this parameter is specified in the linear unit of measurement defined by the spatial reference of the instance on which the method is called. For example, the radius of the buffer zone for any instance of the geography datatype defined with SRID 4296 must be specified in meters. When used on items of geometry data, the `distance` parameter is specified in the same unit of measure as the coordinate values of the geometry itself.

Tip To check the linear unit of measure used by any geographic spatial reference system, you can look at the value in the `unit_of_measure` column of the `sys.spatial_reference_systems` table.

The result of the `STBuffer()` method is a geometry or geography instance containing all those points contained within the buffered geometry, defined using the same datatype and SRID as the instance on which it was invoked.

Tip If the `distance` parameter passed to the `STBuffer()` method is less than zero, the resulting geometry *removes* those points lying within the stated distance from the interior of the geometry. This is a useful technique to "contract" Polygon instances.

Example

Suppose that you run a pizza delivery business that offers a free home-delivery service to all customers who live within a 5 km distance from the restaurant. You could use the `STBuffer()` method to define the Polygon area within which customers are entitled to receive free delivery. The following code illustrates this example, creating a 5 km buffer around a Point representing a restaurant located at a latitude of 52.6285N, longitude of 1.3033E, using SRID 4326:

```
DECLARE @Restaurant geography
SET @Restaurant = geography::STGeomFromText('POINT(1.3033 52.6285)', 4326)

DECLARE @FreeDeliveryZone geography
SET @FreeDeliveryZone = @Restaurant.STBuffer(5000)

SELECT
  @FreeDeliveryZone AS Shape,
  @FreeDeliveryZone.STAsText() AS WKT
```

Notice that we pass a value of 5000 as a parameter to STBuffer() to specify the radius of the buffer created. Since @Restaurant is a Point geometry defined using EPSG:4326, this value represents 5,000 *meters*—the 5 km radius zone we are interested in. Executing this query gives the following results:

```
POLYGON (( 1.2295142928970047 52.630278502274521, 1.2295472461369081
52.62631339458612, 1.2301545201280502 52.622365323475293, 1.2313314083048608
52.618464739096979, 1.2330689454853565 52.614641764288088, 1.235353942999466
52.610925976416176, 1.2381690623910524 52.607346209120308, 1.2414928988997975
52.603930332522118, 1.2453001225223537 52.60070505640406, 1.2495616038691015
52.597695712042189, 1.2542446053638217 52.594926048495672, 1.2593129907012768
52.59241803211485, 1.2647274718221007 52.590191652343655, 1.2704458908315621
52.588264740451066, 1.276423533941953 52.586652794489723, 1.2826134891460548
52.5853688336417, 1.2889670141086018 52.584423258904152, 1.2954339565898492
52.583823754630856, 1.3019631745506761 52.583575201434648, 1.3085029759219142
52.5836796202181, 1.3150015842393175 52.584136163917918, 1.3214075741984501
52.584941114208341, 1.3276703389598035 52.586087924183644, 1.3337405093551296
52.587567291244341, 1.3395703743341678 52.589367251335212, 1.3451142811293637
52.591473303153705, 1.35032899900776161 52.593868545985828, 1.3551739986447167
52.596533854416577, 1.359611855301528 52.599448044708396, 1.3636084040820327
52.602588071945057, 1.3671330192767153 52.605929227306689, 1.3701588001322242
52.609445339213529, 1.3726627175069714 52.6131089868275, 1.3746257639239476
52.616891705353538, 1.3760330440132762 52.620764192248004, 1.3768738577368669
52.624696525439845, 1.3771417561338952 52.6286583654713, 1.3768345809834786
52.6326191597911, 1.3759544721020014 52.63654836107424, 1.3745078711610552
52.640415622543479, 1.3725054902541798 52.644191014018475, 1.3699622689756672
52.647845233385752, 1.3668973005117668 52.651349810357452, 1.3633337327618695
52.654677324100653, 1.3592986521212407 52.657801619601393, 1.3548229107552452
52.660698004822308, 1.3499409480849405 52.66334347138249, 1.3446905555854709
52.665716895922813, 1.3391126095996737 52.667799221810554, 1.3332507731399534
52.669573652798491, 1.3271511624733332 52.671025810256374, 1.3208619546920446
52.6721438825306, 1.3144330073043538 52.672918747233538, 1.3079154101664321
52.673344063969346, 1.3013610473723278 52.673416351600068, 1.2948221297653622
52.673135015438916, 1.2883507323915036 52.672502360668375, 1.281998328174371
52.671523568729668, 1.2758153288993523 52.670206631571865, 1.2698506457504053
52.668562278677896, 1.2641512876131587 52.666603855018472, 1.258761959943709
52.664347188739917, 1.2537247188323759 52.661810431379195, 1.2490786505455085
52.659013885997183, 1.2448595929469528 52.655979818116634, 1.2410998878660526
52.652732251312585, 1.2378281815420607 52.649296770686853, 1.2350692475858318
52.64570030786836, 1.2328438521332565 52.641970925731265, 1.2311686394161832
52.638137614778167, 1.230056056050834 52.634230072684112, 1.2295142928970047
52.630278502274521))
```

This represents an approximately circular Polygon of radius 5 km, centered on the location of the restaurant.

Creating a Simpler Buffer

In the last example, I showed you how to use the STBuffer() method to create a "circular" Polygon around a Point geometry. The resulting geometry contained a lot of points . . . too many to count. Fortunately, we don't need to count them—we can use the STNumPoints() method to do that for us, by adding the following line to the end of the code:

```
SELECT @FreeDeliveryZone.STNumPoints()
```

The result is as follows:

72

The buffer object created by the STBuffer() method centered around the restaurant contains 72 points, defining a regular 71-sided polygon (a *heptacontakaihenagon*!). If you needed to maintain the maximum accuracy of the buffer around any geometry, you could use this Polygon shape definition as it is. However, for the particular application demonstrated in this example— offering free pizza delivery—we probably don't need our buffer to be that accurate. Performing computations on a complex geometry takes more processing resources than doing so on a simpler shape with fewer point definitions, and in this case we could obtain sufficient accuracy of the free delivery zone from a "circular" Polygon containing many fewer points. To create a simpler Polygon, you can use the BufferWithTolerance() extended method instead.

Supported Datatypes

The BufferWithTolerance() method can be used on any instance of the following datatypes:

- geometry
- geography

Usage

You can use the BufferWithTolerance() method on any item of geography or geometry data as follows:

```
Instance.BufferWithTolerance(distance, tolerance, relative)
```

This method requires the following three parameters:

The distance parameter is a floating-point value defining the radius of the created buffer, measured using the linear unit of measure for the spatial reference system of a geography instance, or using the same unit of measure as the coordinate values of a geometry instance. This is the same as required by the distance parameter of the STBuffer() method.

The tolerance parameter is a floating-point value that specifies the maximum variation allowed between the "true" buffer distance as calculated by STBuffer() and the simpler approximation of the buffer returned by the BufferWithTolerance() method. In the example of creating a buffer around a Point, this represents how closely the buffer created by BufferWithTolerance() resembles a true circle around the point. The smaller the tolerance value, the more closely the resulting Polygon buffer will resemble the actual buffer, but also the more complex it will be.

The relative parameter is a bit value specifying whether the supplied tolerance parameter of the buffer is relative or absolute. If relative is 'true' (or 1), then the tolerance of the buffer is determined relative to the extent of the geometry in question. For the geometry datatype, this means that the buffer is calculated from the product of the tolerance parameter and the diameter of the bounding box of the instance. For geography instances (where bounding boxes do not apply), the relative tolerance is instead calculated as the product of the tolerance parameter and the angular extent of the object multiplied by the equatorial radius of the reference ellipsoid of the spatial reference system. If relative is 'false' (or 0), then the value of the tolerance parameter is treated as an absolute value and applied uniformly as the maximum tolerated variation in the buffer created.

The tolerance and relative parameters, used together, define the acceptable level of tolerance by which the simpler buffer created by BufferWithTolerance() may deviate from the "true" buffer as created by STBuffer(). Only those points that deviate by more than the accepted tolerance are included in the resulting geometry. When determining an appropriate tolerance value to use with the BufferWithTolerance() method, remember that a greater tolerance will lead to a simpler geometry, but will result in a loss of accuracy. A lower tolerance will lead to a more complex, but more accurate, buffer.

Tip The results obtained from the STBuffer() method are the same as those obtained from BufferWithTolerance() when tolerance = 0.001 and relative = 'false'.

Example

Let's recalculate the free delivery area from the STBuffer() example using the BufferWithTolerance() method instead. We will keep the same buffer radius of 5 km around the restaurant, but this time we will specify an absolute tolerance of 250 m—any points from the "true" buffer that deviate by less than this accepted tolerance will not be included in the resulting geometry.

```
DECLARE @Restaurant geography
SET @Restaurant = geography::STGeomFromText('POINT(1.3033 52.6285)', 4326)

DECLARE @FreeDeliveryZone geography
SET @FreeDeliveryZone = @Restaurant.BufferWithTolerance(5000, 250, 'false')
```

```
SELECT
  @FreeDeliveryZone AS Shape,
  @FreeDeliveryZone.STAsText() AS WKT
```

The result of the BufferWithTolerance() method is represented by the following WKT:

```
POLYGON ((1.2334543773235034 52.643101987508025, 1.2326771600618118
52.615398722587805, 1.2587899052764109 52.592656472913966, 1.3019631745506761
52.583575201434648, 1.3456512565414915 52.591700111972628, 1.3730991328944517
52.613856932852478, 1.3739649138786716 52.641559233221273, 1.3478827812694929
52.664326601552453, 1.3046395628365322 52.673424441701911, 1.2608777319130935
52.665284532304732, 1.2334543773235034 52.643101987508025))
```

It is clear from examining the WKT representation that the Polygon geometry created using BufferWithTolerance() contains fewer points than the geometry previously created using the STBuffer() method. Using STNumPoints(), we can confirm that the simpler geometry created using BufferWithTolerance() actually contains 11 points, which define a regular decagon. This still provides a reasonable approximation of the circular zone we are interested in for this application, and has the benefit that any methods that operate on the simpler resulting geometry will perform more efficiently.

Figure 12-12 compares the results obtained using the STBuffer() and BufferWithTolerance() methods.

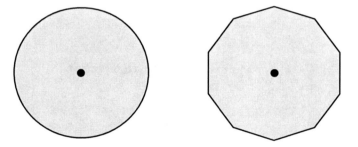

Figure 12-12. *Comparing the Polygons created by STBuffer() (left) and BufferWithTolerance() (right) on a Point geometry*

Creating the Convex Hull of a Geometry

The STConvexHull() method returns the smallest *convex* geometry that contains all the points in an instance. A convex geometry is one in which no interior angle is greater than 180 degrees, so that the sides do not ever bend inwards or contain indentations. Figure 12-13 illustrates the convex hull of a variety of different types of geometries.

■**Tip** To help visualize the convex hull of a geometry, consider an elastic band stretched around the outside of all the points in the geometry. When the elastic band is released, the shape that it snaps back into represents the convex hull of the geometry.

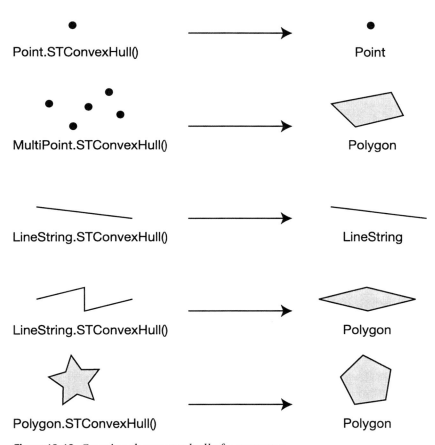

Figure 12-13. *Creating the convex hull of a geometry*

Supported Datatypes

The STConvexHull() method can be used on instances of the following datatype:

- geometry

Usage

The STConvexHull() method can be used with any instance of the geometry datatype as follows:

```
Instance.STConvexHull()
```

The type of geometry returned by STConvexHull() will be the smallest convex geometry that contains all of the points contained in an instance. If used on a single Point instance, the convex hull will be the Point itself. If used on a series of Points or LineStrings that lie in a straight line, the convex hull will be a LineString. In all other cases, the convex hull of a geometry will be a Polygon.

Example

Creating and analyzing the convex hull of a geometry can be a useful way of examining the geographical spread of an object on the earth. To demonstrate this, the following example creates a convex hull representing the spread of reported cases of the H5N1 virus (commonly known as "bird flu").

In this example, @H5N1 is a MultiPoint geometry, containing a point at each location where a case of the H5N1 virus had been recorded in humans, as reported by the World Health Organization between September 2003 and December 2004. We then use the STConvexHull() method of the geometry datatype to create the convex hull Polygon that encompasses each point in the MultiPoint geometry to describe the overall spread of the disease.

```
DECLARE @H5N1 geometry
SET @H5N1 = geometry::STMPointFromText(
  'MULTIPOINT(
    105.968 20.541, 105.877 21.124, 106.208 20.28, 101.803 16.009, 99.688 16.015,
    99.055 14.593, 99.055 14.583, 102.519 16.215, 100.914 15.074, 102.117 14.957,
    100.527 14.341, 99.699 17.248, 99.898 14.608, 99.898 14.608, 99.898 14.608,
    99.898 14.608, 100.524 17.75, 106.107 21.11, 106.91 11.753, 107.182 11.051,
    105.646 20.957, 105.857 21.124, 105.867 21.124, 105.827 21.124, 105.847 21.144,
    105.847 21.134, 106.617 10.871, 106.617 10.851, 106.637 10.851, 106.617 10.861,
    106.627 10.851, 106.617 10.881, 108.094 11.77, 108.094 11.75, 108.081 11.505,
    108.094 11.76, 105.899 9.546, 106.162 11.414, 106.382 20.534, 106.352 20.504,
    106.342 20.504, 106.382 20.524, 106.382 20.504, 105.34 20.041, 105.34 20.051,
    104.977 22.765, 105.646 20.977, 105.646 20.937, 99.688 16.015, 100.389 13.927,
    101.147 16.269, 101.78 13.905, 99.704 17.601, 105.604 10.654, 105.817 21.124,
    106.162 11.404, 106.362 20.504)',
  4326)

SELECT
  @H5N1 AS Shape
UNION ALL SELECT
  @H5N1.STConvexHull() AS Shape
```

The resulting convex Polygon of the STConvexHull() method and the MultiPoint @H5N1 geometry from which it was created are illustrated in Figure 12-14, overlaid onto a map illustrating the affected area.

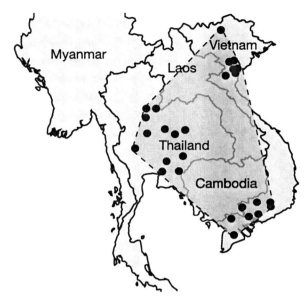

Figure 12-14. *Using a convex hull to illustrate the spread of the H5N1 virus*

One interesting fact to note is that, for the period of time that this data represents (from September 2003 through December 2004), individual cases of H5N1 had only been reported in the countries of Thailand and Vietnam. However, the convex hull encompassing the points at which those cases were discovered covers most of Laos and Cambodia, which indicates that they lie within the area that the disease had spread to, and were at significant risk of exposure to the disease. Indeed, by January 21, 2005, less than one month after the last date recorded in this dataset, the first case of human infection of H5N1 had been reported in Cambodia itself.

Summary

In this chapter, I showed you the methods that SQL Server 2008 provides to combine, subtract from, or modify existing geometries. Table 12-1 provides a summary of the methods introduced in this chapter.

Table 12-1. *Methods That Modify Spatial Objects*

Method	Description	geometry	geography
MakeValid()	Modifies a geometry to ensure it is valid	•	
STUnion()	Defines an instance from the union of two geometries	•	•
STIntersection()	Defines an instance from the intersection of two geometries	•	•
STDifference()	Defines an instance from the points representing the difference between two geometries	•	•
STSymDifference()	Defines an instance from the symmetric difference of two geometries	•	•
Reduce()	Defines a simplified instance of a Polygon or LineString	•	•
STBuffer()	Defines a Polygon or MultiPolygon representing the buffer zone around an object	•	•
BufferWithTolerance()	Defines a Polygon or MultiPolygon representing the buffer zone around an object with a given tolerance	•	•
STConvexHull()	Creates a Polygon representing the convex hull of a geometry	•	

CHAPTER 13

Testing Spatial Relationships

When analyzing spatial information, we frequently want to understand the relationship between two or more features on the earth—for example, how far is it from *a* to *b*? Does the route between *x* and *y* pass through *z*? Does *p* share a common border with *q*? In this chapter, I'll introduce the methods that you can use to answer these questions, by comparing different aspects of the relationship between two items of spatial data. The general syntax of all of these methods is that the instance to which a comparison is being made is provided as a parameter to a method acting upon the first instance; for example:

```
Instance1.Method(Instance2)
```

While the geometry and geography datatypes both implement a core set of methods used to compare spatial relationships, including STEquals(), STIntersects(), and STDistance(), the geometry datatype implements a more extensive range of methods that can be used to test specific relationships, such as STOverlaps(), STCrosses(), and STContains(). For each method introduced in this chapter, I'll tell you which datatypes it can be used with. You should be aware that, even in cases where a method is implemented by both the geography and geometry datatypes, you cannot use a method to compare instances of different types. For example, the STDistance() method can be used to calculate the distance between two geography instances, or between two geometry instances, but cannot be used to determine the distance between a geometry and geography instance.

Testing the Equality of Two Geometries

The STEquals() method tests whether two spatial instances are *equal*—that is, they comprise exactly the same set of points. Note that the geometries do not necessarily have to be the same type of geometry to be considered equal. For example, the single LineString geometry joining three points, LINESTRING(0 0, 2 2, 5 2), is equal to the MultiLineString geometry MULTILINESTRING((0 0, 2 2), (2 2, 5 2)), since they represent the same set of points.

Supported Datatypes

The STEquals() method can be used on instances of the following datatypes:

- geometry

- geography

Usage

The STEquals() method can be used as follows to test whether two instances of the geography or geometry datatype are equal:

```
Instance1.STEquals(Instance2)
```

The result of the method is a bit value—the value 1 represents true, indicating the instances do contain the same set of points, and 0 represents false, indicating the instances do not contain the same set of points.

Example

To demonstrate the STEquals() method, the following example creates three geography instances, each representing a single Point located at the coordinates of latitude 40.689235°N, longitude 74.044480°W. This is the approximate location of the Statue of Liberty in New York City. Each instance is created using a different static method, yet they all create a Point based on the same coordinates, defined using the same SRID. The STEquals() method is then used to test whether the instances created by each method are the same.

```
DECLARE @a geography
DECLARE @b geography
DECLARE @c geography

SET @a = geography::STPointFromText(
  'POINT(-74.044480 40.689235)',
  4326)

SET @b = geography::STPointFromWKB(
  0x0101000000DE54A4C2D88252C018213CDA38584440,
  4326)

SET @c = geography::GeomFromGml(
  '<Point xmlns="http://www.opengis.net/gml">
    <pos>40.689235 -74.044480</pos>
  </Point>',
  4326)

SELECT
  @a.STEquals(@b) AS 'a = b',
  @b.STEquals(@c) AS 'b = c'
```

The results are as follows:

```
a = b    b = c
1        1
```

This confirms that, in this simple case, the three Point instances created from the WKT, WKB, and GML representations are all equal. However, you should exercise caution when using the STEquals() method, because sometimes seemingly "equal" instances are not as equal as you might expect. For example, remember from Chapter 12 that the intersection of a Point and a Polygon is the same as the Point itself, and that you can obtain the result of the intersection of two geometries using the STIntersection() method. With this in mind, consider the following code:

```
DECLARE @Point geometry
SET @Point = geometry::Point(1.1,50,0)

DECLARE @Polygon geometry
SET @Polygon = geometry::STPolyFromText(
  'POLYGON((0 0, 100 0, 100 100, 0 100, 0 0))', 0)

DECLARE @Intersection geometry
SET @Intersection = @Point.STIntersection(@Polygon)

SELECT @Point.STEquals(@Intersection)
```

You would expect @Intersection, the geometry created from the intersection of @Point and @Polygon, to be equal to @Point. Therefore, the STEquals() method will return 1, right? Actually, when you execute the query, you get the following result instead:

0

In this particular case, the x coordinate value at which @Point is defined, 1.1, cannot be exactly expressed in floating-point binary format. The results of the STIntersection() method therefore introduce a fractional error in the coordinate value returned, because they are based on a calculation using the closest possible approximation of that value (I discussed this in more detail in the sidebar in Chapter 12 called "The Imprecision of Floating-Point Spatial Data"). You can confirm this error by adding the following line to the end of the code:

```
SELECT
  @Point.STAsText(),
  @Intersection.STAsText()
```

You will see the following results:

POINT (1.1 50)   POINT (1.1000003814697266 50)

Notice the slight discrepancy in the x coordinate values, which means that these two instances are not considered equal. These slight changes in coordinate values may occur in any operations in SQL Server that involve floating-point binary arithmetic, and they cannot be avoided. The magnitude of any errors is rarely large enough to affect the accuracy of any one item of data. However, such errors can have significant impact on the results from precise comparative methods, such as STEquals().

Calculating the Distance Between Geometries

The STDistance() method can be used to calculate the shortest distance between the points of any two geometries. When used on instances of the geometry datatype, this is the length of the shortest straight line that can be drawn between the two instances. For the geography datatype, this is the length of the shortest great elliptic arc drawn between any two points contained in the two geometries, following the surface of the reference ellipsoid between them. Figure 13-1 illustrates the results of the STDistance() method when performed on a variety of different types of geometries.

Point.STDistance(Point) = *d*

Point.STDistance(LineString) = 0

Point.STDistance(Polygon) = 0

LineString.STDistance(Point) = *d*

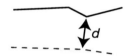

LineString.STDistance(LineString) = *d*

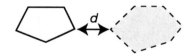

Polygon.STDistance(Polygon) = *d*

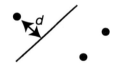

MultiPoint.STDistance(LineString) = *d*

Figure 13-1. *Calculating the shortest distance between different types of geometries by using STDistance()*

Supported Datatypes

The STDistance() method can be used on instances of the following datatypes:

- geometry
- geography

Usage

The STDistance() method can be applied to calculate the distance between two instances of the geometry or geography datatype as follows:

```
Instance1.STDistance(Instance2)
```

The result is a floating-point numeric value, representing the distance between the two geometries. When used to calculate the distance between two instances of the geography datatype, the result is expressed in the corresponding unit of measurement for the spatial reference system in which the instances were defined. For the geometry datatype, the result is expressed in the same unit of measurement as the coordinate values themselves.

Note While the STDistance() method can be applied to either the geography or the geometry datatype, you cannot use it to calculate the distance between instances of different types. In the usage example provided, Instance1 and Instance2 must be of the same datatype, defined using the same SRID.

Example

One common application of spatial data is to identify the feature that lies closest to a given location or, in the more general case, to find the nearest *n* features to a location. This type of query is commonly called a *nearest-neighbor* query. To demonstrate how you might use the STDistance() method to perform a nearest-neighbor query in SQL Server, let's suppose that you are operating a disaster response service based in the state of Massachusetts. When you are notified of a major fire, you need to identify and contact the closest fire station to the incident so that it can send out a response unit. For this example, let's first create a table containing details of fire stations in the state of Massachusetts:

```
CREATE TABLE MA_Firestations (
  Name varchar(255),
  Address varchar(255),
  City varchar(255),
  Location geometry
)
```

To populate this table, let's create a few records representing individual fire stations. The location of each fire station is represented by a Point geometry defined using the Massachusetts State Plane Coordinate System, which is a projected coordinate system denoted by the SRID 26986:

```
INSERT INTO MA_Firestations VALUES
('SANDWICH FIRE DEPARTMENT',
'115 Rt. 6A',
'SANDWICH',
0x6A690000010C07095104C54C114113471EC8467D2941),

('BROCKTON FIRE DEPARTMENT',
'560 West Street',
'BROCKTON',
0x6A690000010C9DD497310C050D4113471EC8A4852A41),
```

```
('SWANSEA FIRE DEPARTMENT',
'50 New Gardner Neck Road',
'SWANSEA',
0x6A690000010C4145D53B43770B4113471EC89F5C2941),

('ASHLAND FIRE DEPARTMENT',
'70 Cedar Street',
'ASHLAND',
0x6A690000010C22C0E9E9A0090941C18EFF4247252B41)
```

Although you can use just these four examples, it is easier to see how a nearest-neighbor query works when selecting records from a dataset containing thousands, or millions, of records. If you want to add more records to the MA_Firestations table, you can download the full dataset of fire stations in Massachusetts as part of the code archive accompanying this book, available in the Source Code/Download area of the Apress web site (http://www.apress.com).

For this example, suppose that we have been informed of a fire at coordinates (210000,890000) defined, like the location of the fire stations, using EPSG:26986. These coordinates relate to a point about 15 miles southwest of Boston. Now that we have a sample table set up, and we know the location of the fire, how do we go about identifying the nearest fire station? Let's look at three possible methods, in turn.

Finding Nearest Neighbors: Basic Approach

The simplest method of identifying the nearest neighbor is to use the STDistance() method in the ORDER BY clause of a SELECT statement. Once the records have been sorted in ascending order of distance, you can use the SELECT TOP n syntax to return only the top *n* nearest neighbors, as demonstrated in the following query:

```
DECLARE @Fire geometry
SET @Fire = geometry::STPointFromText('POINT (210000 890000)', 26986)

SELECT TOP 1
  Name,
  Address,
  City,
  Location.STDistance(@Fire) AS Distance
FROM
  MA_Firestations
ORDER BY
  Location.STDistance(@Fire) ASC
```

The result is as follows:

Name	Address	City	Distance
ASHLAND FIRE DEPARTMENT	70 Cedar Street	ASHLAND	4916.60186549405

Although this query will correctly identify and return the nearest *n* neighboring fire stations to the fire (in this example, just the single nearest fire station is chosen), this approach is not very efficient. STDistance() is a precise method, but it is computationally expensive. Because SQL Server has to execute the STDistance() method on every row in the table and then order the records by the resulting distance to identify and return the top row only, when performing a nearest-neighbor search like this on a table containing millions of rows, processing the results could take a very long time. This approach is therefore most useful if you have only a limited amount of data within which to search for nearest neighbors (hundreds, or thousands, of rows, for example).

Finding Nearest Neighbors Within a Fixed Search Zone

This solution modifies the basic nearest-neighbor query described previously into a two-stage approach. The first stage is to identify a set of likely nearest-neighbor candidates, by selecting only those records that lie within a predetermined radius of the feature in question. To do this, a search area is created around the feature, using the STBuffer() method. The size of the buffer is chosen to be large enough to ensure that it contains the required number of nearest neighbors, but not so large that it includes lots of additional rows of data that exceed the desired number of results. Features lying within the search area are selected using the efficient Filter() method, which uses the spatial index to create a table of possible candidates. The STDistance() method is then used to calculate the associated distance of only those candidate records, rather than processing and sorting the whole table. The following listing demonstrates this approach, using STBuffer(25000) to identify candidate records that lie within a 25 km search area around the fire in which to search for nearest neighbors:

```
DECLARE @Fire geometry
SET @Fire = geometry::STPointFromText('POINT (210000 890000)', 26986)

DECLARE @SearchArea geometry
SET @SearchArea = @Fire.STBuffer(25000)

DECLARE @Candidates TABLE (
  Name varchar(255),
  Address varchar(255),
  City varchar(255),
  Distance float
)

INSERT INTO @Candidates
SELECT
  Name,
  Address,
  City,
  Location.STDistance(@Fire) AS Distance
FROM
  MA_Firestations
WHERE
  Location.Filter(@SearchArea) = 1

SELECT TOP 1 * FROM @Candidates ORDER BY Distance
```

As in the last example, the fixed search zone approach correctly identifies the closest fire station in this case as follows:

Name	Address	City	Distance
ASHLAND FIRE DEPARTMENT	70 Cedar Street	ASHLAND	4916.60186549405

The advantage of the fixed search zone approach is that, by first filtering the set of candidate results to only those that lie within the vicinity of the feature, the number of times that STDistance() needs to be called is reduced, and the dataset requiring sorting is much smaller, making the query significantly faster than the basic nearest-neighbor approach described previously. However, the problem with this approach is that you must choose an appropriate fixed value to pass to the STBuffer() method to set the radius of the search area. If you set the value too high, then there will be too many possible candidates returned and the filter will not be efficient. If you set the buffer size too small, then there is a risk that the search area will not contain any candidates, resulting in the query failing to identify any nearest neighbors at all. For example, the fire station located the shortest distance from the fire is Ashland Fire Department, which lies 4.9 km away. If we had narrowed our nearest-neighbor query to only search for candidates lying within a 4 km distance, using SET @SearchArea = @Fire.STBuffer(4000), the query would not have returned any results.

The fixed search zone approach is most useful in situations where you are able to reliably set an appropriate buffer size in which to select a set of candidate nearest neighbors. This might be based on known, uniform distribution of your data—for example, you know that any item in the dataset will never lie more than 25 km from its nearest neighbor. Alternatively, you might want to obtain nearest neighbors within a particular distance constraint—for example, to answer the query "Show me the three closest gas stations to this location, but only if they are within 10 miles."

■**Tip** To ensure that the Filter() method acts as an efficient primary filter to identify nearest-neighbor candidates, make sure that your table contains a spatial index. For more information on spatial indexes, see Chapter 14.

Finding Nearest Neighbors Within an Expanding Search Zone

This approach, like the previous one, uses a two-stage approach to identify nearest neighbors, where a set of possible nearest-neighbor candidates is first identified before selecting the actual nearest neighbor from the set of possible candidates. Rather than using the Filter() method to identify candidates that lie within a fixed search zone (which faces the risk of failing to identify any nearest neighbors at all), this approach creates a series of expanding search ranges, which is ultimately guaranteed to find the nearest neighbor.

To create the expanding ranges, this approach requires the use of a *numbers table* (sometimes called a tally table), which is a simple table containing a single column of consecutive integers. Although, at first, it may seem unnecessary to create a table containing nothing more than a sequential list of numbers, numbers tables can prove very useful when it comes to solving certain problems in a set-based environment, as you will see in this example. To create and populate a numbers table with the integers between 0 and 1,000, execute the following code:

```
CREATE TABLE Numbers (
  Number int PRIMARY KEY CLUSTERED
)
DECLARE @i int = 0
WHILE @i <= 1000
BEGIN
  INSERT INTO Numbers VALUES (@i)
  SET @i = @i + 1
END
```

The Numbers table will be joined to the MA_Firestations table to create a series of expanding search ranges. The distance to which each successive search extends increases exponentially until a search area of sufficient size is found that contains the requisite number of nearest neighbors. All of the features in this search area are returned as candidates, and then the TOP 1 is selected as the true nearest neighbor. This approach is demonstrated in the following code:

```
DECLARE @Fire geometry
SET @Fire = geometry::STPointFromText('POINT (210000 890000)', 26986)

DECLARE @Candidates TABLE (
  Name varchar(255),
  Address varchar(255),
  City varchar(255),
  Distance float,
  Range int
)

INSERT INTO @Candidates
SELECT TOP 1 WITH TIES
  Name,
  Address,
  City,
  Location.STDistance(@Fire) AS Distance,
  1000*POWER(2, Number) AS Range
FROM
  MA_Firestations
  INNER JOIN Numbers
  ON MA_Firestations.Location.STDistance(@Fire) < 1000*POWER(2,Numbers.Number)
  ORDER BY Number

SELECT TOP 1 * FROM @Candidates ORDER BY Range DESC, Distance ASC
```

The result obtained is as follows:

Name	Address	City	Distance	Range
ASHLAND FIRE DEPARTMENT	70 Cedar Street	ASHLAND	4916.60186549405	8000

Remember that the Numbers table contains consecutive integers, starting at zero. So, the condition MA_Firestations.Location.STDistance(@Fire) < 1000*POWER(2,Numbers.Number) specifies that the initial criterion for a feature to be considered a nearest-neighbor candidate is that the result of STDistance() is less than 1000 * 2^0 for that feature. Since the EPSG:26986 spatial reference system defines distances in meters, this equates to a 1 km search area—if you want to specify an alternative starting search radius, you may do so by changing the value of 1000 to another value (remember to use the unit of measure appropriate to the datatype and SRID of the data in question). If the requisite number of neighbors (in this case, we are searching only for the TOP 1) are not found within the specified distance, then the search range is increased in size. Successive search ranges are obtained by raising 2 to the power of the next number in the Numbers table. Thus, the first range extends to 1 km around the fire, the second range extends to 2 km, then 4 km, 8 km, 16 km, and so on. By adopting an exponential growth model, this method is guaranteed to find the nearest neighbor within a relatively short number of iterations, however dispersed the distribution of the underlying features is.

Once the search range has been sufficiently increased to contain at least the required number of candidate nearest neighbors, all of the features lying within that range are selected as candidates, by using a SELECT statement with the WITH TIES argument. Finally, the candidates are sorted by ascending distance from the fire, and the TOP 1 record is selected as the true nearest neighbor. The Range column returned in the results states the distance to which the search range was extended to find the nearest neighbor—in this case, 8 km.

While it is slightly more complex, this approach provides the most flexible solution for implementing a nearest-neighbor query. It is significantly faster than the basic approach and, although it is not quite as fast as the fixed search area technique, it does not suffer from the limitations associated with having to specify a fixed search radius.

■**Tip** You can improve the performance of this query by adding a spatial index to the MA_Firestations table.

MEASURING THE DISTANCE BETWEEN POINTS IN THREE DIMENSIONS

Like all of the spatial functionality in SQL Server 2008, the STDistance() function only operates in two dimensions—describing the distance along the flat surface of the geometry datatype, or the curved ellipsoidal surface used by the geography datatype. Although you may define the points of a geometry using z coordinate values that describe their elevation (or depth) relative to this surface, these z coordinates are not considered by any of the in-built spatial methods. However, you can extend the spatial functionality provided in SQL Server by creating your own user-defined functions that take into account the z coordinates defined by a geometry.

The following code demonstrates how to create a user-defined function that calculates the distance in three-dimensional space between any two geometry Points, taking into consideration the x, y, and z coordinate values:

```
CREATE FUNCTION dbo.DistanceXYZ (
  @Point1 geometry,
  @Point2 geometry )
RETURNS float
AS
```

```
BEGIN
  RETURN
  SQRT(
    SQUARE(@Point2.STX - @Point1.STX) +
    SQUARE (@Point2.STY - @Point1.STY) +
    SQUARE (COALESCE(@Point2.Z,0) - COALESCE(@Point1.Z,0))
  )
END
GO
```

You may use this function to calculate the distance between two Points of the geometry datatype, as follows:

```
DECLARE @a geometry = geometry::STPointFromText('POINT(3 4 12)',0)
DECLARE @b geometry = geometry::STPointFromText('POINT(0 0 0)',0)
SELECT
  @a.STDistance(@b) AS STDistance,
  dbo.DistanceXYZ(@a, @b) AS DistanceXYZ
```

The results demonstrate the comparison between the two-dimensional distance between the Points located at (0,0,0) and (3,4,12) along the plane, obtained using STDistance(), and the distance between the Points calculated in three dimensions using the new DistanceXYZ function, as follows:

STDistance	DistanceXYZ
5	13

Note that, although STDistance() can calculate the shortest distance between *any* two geometries in two-dimensional space, the DistanceXYZ function described here can only calculate the distance between two Point geometries.

Testing Whether Two Geometries Intersect

Two geometries are said to *intersect* if they share at least one point in common. Those common points may lie on the boundary of the geometries concerned, or may be contained in their interior. Testing to see whether one geometry intersects another is one of the most commonly used methods to identify objects that have some generalized spatial relationship to each other—for instance, identifying all the features related to a particular area of interest.

Note In addition to providing a method to test for the general case of whether two geometries intersect, the geometry datatype provides additional methods that allow you to test for specific *types* of intersection—whether two geometries touch, cross, or overlap each other, or whether a geometry contains or is contained by another geometry. These specific types of intersections are discussed later in this chapter.

The STIntersects() method tests whether two instances have at least one point in common. Figure 13-2 illustrates the results of the STIntersects() method when used to test the case of intersection between different types of geometries.

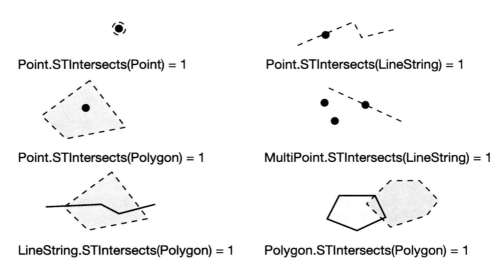

Point.STIntersects(Point) = 1   Point.STIntersects(LineString) = 1

Point.STIntersects(Polygon) = 1   MultiPoint.STIntersects(LineString) = 1

LineString.STIntersects(Polygon) = 1   Polygon.STIntersects(Polygon) = 1

Figure 13-2. *Results of the STIntersects() method when testing the intersection of different geometries*

Note STIntersects() is used to test *whether* two geometries intersect. If you want to return the shape created by the intersection of two geometries, you should use the STIntersection() method instead.

Supported Datatypes

The STIntersects() method can be used on instances of the following datatypes:

- geometry
- geography

Usage

The STIntersects() method can be used to test whether two instances of the geometry or geography datatype intersect as follows:

Instance1.STIntersects(Instance2)

The result of the method is 1 if the instances share any point in common, or 0 if they do not.

Example

The following example code creates a table containing a Point, a LineString, and a Polygon geometry representing three well-known landmarks in Sydney: the Sydney Opera House, the

Sydney Harbour Bridge, and the Royal Botanic Gardens. It then defines a Polygon geometry representing a 1 km square area of interest in the center of the city, and determines which of the geometries in the table intersect that area.

```
DECLARE @SydneyFeatures TABLE (
  Name varchar(32),
  Shape geometry
  )

INSERT INTO @SydneyFeatures VALUES
('Sydney Opera House', geometry::STPointFromText('POINT(334900 6252300)', 32756)),
('Sydney Harbour Bridge', geometry::STLineFromText(
  'LINESTRING(334300 6252450, 334600 6253000)', 32756)),
('Royal Botanic Garden', geometry::STPolyFromText('
  POLYGON ((334750 6252030, 334675 6251340, 335230 6251100, 335620 6251700,
  335540  6252040,335280 6251580, 335075 6251650, 335075 6251960, 334860 6252120,
  334750 6252030))', 32756))

DECLARE @AreaOfInterest geometry = geometry::STPolyFromText('
  POLYGON((334400 6252800, 334400 6251800, 335400 6251800, 335400 6252800,
  334400 6252800))', 32756)

SELECT
  Name
FROM
  @SydneyFeatures
WHERE
  Shape.STIntersects(@AreaOfInterest) = 1
```

This produces the following results:

```
Sydney Opera House
Sydney Harbour Bridge
Royal Botanic Garden
```

Note that the features do not need to be completely contained within the area of interest to be included in the results of the STIntersection() method—they simply need to intersect some part of it (this contrasts with the STWithin() method, discussed later in this chapter). To visualize the relationship between the particular features in this example, add the following code immediately after the end of the previous query:

```
SELECT Shape FROM @SydneyFeatures
UNION ALL SELECT @AreaOfInterest
```

After executing the query, click the Spatial Results tab and you will see the results illustrated in Figure 13-3.

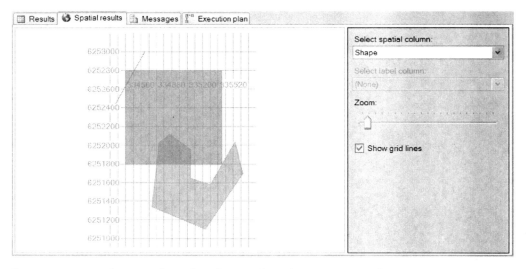

Figure 13-3. *Using the Spatial Results tab to confirm the intersection of geometries representing features in Sydney*

The large square geometry represents the area of interest. The Point geometry representing Sydney Opera House is fully contained and located roughly in the center of the square. The LineString geometry representing the Sydney Harbour Bridge crosses the northwest corner of the area, and the Polygon geometry representing the Royal Botanic Gardens overlaps on the south side. All three geometries intersect the area of interest in some way, and are therefore included in the results returned by the condition STIntersects(@AreaOfInterest) = 1.

Performing a Quick Test of Intersection Between Two Geometries

The Filter() method provides similar functionality to the STIntersects() method—it too is used to test for any kind of intersection between two geometry or geography instances, returning the value 1 if intersection occurs, or 0 if no intersection occurs. However, instead of directly testing the two geometries in question to establish whether they share any points in common (as STIntersects() does), the Filter() method searches the spatial index to find those objects that occupy index grid cells that intersect the parameter object. Depending on the complexity of the geometries in question, this means that testing for the intersection between two geometries using the Filter() method can be much faster than using the STIntersects() method.

Note Spatial indexes store a record of the generalized extent of a geometry by creating a grid of cells, and recording those cells in the grid required to completely cover each geometry. This can be used as a primary filter to approximate the area of space that the geometry itself occupies. The topic of spatial indexes is covered fully in the next chapter.

Although Filter() may be faster than STIntersects(), one disadvantage of the Filter() method is that there is a risk of returning false positive results—that is, the Filter() method might return the value 1 in some cases where the cell in the spatial index intersects the geometry in question, even if the instance itself does not. One of the most useful applications of the Filter() method is therefore when you want to perform a fast, approximate query of all those geometries that intersect a particular instance, in order to pass the data to a client that can then perform more detailed analysis of those geometries.

Supported Datatypes

The Filter() method can be used on instances of the following datatypes:

- geometry
- geography

Usage

The Filter() method can be applied to perform a quick test of intersection between two instances of the geometry or geography datatype as follows:

```
Instance1.Filter(Instance2)
```

Like the STIntersects() method, the Filter() method returns the value 1 if the instances intersect, or 0 if they do not. However, you should remember that, since this result is based on an approximate test of intersection from the cells occupied by each instance in the spatial index, the Filter() method might not return the same result as the STIntersects() method.

Note If used on a column of data that does not have a defined spatial index, the Filter() method defaults to exactly the same behavior as the STIntersects() method.

Example

To demonstrate the Filter() method, you first need to create a table with a spatial index as follows:

```
CREATE TABLE #Geometries (
  id int IDENTITY(1,1) PRIMARY KEY,
  geom geometry
  )
CREATE SPATIAL INDEX [idx_Spatial]
  ON [dbo].[#Geometries] ( geom )
  USING GEOMETRY_GRID
  WITH (
    BOUNDING_BOX =(-180, -90, 180, 90),
    GRIDS =(
      LEVEL_1 = MEDIUM,
      LEVEL_2 = MEDIUM,
```

```
    LEVEL_3 = MEDIUM,
    LEVEL_4 = MEDIUM),
 CELLS_PER_OBJECT = 4 )
```

Then, execute the following code to insert two Polygon geometries into the table:

```
INSERT INTO #Geometries (geom) VALUES
(geometry::STPolyFromText(
 'POLYGON((52.09 -2.14, 51.88 -2.15, 51.89 -1.89,52.12 -1.99, 52.09 -2.14))', 0)),
(geometry::STPolyFromText(
 'POLYGON((52.1 -2, 52.05 -2.01, 51.9 -1.9, 52.11 -2.15, 52.15 -1.9, 52.1 -2))', 0))
```

Let's try to find out which Polygons in the #Geometries table intersect a Point geometry at coordinates (52.07,–2). First, we'll try using the Filter() method:

```
DECLARE @Point geometry = geometry::STGeomFromText('POINT(52.07 -2)', 0)
SELECT id
FROM #Geometries
WITH(INDEX(idx_Spatial))
WHERE geom.Filter(@Point) = 1
```

The results of the Filter() method suggest that both Polygons contain this Point:

```
1
2
```

Now let's ask the same question using the STIntersects() method:

```
SELECT
id
FROM #Geometries
WITH(INDEX(idx_Spatial))
WHERE geom.STIntersects(@Point) = 1
```

In this case, the result of STIntersects() accurately shows that the Point is contained only in the first Polygon:

```
1
```

Why do the Filter() method and the STIntersects() method give different results in this example? Although the Point @Point lies very close to the edge of the second Polygon, it is not contained within it. However, the index grid cells provide only a loose fit around the shape of the Polygon, so they *do* contain the Point in question. Since the Filter() method obtains an approximate answer based on the cells representing each object in the spatial index, it can therefore create false positive results, as in this case.

Note The degree of accuracy with which the results of the Filter() method represent the actual intersection between two geometries depends on the properties of the spatial index that it uses.

Testing Whether Two Geometries Are Disjoint

The STDisjoint() method tests whether two instances are disjoint—that is, they have no points in common. STIntersects() and STDisjoint() are complementary methods, which means that for any given pair of geometries, if the result of one of these methods is true, the result of the other must be false. A.Intersects(B) = 1 is logically equivalent to A.Disjoint(B) = 0.

Tip STIntersects() returns 1 if the instances intersect, and 0 if they are disjoint. STDisjoint() returns 1 if the instances are disjoint, and 0 if they intersect.

Figure 13-4 illustrates the results of the STDisjoint() method when used to test whether various geometries are disjoint.

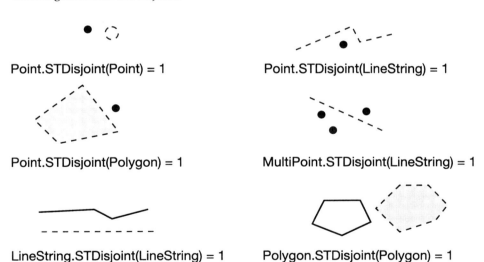

Point.STDisjoint(Point) = 1   Point.STDisjoint(LineString) = 1

Point.STDisjoint(Polygon) = 1   MultiPoint.STDisjoint(LineString) = 1

LineString.STDisjoint(LineString) = 1   Polygon.STDisjoint(Polygon) = 1

Figure 13-4. *Testing whether different geometries are disjoint by using STDisjoint()*

Supported Datatypes

The STDisjoint() method can be used on instances of the following datatypes:

- geometry
- geography

Usage

The STDisjoint() method can be used on any two instances of the geometry or geography datatype as follows:

```
Instance1.STDisjoint(Instance2)
```

If the two instances are disjoint (share no points in common), then the result of the STDisjoint() method is 1. If the two instances have any point in common, the STDisjoint() method returns the value 0.

Example

In order to protect and preserve the natural environment, many countries designate specific areas of outstanding natural beauty, such as national parks, which are governed by special planning restrictions that prevent industrial development in those areas. In this example, a Polygon geometry is created representing an area of protected countryside in Dorset, England. A LineString geometry is then defined representing the proposed route of a new road being developed in the area. The STDisjoint() method is used to test whether the road avoids the designated area of countryside.

```
DECLARE @Countryside geography
SET @Countryside = geography::STPolyFromText(
  'POLYGON((-2.66 50.67, -2.47 50.59, -2.39 50.64, -1.97 50.58,
  -1.94 50.66, -2.05 50.69, -2.02 50.72, -2.14 50.75, -2.66 50.67))',  4326)

DECLARE @Road Geography
SET @Road = geography::STGeomFromText(
  'LINESTRING(-2.44 50.71, -2.46 50.66, -2.45 50.61 )', 4326)

SELECT
@Road.STDisjoint(@Countryside)
```

The result of the STDisjoint() method is as follows:

```
0
```

This indicates that, in this case, the road is *not* disjoint to the protected countryside area, and the road plan must be reconsidered.

Finding Out Whether One Geometry Crosses Another

The STCrosses() method can be used to test the specific case of intersection where one geometry *crosses* another. In spatial terms, geometry A crosses geometry B when either of the following conditions is met:

- Geometry B is a Polygon and geometry A intersects both the interior and exterior of that Polygon. Note that this only applies when geometry A is a MultiPoint, LineString, or MultiLineString—if geometry A is also a Polygon, then this condition would be described as the Polygons *overlapping*.

- Geometry A and geometry B are both either LineStrings or MultiLineStrings, and the geometry created by their intersection occupies zero dimensions (i.e., the LineStrings intersect each other at a single point, or at multiple points, but not along a continuous stretch of LineString).

■Note A Point cannot cross any object. Two Polygons cannot cross each other, but they may overlap (for more information, see the discussion of the STOverlaps() method, later in this chapter).

Figure 13-5 illustrates a number of scenarios where one geometry crosses another, as tested by the STCrosses() method.

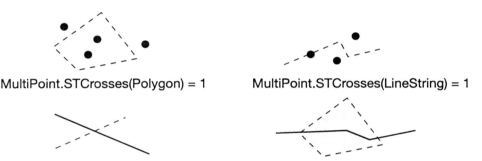

MultiPoint.STCrosses(Polygon) = 1   MultiPoint.STCrosses(LineString) = 1

LineString.STCrosses(LineString) = 1   LineString.STCrosses(Polygon) = 1

Figure 13-5. *Testing cases where one geometry crosses another by using STCrosses()*

■Caution The STCrosses() method is not symmetric. For example, a LineString can cross a Polygon, but a Polygon cannot cross a LineString. Be sure to specify the instances in the correct order when using the STCrosses() method.

Supported Datatypes

The STCrosses() method can be used on instances of the following datatype:

- geometry

Usage

The STCrosses() method can be used to test whether one geometry instance crosses another as follows:

```
Instance1.STCrosses(Instance2)
```

If the two instances satisfy the conditions described previously, then they are deemed to cross, and the STCrosses() method returns the value 1. Otherwise, the STCrosses() method returns the value 0.

Example

To demonstrate the use of the STCrosses() method, consider the following example based on the London congestion charging zone. In order to reduce traffic in the city, in 2003 the mayor of London introduced a congestion zone covering an area in the center of London. Any vehicles entering the zone between 07:00 and 18:00 on a weekday are subject to a charge.

In this example, we will define a Polygon representing the zone in which the charge applies. Then we will create a LineString representing the route that a delivery van takes across the city, such as might be recorded by a GPS tracking system. We will then use the STCrosses() method to determine whether the route taken by the vehicle crosses the congestion zone and thus is subject to the charge.

```
DECLARE @LondonCongestionZone geometry
SET @LondonCongestionZone = geometry::STPolyFromText(
   'POLYGON ((-0.12367 51.48642, -0.07999 51.49773, -0.07256 51.51593,
               -0.08115 51.52472, -0.10977 51.53168, -0.17644 51.51512,
               -0.21495 51.52631, -0.22672 51.51943, -0.18149 51.48174,
               -0.12367 51.48642))',
   4326)

DECLARE @DeliveryRoute geometry
SET @DeliveryRoute = geometry::STLineFromText(
   'LINESTRING(
     -0.1428  51.5389, -0.1209 51.5190, -0.1171 51.5129, -0.1187 51.5112,
     -0.1136 51.5047, -0.1059 51.4983, -0.1043 51.4986, -0.1003 51.4946,
     -0.0935 51.4850, -0.0945 51.4827, -0.0929 51.4713
   )', 4326)

SELECT
@DeliveryRoute.STCrosses(@LondonCongestionZone)
```

The result of the STCrosses() method, confirming that the route does cross the congestion charging zone, is as follows:

1

To illustrate the geometries used in this example, you can add the following statement to the end of the query:

```
SELECT @LondonCongestionZone
UNION ALL SELECT @DeliveryRoute
```

Switching to the Spatial Results tab displays the illustration shown in Figure 13-6. The LineString representing the route taken by the delivery van clearly crosses the Polygon representing the congestion charging zone.

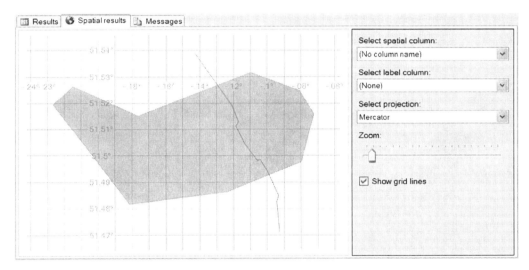

Figure 13-6. *Illustrating the results of the STCrosses() method for a route across the London congestion charging zone*

Finding Out Whether Two Geometries Touch

In order for two geometries to *touch* each other, the intersection between them must contain at least one point from the boundary of the geometries in question, but no interior points. You can test whether two geometries touch each other by using the STTouches() method. Figure 13-7 illustrates some examples of touching geometries.

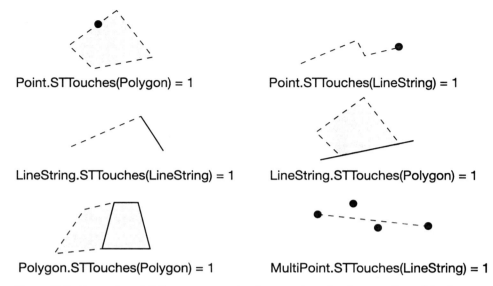

Figure 13-7. *Examples of different geometries that touch each other, confirmed by the STTouches() method*

Supported Datatypes

The STTouches() method can be used on instances of the following datatype:

- geometry

Usage

The STTouches() method can be used to test whether two instances of the geometry **datatype** touch each other as follows:

Instance1.STTouches(Instance2)

If Instance1 touches Instance2, then the STTouches() method returns the value 1, otherwise it returns the value 0.

■Note STTouches() is a symmetric method—that is, for any two given instances, Instance1.STTouches(Instance2) = Instance2.STTouches(Instance1).

Example

Metropolitan France is divided into 21 administrative *regions* (not including the island of Corsica). In this example, Polygon geometries are created to represent the regions of Aquitaine and Limousin. The STTouches() method is then used to test whether the two regions touch each other.

```
DECLARE @Aquitaine geometry
SET @Aquitaine = geometry::STPolyFromText(
  'POLYGON ((-1.785 43.357, -0.389 42.809, 0.006 43.371, -0.280 43.603,
  -0.226 43.947, 0.871 44.162, 0.981 44.580, 1.340 44.780, 1.422 45.037,
  1.116 45.580, 0.737 45.624,0.532 45.649, -0.061 45.113, -1.127 45.512,
  -1.442 43.661, -1.785 43.357))', 4326)

DECLARE @Limousin geometry
SET @Limousin = geometry::STPolyFromText(
  'POLYGON ((0.737 45.624, 1.116 45.580, 1.422 45.037, 1.798 44.937, 2.122 45.003,
  2.481 45.396, 2.402 45.855, 2.597 46.020, 2.286 46.390, 1.825 46.437, 0.888
  46.297, 0.952 45.966, 0.737 45.624 ))', 4326)

SELECT @Aquitaine.STTouches(@Limousin)
```

The result of the STTouches() method is as follows:

1

The geometries created in this example do touch each other, and are illustrated in Figure 13-8, shown in relation to an outline map of France.

Figure 13-8. *Illustrating the touching French regions of Limousin and Aquitaine*

Testing Whether One Geometry Overlaps Another

Two geometries, A and B, are considered to *overlap* if the following criteria are all met:

- Both A and B are the same type of geometry.

- A and B share some, but not all, interior points in common.

- The geometry created by the intersection of A and B occupies the same number of dimensions as both A and B themselves.

You can use the STOverlaps() method to determine whether two instances of the geometry datatype meet these criteria, as illustrated in Figure 13-9.

MultiPoint.STOverlaps(MultiPoint) = 1   LineString.STOverlaps(LineString) = 1

Polygon.STOverlaps(Polygon) = 1

Figure 13-9. *Examples of geometries that overlap one another, as confirmed using the STOverlaps() method*

Supported Datatypes

The STOverlaps() method can be used on instances of the following datatype:

- geometry

Usage

The STOverlaps() method can be used to test whether two instances of the geometry datatype overlap as follows:

```
Instance1.STOverlaps(Instance2)
```

The result is a value of 1 (true) if the instances do overlap, or 0 (false) if they do not.

Example

The US states of Arizona, Colorado, Utah, and New Mexico all meet at a single point, known as the Four Corners. The Four Corners Monument is located at this spot, marked by a large circular bronze disk that lies partially in each of the four states—the only location in the United States where it is possible to do so. In this example, Polygon instances are created to represent each of the four states, and a further Polygon instance is used to represent the circular Four Corners Monument (defined by creating a buffer of radius 1 meter about a Point). The STOverlaps() method is then used to test whether the monument overlaps each of the states.

```
DECLARE @States TABLE (
  Name varchar(32),
  Shape geometry
)
INSERT INTO @States(Name, Shape) VALUES
  ('Arizona', geometry::STPolyFromText('POLYGON((500000 4094872, 54963 4106576,
   -45243 3610718, 309650 3464577, 500000 3462850, 500000 4094872))', 9999)),
  ('Colorado', geometry::STPolyFromText('POLYGON((500000 4094872, 1123261 4117851,
   1088915 4562422, 500000 4538757, 500000 4094872))', 9999)),
  ('Utah', geometry::STPolyFromText('POLYGON((500000 4094872,500000 4538757,
    331792 4540684, 334361 4651711,85856 4661884, 54963 4106576, 500000 4094872))',
    9999)),
  ('New Mexico', geometry::STPolyFromText('POLYGON((500000 4094872, 500000 3462850,
    576134 3463126, 575729 3518546, 736683 3520990, 726722 3542965, 1067189 3556209,
    1034127 4111739, 500000 4094872))', 9999))

DECLARE @Monument geometry
  SET @Monument = geometry::STPointFromText('POINT(500000 4094872)',
  9999).STBuffer(1)

SELECT
Name FROM @States WHERE
@Monument.STOverlaps(Shape) = 1
```

The result confirms that the monument does overlap all four states:

```
Arizona
Colorado
Utah
New Mexico
```

☐**Note** It is not easy to choose a projection that can accurately portray the combined areas of Arizona, Colorado, Utah, and New Mexico—they do not lie in the same UTM zone, and each one has its own state plane projection system. In order to portray the four states with the least amount of distortion, the coordinates in this example are based on a transverse Mercator projection as used in the UTM system, but centered on a central meridian of 109° west longitude—the line of longitude on which the Four Corners point itself lies. This projection lies between UTM Zones 12N and 13N and is not a recognized EPSG spatial reference system. Since it does not have an associated spatial reference identifier, the SRID 9999 is used instead.

Testing Whether a Geometry Is Contained Within Another Geometry

Geometry A is said to be *within* geometry B if the interior of A is completely contained within B. Specifically, the two geometries must meet the following criteria:

- The interiors of both geometries must intersect.

- No point from geometry A may lie in the exterior of geometry B (although points from geometry A may lie in the boundary of B).

You can use the STWithin() method to test whether one geometry is contained within another geometry, as illustrated in the examples in Figure 13-10.

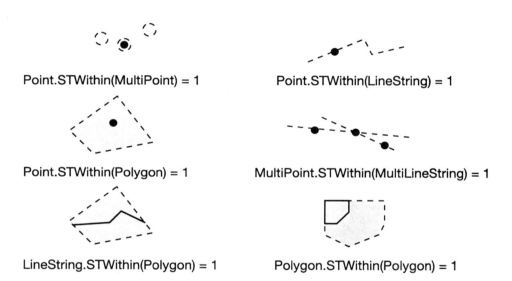

Point.STWithin(MultiPoint) = 1   Point.STWithin(LineString) = 1

Point.STWithin(Polygon) = 1   MultiPoint.STWithin(MultiLineString) = 1

LineString.STWithin(Polygon) = 1   Polygon.STWithin(Polygon) = 1

Figure 13-10. *Examples of geometries contained within other geometries, as tested using the STWithin() method*

Supported Datatypes

The STWithin() method can be used on instances of the following datatype:

- geometry

Usage

The STWithin() method can be used to test whether an instance of the geometry datatype, Instance1, is contained within another instance, Instance2, as follows:

Instance1.STWithin(Instance2)

The result is the value 1 if Instance1 lies within Instance2, or 0 if it does not.

Example

The following example creates a Polygon geometry representing the political ward of Stormont, Belfast. A ward is a district of local government in Northern Ireland, for which an individual councilor is elected. The points of the Polygon representing the Stormont ward are defined using the Irish National Grid system (SRID 29901). The example then demonstrates how the STWithin() method can be used to test whether the residents of a particular house represented by a Point, @Constituents, are constituents within that ward.

```
DECLARE @Stormont geometry
SET @Stormont = geometry::STPolyFromText('POLYGON ((338109 373760, 341057
373912, 341208 375079, 338560 376107, 338109 373760))', 29901)

DECLARE @Constituents geometry
SET @Constituents = geometry::STPointFromText('POINT(340275 375032)', 29901)

SELECT @Constituents.STWithin(@Stormont)
```

The result indicates that the Point representing the house does lie within the Polygon geometry representing the Stormont ward. The residents of that house are therefore constituents of that ward.

1

Testing Whether a Geometry Contains Another Geometry

STContains() can be used to test whether one geometry *contains* another geometry. Geometry A contains geometry B if the following criteria are met:

- The interior of both geometries intersects.
- None of the points of geometry B lies in the exterior of geometry A.

The STContains() method provides complementary functionality to the STWithin() method, such that a.Contains(b) is logically equivalent to b.Within(a).

Caution In order for geometry A to contain geometry B, it is not sufficient that no point of geometry B lies outside of geometry A—to suffice, *at least one point* of the interior of B must lie in the interior of A. For example, the LineString geometry that defines the exterior ring of a Polygon is not contained within that Polygon, since none of the points in the LineString lies in the interior of the Polygon—only in its boundary.

Figure 13-11 illustrates a variety of examples of spatial objects that contain other objects.

MultiPoint.STContains(Point) = 1

LineString.STContains(Point) = 1

Polygon.STContains(Point) = 1

MultiLineString.STContains(MultiPoint) = 1

Polygon.STContains(LineString) = 1   Polygon.STContains(Polygon) = 1

Figure 13-11. *Testing whether one object contains another by using STContains()*

Supported Datatypes

The STContains() method can be used on instances of the following datatype:

- geometry

Usage

The STContains() method can be used to test whether one instance of the geometry datatype, Instance1, contains another, Instance2, as follows:

Instance1.STContains(Instance2)

The result is the value 1 if Instance1 contains Instance2, or 0 if it does not.

Example

The following example creates a Polygon geometry representing the jurisdiction of the Oxfordshire Local Education Authority (LEA)—the local authority responsible for education and library services within the county of Oxfordshire, England. It then creates a Point geometry representing a school, and uses the STContains() method to determine whether or not the area for which the LEA has responsibility contains the school.

```
DECLARE @OxfordshireLEA geometry
SET @OxfordshireLEA = geometry::STPolyFromText('POLYGON ((478150 178900, 446250
252400, 419900 209050, 428200 180250, 478150 178900))', 27700)

DECLARE @School geometry
SET @School = geometry::STPointFromText('POINT(431400 214500)', 27700)

SELECT @OxfordshireLEA.STContains(@School)
```

The result of the STContains() method, confirming that the area for which the authority is responsible does contain the school, is as follows:

1

Testing Custom Relationships Between Geometries

The STRelate() method allows you to test for an explicit set of intersections between two geometry instances, using a Dimensionally Extended 9-Intersection Model (DE-9IM) pattern. The DE-9IM model is a mathematical matrix that represents each of the possible intersections that can occur between the points located in the interior, boundary, and exterior of two geometries. A pattern from the DE-9IM model allows you to specify a relationship between two geometries based on whether an intersection occurs between the geometries at each possible intersection and, if so, what the dimension of the resulting intersection is. By using one or more of these patterns, it is possible to reproduce the functionality of any of the other methods introduced in this chapter, as well as define your own custom relationships.

For the majority of spatial applications, all of the necessary functionality to compare the relationships between two items of spatial data can be provided using the predefined SQL Server methods already discussed: STIntersects(), STContains(), STCrosses(), and so on. However, some applications require you to establish specific, custom spatial relationships between two instances, which are not catered for by existing methods. In these circumstances, you can explicitly state the relationship to be tested in a DE-9IM pattern supplied to the STRelate() method.

Supported Datatypes

The STRelate() function can be used on instances of the following datatype:

- geometry

Usage

The syntax for the STRelate() method is as follows:

```
Instance1.STRelate(Instance2, Pattern)
```

Instance1 and Instance2 are instances of the geometry datatype. Pattern is a nine-character string pattern from the DE-9IM model that describes the relationship that you want to test. Each character in the Pattern string represents the type of intersection allowed at one of the nine possible intersections between the interior, boundary, and exterior of the two geometries. The values used in the pattern are as follows:

- T: An intersection must occur between the geometries.

- F: An intersection must not occur.

- 0: An intersection must occur that results in a zero-dimensional geometry (i.e., a Point or MultiPoint).

- 1: An intersection must occur that results in a one-dimensional geometry (i.e., a LineString or MultiLineString).

- 2: An intersection must occur that results in a two-dimensional geometry (i.e., a Polygon or MultiPolygon).

- *: It does not matter whether an intersection occurs or not.

To demonstrate how to construct a DE-9IM pattern for use with the STRelate() method, consider the intersections between two geometries that must exist for the STWithin() method to return true:

- The interior of geometry 1 must intersect the interior of geometry 2. It does not matter what the dimensions of this intersection are.

- Neither the interior nor the boundary of geometry 1 is allowed to intersect the exterior of geometry 2.

Using the DE-9IM model, this relationship can be represented by the matrix shown in Table 13-1.

Table 13-1. *DE-9IM Matrix Representing the STWithin() Method*

	Geometry 2 Interior	Geometry 2 Boundary	Geometry 2 Exterior
Geometry 1 Interior	T	*	F
Geometry 1 Boundary	*	*	F
Geometry 1 Exterior	*	*	*

In order to use the relationship stated in this matrix in combination with the STRelate() method, we first need to express the values contained in the cells of the matrix as a nine-character string. To do this, start at the top-left cell of the matrix and read the values from left to right and from top to bottom. For the relationship shown in the matrix in Table 13-1, this produces the pattern: T*F**F***.

You can test two geometries to see if they exhibit the relationship specified by supplying this pattern to the STRelate() method as follows:

```
Instance1.STRelate(Instance2, 'T*F**F***')
```

If the relationship of the two geometries meets the criteria specified in the pattern, then the STRelate() method returns 1. Otherwise, the method returns 0. In this example, since the pattern T*F**F*** represents the intersections that must exist for one geometry to be contained within another, Instance1.STRelate(Instance2, 'T*F**F***') is equivalent to Instance1.STWithin(Instance2).

Example

Suppose that we want to define and test for a specific type of intersection between two geometries—whether two instances are "connected," let's say. We'll define the conditions for two geometries to be connected as follows:

- The boundaries of the two geometries must intersect at one or more points, but they must not share a common side. In other words, the intersection between the two boundaries must be zero-dimensional.

- No parts of the interior of either geometry may intersect the other.

This relationship can be expressed in the DE-9IM matrix shown in Table 13-2.

Table 13-2. *DE-9IM Matrix to Determine Whether Two Geometries Are Connected*

	Geometry 2 Interior	Geometry 2 Boundary	Geometry 2 Exterior
Geometry 1 Interior	F	F	*
Geometry 1 Boundary	F	0	*
Geometry 1 Exterior	*	*	*

From this matrix, we can obtain the following DE-9IM pattern: FF*F0****. To demonstrate a situation in which you might use this pattern to test whether two geometries are connected, the following example creates two LineString geometries representing gas pipelines in Utah, expressed using the UTM Zone 12N projection based on the NAD 83 datum (SRID 26912). By supplying the DE-9IM pattern FF*F0**** to the STRelate() method, the example then checks whether the two pipelines are connected or not.

```
DECLARE @Pipe1 geometry
SET @Pipe1 = geometry::STLineFromText('LINESTRING (446683 4441938, 446878 4442269,
  447236 4447851, 448057 4448802, 448060 4449019, 447303 4450244, 446746 4450760)',
  26912)

DECLARE @Pipe2 geometry
SET @Pipe2 = geometry::STLineFromText('LINESTRING (437751 4438849, 443022 4438830,
  444164 4439588, 445240 4439580, 446683 4441938)', 26912)

SELECT
  @Pipe1.STRelate(@Pipe2,'FF*F0****')
```

The result of the STRelate() method is as follows:

1

This result confirms that, in this case, the two geometries satisfy the conditions specified by the pattern FF*F0****: the boundaries of both geometries intersect each other, leading to a zero-dimensional (Point) geometry, and neither the interior nor the boundary of either geometry intersects the interior of the other.

Summary

In this chapter, I introduced a number of methods to query the possible relationships that exist between spatial objects, including intersection- and proximity-based queries. Table 13-3 contains a summary of the methods introduced in this chapter.

Table 13-3. *Supported Methods to Test Relationships Between Spatial Instances*

Method	Description	geometry	geography
STEquals()	Determines whether two instances are composed of the same point set	•	•
STDistance()	Calculates the shortest distance between two instances	•	•
STIntersects()	Determines whether two instances intersect	•	•
Filter()	Tests whether two instances intersect based on a spatial index	•	•
STDisjoint()	Determines whether two instances are disjoint	•	•
STTouches()	Tests whether two instances touch	•	
STOverlaps()	Determines whether two instances overlap	•	
STCrosses()	Determines whether one instance crosses another	•	
STWithin()	Determines whether one instance is within another	•	
STContains()	Determines whether one instance contains another	•	
STRelate()	Tests whether two instances exhibit a given relationship specified using the DE-9IM model	•	

Remember that, even if a method is implemented in both the geography and geometry datatypes, you cannot test the spatial relationships between objects using a different datatype. For example, the STDistance() method can be used to calculate the distance between two geometry instances, or between two geography instances, but not the distance between a geography and a geometry instance.

Ensuring Spatial Performance

Designing a database that performs effectively and efficiently is important for any application, but becomes particularly important when dealing with complex spatial data. Poor design may lead to slow, inefficient queries that use a lot of resources, causing unnecessary load on the database. This part looks at the key topic of spatial indexes and how they can be used to improve the performance of operations using geometry and geography data.

CHAPTER 14

Indexing

Effective indexing is the key to making database applications find the results you want quickly and efficiently. In addition to the geometry and geography datatypes designed to store spatial data, SQL Server 2008 also includes a new type of index for use with spatial data, called (perhaps unsurprisingly) a *spatial index*. In this chapter I'll explain how spatial indexes work, and how you can use a spatial index to improve the speed of your spatial queries.

Note SQL Server's generic clustered and nonclustered index types can be used to index many different types of data, including values stored using the `int`, `char`, and `datetime` datatypes. However, a spatial index can only be created on a column of the `geography` or `geometry` datatype, and columns of these datatypes can only be added to a spatial index.

What Does a Spatial Index Do?

Spatial operations can be complex, and performing them requires a significant amount of processing power. This is particularly true when using methods that compare the relationships between two geometries, such as those discussed in Chapter 13, because the results of these methods may depend on evaluating the relationship between each point contained in the point set of both geometries. If used to compare two complex geometries, these methods may involve performing thousands of individual calculations before obtaining the desired result. Therefore, wherever possible, you should try to find a way to reduce the number of times that expensive methods such as `STIntersects()`, `STDistance()`, and `STContains()` are called.

For example, suppose that you have a table, `Vineyards`, containing Polygon geometries representing the plantations of vineyards of the world. If you wanted to identify those vineyards that were located in the Champagne region of France, you could write a query using the `STWithin()` method as follows (assuming that the variable `@Champagne` is a Polygon geometry representing the Champagne region):

```
Select
  VineyardName
FROM
  Vineyards
WHERE
  VineyardGeometry.STWithin(@Champagne)
```

This query would have to test every row of data in the Vineyards table, using the computationally expensive STWithin() method to compare the Polygon geometry representing every vineyard against the Polygon representing the Champagne region. Now, we know that some vineyards, such as Jacob's Creek in the Barossa Valley of Australia and those in the Napa Valley of California, *clearly* don't lie within the Champagne region of France, since they don't even lie within the country of France itself. However, SQL Server can't apply common sense like this, so it must use STWithin() to evaluate the intersection of every geometry to see if it should be included in the results. This is a lot of effort, and will be a slow, laborious query to execute.

To make this query perform more efficiently, we need some way of initially narrowing down the dataset, so that we call the STWithin() method only on those rows that we know are approximately located within the correct area—"in the right ballpark," so to speak. This is where the spatial index comes in.

When you execute a spatial query, SQL Server uses a two-stage query model, which involves two filters, to obtain the results:

- *Primary filter*: A fast, approximate query to select a set of potential candidate records to pass to the secondary filter. The set of candidates returned by the primary filter is a superset of the actual result set—that is, while it is guaranteed to include all of the values that should be present in the result, it may also include additional records. These "false positive" results must be filtered out by the secondary filter.

- *Secondary filter*: An accurate, but computationally expensive (and therefore slow to perform) query that acts upon the candidate results identified by the primary filter to refine them into the true result set required by the query in question.

When a spatial index exists on a column of spatial data, that index can be used as the primary filter for a spatial query—designed to quickly identify an approximate set of candidate results. In doing so, it reduces the number of matching records that must be tested by the slower, more accurate secondary filter. The results of the secondary filter, which returns the precise result required by the query, are obtained by executing the relevant spatial method on the set of records identified by the primary filter. In the example given previously, the secondary filter is therefore the STWithin() method.

When a spatial index does not exist, no primary filtering of the dataset can occur, so the slower secondary filter must be applied to every row in the source dataset, which can be very costly.

How Do Spatial Indexes Work?

Now that you understand the purpose of a spatial index, you may be wondering how you go about indexing spatial data to perform a primary filter on the results of a query. The entries in a spatial index (as with any type of index) must be sorted in some logical order, so that we can quickly identify and access the set of candidate results for a particular spatial query. Consider how values from the following datatypes can be ordered in an index:

- Values stored using the int, money, decimal, or float datatype can be sorted in *numerical* order.

- Values stored using the char or varchar datatype (or the nchar or nvarchar Unicode equivalent, respectively) can be sorted in a collating sequence (usually *alphabetical*) order.

- Values stored using the datetime datatype can be sorted in *chronological* order.

How, then, do we create an index of values of the geometry and geography datatypes, which define the position of objects in space? The solution used by SQL Server 2008 (and in several other spatial databases) is to define a grid system that extends over the area of space in which geometries lie, where each feature to be indexed intersects one or more cells in the grid. The grid cells are logically arranged and ordered, and the index entry describing each feature stores a reference to the grid cells that that geometry intersects. To explain this concept in more detail, I'll demonstrate how it works with a simple example—consider the geometry illustrated in Figure 14-1.

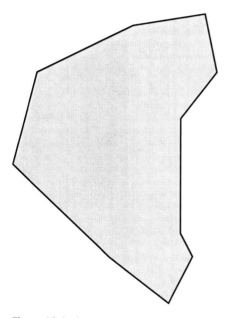

Figure 14-1. *A geometry*

Now suppose we were to overlay a regular 4 × 4 grid of cells on top of this geometry. The cells of the grid are numbered sequentially from left to right and top to bottom, starting with cell 1 in the top left corner and increasing to cell 16 in the bottom right corner. This is illustrated in Figure 14-2.

Note The actual numbering system used by SQL Server to allocate a reference to each grid cell is more complicated than described here, and is based on the Hilbert curve model. However, for the purposes of illustration, I will use simple incremental numbering.

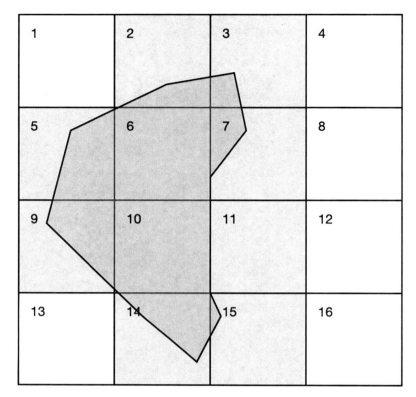

Figure 14-2. *A low-resolution grid index*

The geometry illustrated in Figure 14-2 intersects the shaded grid cells 2, 3, 5, 6, 7, 9, 10, 11, 14, and 15—that's ten cell references that form a spatial index entry describing the extent of this geometry. Even based on this simple example, this list of grid cells provides sufficient information to create an index that can be used as a primary filter for certain types of spatial query. For instance, suppose that you want to use the STWithin() method to determine whether a Point located somewhere in cell 12 is contained within the geometry illustrated (remember from Chapter 13 that for geometry B to be contained within geometry A, no part of B may lie in the exterior of A). Although the cell references contained in the index entry don't fit the shape of geometry A exactly, they do cover the whole of the geometry. Therefore, if a given cell *doesn't* appear in the index entry for a geometry, we know that the space contained by that cell must lie in that geometry's exterior. When you come to execute a query, SQL Server first applies a primary filter by examining the cells contained in the index. Because the index entry for the geometry illustrated does not contain cell 12, SQL Server can immediately determine that the Point in question cannot be contained within the geometry, and the query can return the appropriate result using the primary filter without ever needing to examine any further conditions required by the STWithin() method.

Note The spatial index entry for a geometry contains those cells that are *touched*, *partially covered*, or *fully covered* by the geometry.

While functional, the index entry obtained using this method is not very precise—although cells 6 and 10 are fully occupied by the geometry, some cells, such as cell 15, hardly contain any of the geometry at all. Cell 11 only touches the geometry, but since touching objects still inter- sect each other, this cell must be included in the index too. Including these cells in the index will lead to more results being returned by the primary filter, therefore generating more work for the secondary filter. For example, suppose that we want to determine whether a Point lying in the center of cell 15 is contained within the geometry. Using the spatial index, we know that cell 15 is partially occupied by the geometry, but based on the index alone we cannot conclude whether or not the geometry contains the Point in the center of that cell. In this case, the secondary filter has to be applied to determine the result.

In order to make the index more precise, we could increase the resolution of the grid, by dividing the space into 64 cells arranged in an 8 × 8 grid instead, as shown in Figure 14-3.

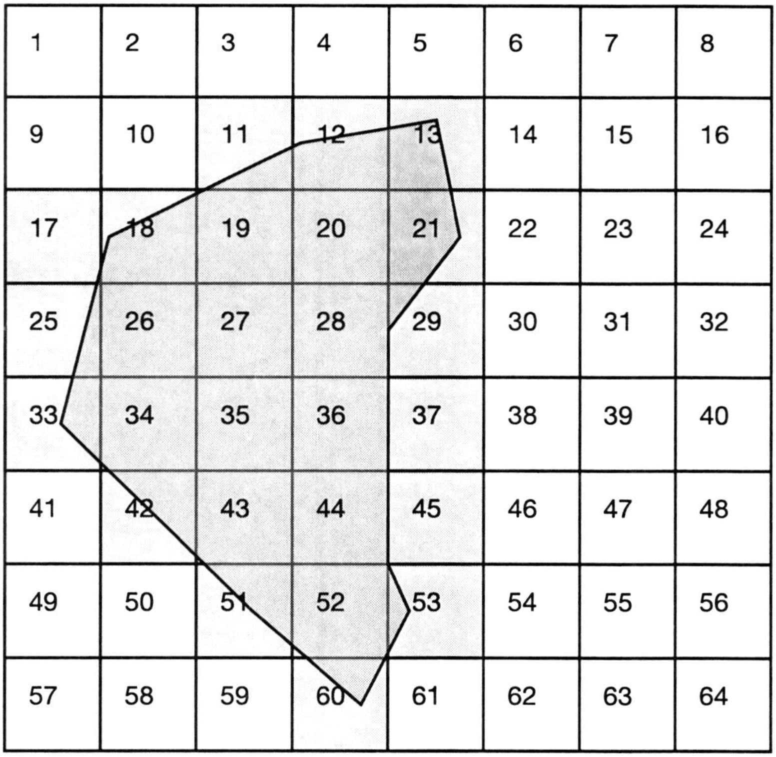

Figure 14-3. *A medium-resolution grid index*

By increasing the resolution of the grid to contain a total of 64 cells, we obtain a closer fit around the shape of the geometry. Since the index is more precise, this means that a primary filter based on this index will be more selective and return fewer candidate geometries that have to be evaluated by the secondary filter. However, this introduces a problem—to be able to describe the geometry, the index must now contain the following grid cells: 11, 12, 13, 17, 18, 19, 20, 21, 25, 26, 27, 28, 29, 33, 34, 35, 36, 37, 42, 43, 44, 45, 51, 52, 53, and 60. The index entry now contains 26 cell values for this geometry—nearly three times as many as in the original index. The increase in precision therefore comes at the expense of a larger index, which has an associated performance cost.

We can extend this approach even further by declaring a high-resolution grid index—a 16 × 16 grid containing a total of 256 cells, as shown in Figure 14-4.

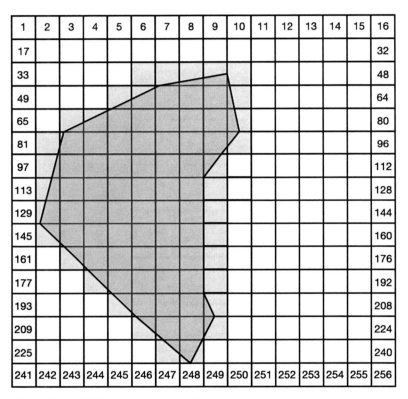

Figure 14-4. *A high-resolution grid index*

The area described by the cells occupied in Figure 14-4 gives the closest approximation of the true area occupied by the shape, which will optimize the accuracy of the results of the primary filter, but it also leads to the most complex index entry. The larger an index grows, the more unwieldy and slow it becomes, to the point where using an index can actually degrade the performance of a query rather than improve it.

In order to be an efficient primary filter, our index needs to be accurate, but it also needs to be small. So, what is the best compromise between these approaches? The solution used by SQL Server is to not actually use a single grid as in these examples, but rather to define a multilevel grid. The multilevel grid consists of four levels of grid, nested within one another. For example, the first, level 1, grid divides the space into 64 cells. The next, level 2, grid then subdivides each of these level 1 cells into a further 64 cells. The third grid subdivides each of those level 2 cells into 64 subcells, and so on until level 4. The multilevel grid is illustrated in Figure 14-5.

The numbering convention illustrated in Figure 14-5 expresses the cell reference as you drill down into each subsequent level of the grid, in the format Level1.Level2.Level3.Level4. For example, the cell 3.9.12.1 refers to the first level 4 cell that is located within cell 3 of the level 1 grid, in cell 9 of that level 2 grid, and within cell 12 of that level 3 grid.

The number of cells contained at each level of the grid may be set independently, to one of three predetermined resolutions:

- LOW resolution grids correspond to a 4 × 4 grid, containing a total of 16 cells (as illustrated in Figure 14-2).

- MEDIUM resolution grids correspond to an 8 × 8 grid, containing a total of 64 cells (as illustrated in Figure 14-3).

- HIGH resolution grids correspond to a 16 × 16 grid, containing a total of 256 cells (as illustrated in Figure 14-4).

The default resolution for each grid level is MEDIUM. This means that the default spatial index containing four grids, each at MEDIUM resolution, contains 64^4 (approximately 16.7 million) level 4 cells. Increasing the grid resolution to HIGH at all four grid levels would result in the maximum of 256^4 cells, which equals approximately 4.3 billion level 4 cells!

Note Each cell in the multilevel grid is subdivided into a complete grid at the next grid level.

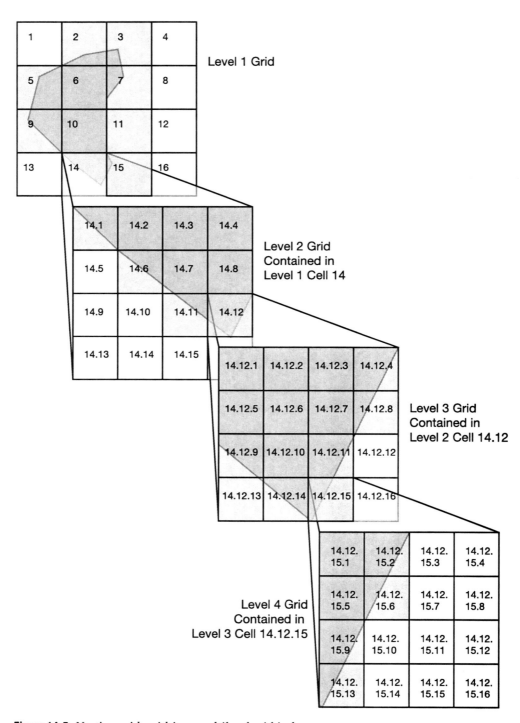

Figure 14-5. *Nesting grids within a multilevel grid index*

Building an Index from the Grid

When using a multilevel grid, you might wonder how we go about creating the index entries representing spatial features covered by that grid. In a simple single-grid index, like that discussed at the beginning of this chapter, each geometry's index entry is constructed from a list of all the individual grid cells intersected by the geometry. We could do the same for a multilevel grid—storing every cell at every level of the grid that intersects the geometry. However, to do so would be inefficient, and fail to take advantage of some of the beneficial properties of a multilevel grid. The spatial index entry describing a geometry may contain cells from different grid levels, but it does not need to contain *every* cell intersected by that geometry at every grid level. To determine those cells that should be included in a spatial index, SQL Server applies three rules:

- The covering rule

- The deepest-cell rule

- The cells-per-object rule

The purpose of these rules is to ensure that each index entry only includes the necessary cells to maximize the accuracy by which that index entry describes a geometry, while minimizing the total size of the index entry required to do so. Let's look at how each rule operates, in turn.

Covering Rule

The covering rule states that if a cell at any grid level is *completely* covered by a geometry, that cell should not be further divided into lower grid levels. For example, if a level 1 cell is completely covered by a geometry, we know that every level 2 cell contained within that level 1 cell must, by implication, also be completely covered (as must every subsequent level 3 and level 4 cell). Therefore, performing this subdivision and storing every lower-level subcell would occupy a lot of space in the index while providing no new information. In these cases, only the completely covered cell needs to be stored in the index. Figure 14-6 illustrates how the covering rule can be applied to an example geometry.

Note The covering rule applies only to cells completely covered by a geometry. If a cell is only partially covered, then it will be subdivided to a lower-level grid.

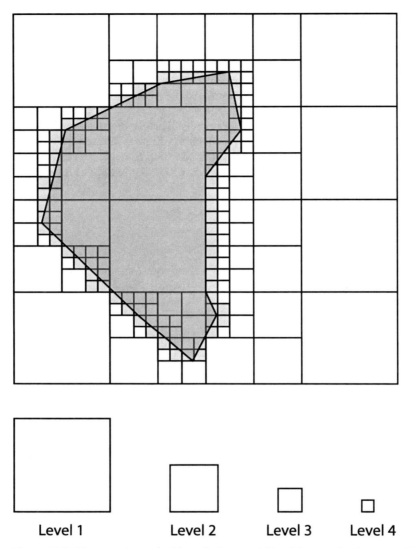

Figure 14-6. *The covering rule. If a cell at any grid level is completely covered by the geometry, it is not further subdivided into lower grid levels. For illustrative purposes, the level 1 grid resolution in this example is 4 × 4 cells, and levels 2, 3, and 4 are all 2 × 2 cells.*

Deepest-Cell Rule

The deepest-cell rules states that when a partially covered cell is subdivided, only the cell or cells that lie at the deepest-level grid at which intersection occurs need to be added to the index—not the cells at any higher grid levels in which those cells are contained. Since every level 4 cell lies within one (and only one) level 3 cell, and that level 3 cell lies within one (and only one) level 2 cell, and so on, once you know that a level 4 cell (the deepest level) intersects the geometry, you know by implication that the cells at each higher grid level in which that level 4 cell lies must also partially intersect the geometry.

To use an example based on the numbering system illustrated in Figure 14-5, if the level 4 cell 10.2.31.5 intersects the geometry, then the cells 10.2.31 (level 3), 10.2 (level 2), and 10 (level 1) must also intersect that geometry, since they contain the stated level 4 cell. As such, only the deepest cell, 10.2.31.5, needs to be added to the index describing that feature. Note that the deepest cell does not always lie in level 4—as shown in Figure 14-6, if a cell at any level is completely covered by the geometry, it is not subdivided further, and the deepest cell lies at the grid level at which the cell is completely covered.

Cells-Per-Object Rule

Even after applying the deepest-cell rule and the covering rule, the index entry necessary to describe a complex geometry might still contain many distinct grid cells. While this maximizes the precision with which an index entry describes the extent of a geometry, it can lead to poor performance of the index. The cells-per-object rule allows you to explicitly set a limit on the maximum number of cells that will be stored for each object. In situations where subdividing a cell would lead to this limit being exceeded, the cell will not be subdivided, and the cell at the current grid level will be included in the index instead (overruling the behavior dictated by the deepest-cell rule). The value of the CELLS_PER_OBJECT parameter on which the cells-per-object rule is based must be specified at the time a spatial index is created, and may be set to any value between 1 and 8,192. The default value is 16 cells per object.

Note The only circumstance in which SQL Server will break the cells-per-object rule is if the number of *level 1* grid cells required to cover a large object exceeds the specified CELLS_PER_OBJECT value. In this case, SQL Server will include as many level 1 grid cells as are necessary to ensure that the object is fully covered.

Applying a Grid to the geography Datatype

Up to this point in the chapter, I have only discussed how a spatial index can be created from a flat, two-dimensional grid system. This sort of system can easily be applied directly to the geometry datatype, but what about creating indexes on geography data?

Like the geometry datatype, spatial indexes on the geography datatype are based on a four-level grid system. However, we cannot directly apply a grid onto the round model of the earth used by the geography datatype—we first need to express any geography features on a flat plane. This involves, you guessed it, *projection*. I have already introduced several types of projection in this book—including the equirectangular, Mercator, Bonne, and Robinson projections used in the Spatial Results tab. To apply an index to geography data, SQL Server uses another projection, created using the following method:

1. Two quadrilateral pyramids are placed over the ends of a model of the earth. The bases of the two pyramids touch at the equator, so that the pyramids fully cover the Northern and Southern Hemispheres, respectively.

2. The features of each hemisphere of the earth are projected onto the sides of the appropriate pyramid.

3. The pyramids are then flattened and joined together to form a single projected image.

> **Note** Data in the `geography` datatype does not itself become projected when you apply a spatial index; rather, merely the representation of that data is projected, so that it can be indexed.

Once the two hemispheres have been projected and combined onto a single plane, the grid can be applied to the resulting data in the same manner as with the `geometry` datatype. This projection process is illustrated in Figure 14-7.

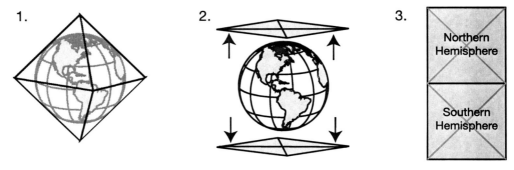

Figure 14-7. *Projecting the geography model onto a flat plane in order to apply a grid index*

This projection process occurs automatically and transparently whenever you create a spatial index on a column of the geography datatype. As a result, you can apply the same basic set of rules for a multilevel grid system to either the `geometry` or `geography` datatype, and SQL Server will handle the rest for you. Unless stated otherwise, you can assume that all of the content discussed in this chapter applies equally to spatial indexes of either type.

> **Note** "Behind the scenes," many of the spatial operations on the `geography` datatype utilize the projection shown in Figure 14-7. Because each hemisphere must be projected separately, this is one of the factors behind the limitation that no `geography` object can span more than a single hemisphere.

Creating a Spatial Index Using T-SQL

Now that I've shown you the mechanics behind how spatial indexes work, let's look at a practical demonstration of how to create a spatial index. To start with, let's create a simple table onto which the index can be applied. We'll create a table containing an `identity` column that increments by one with each row of data added to the table, and columns of both the `geometry` and `geography` datatypes that will record the locations of a series of random Points. To create the table, execute the following query:

```
CREATE TABLE RandomPoints (
  id int identity(1,1),
  geom geometry,
  geog geography
)
GO
```

To populate the table with 10,000 rows of (pseudo) randomly distributed Point data, execute the following:

```
DECLARE @i int = 1
DECLARE @lat float, @long float
WHILE @i < 10000
BEGIN
  SET @lat = (RAND() * 180) - 90
  SET @long = (RAND() * 360) - 180
  INSERT INTO RandomPoints (geom, geog) VALUES (
    geometry::Point(@lat, @long, 4326),
    geography::Point(@lat, @long, 4326)
  )
  SET @i = @i + 1
END
GO
```

Now that we have a table containing a column of spatial data, we can add a spatial index, right? Actually, we can't quite yet. A spatial index identifies grid cells that belong to a geometry by relating those cells to the primary keys of indexed objects. The index itself takes the form of a number of (grid cell id, primary key) pairs. Therefore, you can only add a spatial index to a table that has a defined primary key.

Before we can create a spatial index, we need to add a primary key to the id column of the RandomPoints table, which we can do by using the following query:

```
ALTER TABLE RandomPoints
ADD CONSTRAINT idxCluster PRIMARY KEY CLUSTERED (id ASC)
GO
```

░░**Caution** To create a spatial index on a table, the table must first have a clustered primary key.

Creating a geometry Index

Now that we have a table containing a column of spatial data and a primary key, we can add a spatial index. The way in which indexes are created and used differs slightly between the geometry and geography datatypes. To start with, let's look at how to create a geometry index on the geom column of the RandomPoints table, using the syntax shown in the following query:

```
CREATE SPATIAL INDEX idxGeometry ON RandomPoints ( geom )
USING GEOMETRY_GRID
WITH (
  BOUNDING_BOX = (-180, -90, 180, 90),
  GRIDS = (
    LEVEL_1 = MEDIUM,
    LEVEL_2 = MEDIUM,
    LEVEL_3 = MEDIUM,
    LEVEL_4 = MEDIUM),
  CELLS_PER_OBJECT = 16
)
GO
```

Let's examine each element of this statement in turn:

- The first line uses the standard T-SQL syntax for creating any type of index in SQL Server, `CREATE indextype INDEX indexname ON tablename (columnname)`. In this case, we are creating a `SPATIAL` index called `idxGeometry`, on the geom column of the `RandomPoints` table.

- `USING GEOMETRY_GRID` specifies that we are creating an index based on the `GEOMETRY_GRID` tessellation scheme. Each datatype has a corresponding tessellation scheme—since we are creating an index on a geometry column, we specify the `GEOMETRY_GRID` scheme.

- The `WITH` clause contains a number of parameters that affect how the grid is constructed: `BOUNDING_BOX`, `GRIDS`, and `CELLS_PER_OBJECT`, discussed next in turn.

- `BOUNDING_BOX` specifies the coordinates of the extent over which the grid is overlaid, in the order `xmin`, `ymin`, `xmax`, `ymax`. The geometry datatype describes the position of items on an infinite flat plane. However, the grid that divides that space required for the spatial index can only be applied within the limits of a finite space. You must therefore always specify the extent of the `BOUNDING_BOX` for any spatial index applied to a geometry column. The coordinate values of the bounding box may be any floating-point values, so long as `xmax` is greater than `xmin`, and `ymax` is greater than `ymin`.

▨**Tip** The bounding box specifies the extent of the data to be indexed. Any features lying outside the bounding box will still be included in any spatial queries of the table, but will not be indexed.

- The `GRIDS` property specifies the density of cells contained at each level of the grid. `LOW` resolution represents a 4 × 4 grid, `MEDIUM` resolution corresponds to an 8 × 8 grid, and `HIGH` resolution specifies a 16 × 16 grid. The default resolution at all grid levels is `MEDIUM`.

- The `CELLS_PER_OBJECT` value specifies the maximum number of cells that can be included in the index entry for any individual geometry. Once this limit is reached, the cells will not be further subdivided into lower grid levels. The default value is 16.

The parameters just described relate to properties that are specific to spatial indexes. You can also set a number of additional spatial index options that are generic to all index types in

SQL Server, such as PAD_INDEX and SORT_IN_TEMPDB. For a full list of available index options, consult http://msdn.microsoft.com/en-us/library/bb934196.aspx.

Tip You can check the grid parameters used by any existing spatial index by examining the sys.spatial_index_tessellations table.

Creating a geography Index

The syntax for creating a spatial index on the geography column of a table is very similar to that of the geometry datatype, but with a few differences. To create an index on the geography column of the RandomPoints table, geog, you can use the following code:

```
CREATE SPATIAL INDEX idxGeography ON RandomPoints ( geog )
USING  GEOGRAPHY_GRID
WITH (
  GRIDS = (
    LEVEL_1 = MEDIUM,
    LEVEL_2 = MEDIUM,
    LEVEL_3 = MEDIUM,
    LEVEL_4 = MEDIUM),
  CELLS_PER_OBJECT = 16
)
```

Notice first that, since this index is based on the geography datatype, it must use the appropriate tessellation grid, specified using the USING GEOGRAPHY_GRID clause. Secondly, unlike in the geometry datatype, there is no need to specify a BOUNDING_BOX parameter. All geography index grids are assumed to cover the entire globe, so you do not need to specify an explicit bounding box. Otherwise, the settings used to create a geography index are exactly the same as those specified to create a geometry index.

Creating a Spatial Index in SQL Server Management Studio

If you prefer, you can also add a spatial index by using the menu options provided by SQL Server Management Studio, by following these steps:

1. In the Object Explorer pane, click the + icon to the left of the table to which you want to add the index.

2. From the expanded list of nodes underneath the table, right-click the Indexes tab and select New Index.

3. The New Index dialog box will appear. From the Index Type drop-down list, select Spatial.

4. Type a name for the index.

5. Click Add.

6. You are prompted to select the columns to add to the index. Check the check box corresponding to the appropriate geography or geometry column and click OK.

7. When you are returned to the New Index dialog box, click Spatial under the Select a Page heading.

8. The Spatial index settings page appears, as illustrated in Figure 14-8. Enter the values for the bounding box (if creating a geometry index), cells per object, and the grid resolution at each level of the grid.

9. Once you have entered the appropriate values, click OK.

The index is created, and becomes visible in the Object Explorer pane.

Note You will not be able to drop the primary key from a table once a spatial index has been created, since the spatial index has a dependency on the primary key. To remove the primary key, you must first drop the spatial index.

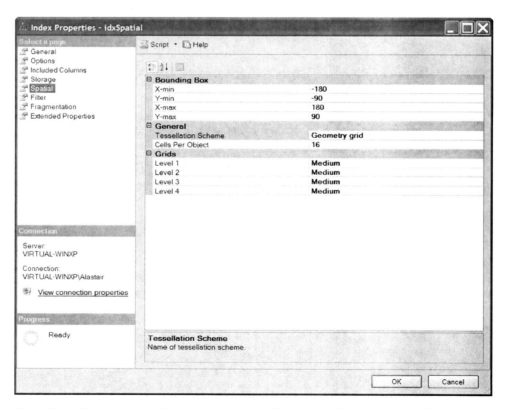

Figure 14-8. *Choosing spatial index settings using SQL Server Management Studio*

Designing Queries to Use a Spatial Index

Spatial indexes can be used to improve the performance of only some spatial queries. Specifically, SQL Server supports the use of a spatial index as a primary filter for the results of the following methods:

- `Filter()`

- `STContains()`

- `STDistance()`

- `STEquals()`

- `STIntersects()`

- `STOverlaps()`

- `STTouches()`

- `STWithin()`

If you are writing a query that selects data based on the result of any of these methods, and you want that query to use a spatial index, you must also ensure that the following criteria are met:

- The method must be used within a condition contained in the `WHERE` clause of the query to filter the result set (i.e., not in the `SELECT`, `HAVING`, or `GROUP BY` clause).

- With the exception of the `STDistance()` method, you must write the condition contained in the `WHERE` clause using the form `A.Method(B) = 1`.

Note To make use of a spatial index, you must write your query conditions using the general syntax `A.Method(B) = 1`. Even though they are logically equivalent, the query `SELECT * FROM TABLE WHERE A.STEquals(B) = 1` can use a spatial index, whereas the query `SELECT * FROM TABLE WHERE 1 = A.STEquals(B)` cannot.

- If you're using the `STDistance()` method, the condition in the `WHERE` clause must be specified using the syntax `A.STDistance(B) < x` or `A.STDistance(B) <= x`.

- The method must be performed on a column of data on which a spatial index has been defined.

Note that, even if all of these criteria are met, the query optimizer might not choose to use the index, depending on the cost of the associated query plan (discussed in more detail in the next section).

Spatial indexes are most effective when used by queries that are highly selective—that is, the window within which intersecting geometries are chosen is relatively small compared to the overall extent of the dataset. As the percentage of rows selected from the underlying table increases, the cost of performing lookups against the spatial index begins to outweigh the cost of performing a full table scan, to the point that using a spatial index actually degrades a query performance rather than improves it.

Figure 14-9 illustrates a graph plotting the time taken to execute a set of queries against the randomly distributed set of points in the RandomPoints table, based on the following general syntax:

```
DECLARE @Window geometry
SET @Window = geometry::STPolyFromText('POLYGON((0 0, 1 0, 1 1, 0 1, 0 0))', 4326)

SELECT *
FROM RandomPoints
WHERE geom.STIntersects(@Window) = 1
```

The query was executed repeatedly, specifying increasing lengths for the sides of the square Polygon @Window. For each size of window, the query was executed with and without a spatial index on the geom column of the RandomPoints table.

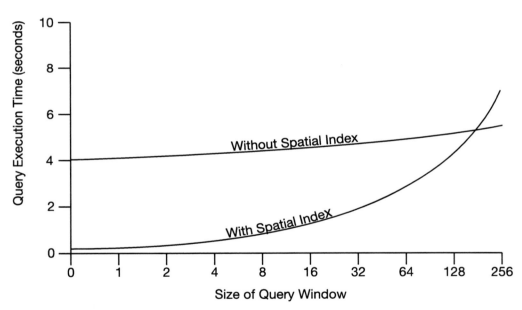

Figure 14-9. *A graph comparing the time taken to execute a query with and without a spatial index*

As can be clearly seen in Figure 14-9, spatial indexes prove most beneficial when the size of a query window is very small. Note that this comparison is based on results obtained from a single, artificial dataset. The relative benefits of using a spatial index in a particular application depend on a number of factors, and may vary from those illustrated here.

Providing a Hint to a Spatial Index

When it comes to executing a query, frequently there is more than one approach that SQL Server can use to locate and return the necessary results from the database. SQL Server employs a *cost-based* query optimizer to decide which particular approach (or *plan*) to use for a particular query. The query optimizer generates a number of alternative query plans, together with an estimated cost of each plan. The actual execution plan chosen for the query is the one that has the lowest estimated cost—that is, the most efficient query.

In general, this process happens automatically and is not something you need to worry about. For example, if an index exists on a table, and the query optimizer estimates that using that index would generate the most efficient query, the index will automatically be used. Spatial indexes make spatial queries more efficient, so, having created a spatial index, you don't need to do anything else—right? Unfortunately, this isn't quite true. In fact, there are two false assumptions in the previous sentence—let's look at each one in turn.

Firstly, using a spatial index does not always lead to a spatial query being more efficient. As you have seen, setting inappropriate values for the GRIDS, CELLS_PER_OBJECT, and BOUNDING_BOX parameters can actually lead to the use of a spatial index being more cumbersome than executing methods directly against spatial data in the source table. In this case, the query optimizer might (correctly) choose not to use the query execution plan that relies on the spatial index, since it has a high associated cost.

The second false assumption relates to the fact that the query optimizer decides on an execution plan based on which plan has the lowest *estimated* cost, and sometimes these estimates aren't accurate. While this can occur with any type of query plan, it is particularly difficult to correctly assign cost estimates to spatial queries. As a result, the optimizer might decide not to choose a query plan that uses the spatial index, because it has failed to accurately estimate the associated cost in comparison to the other plans. As previously, this results in the spatial index not being used—however, in this case, the query optimizer has made an incorrect decision, and the query execution plan chosen is not optimal.

Tip To find out whether the query optimizer has chosen a plan that uses a spatial index, select Query ➤ Include Actual Execution Plan in SQL Server Management Studio before you execute the query. If the execution plan contains a step called Clustered Index Seek (Spatial), then you know that the index was used as part of the query execution. Hovering the mouse cursor over this step provides a window of information that includes the number of rows returned from the primary filter of the index.

If you want to ensure that a spatial index is used to execute a query, you can manually force the query optimizer to use a particular index (or indexes) by adding an *index hint*. You can specify an index hint by using the WITH (INDEX) clause in your query, as highlighted in the following example:

```
DECLARE @Region geometry
SET @Region = geometry::STPolyFromText(
  'POLYGON((0 0, 10 0, 10 10, 0 10, 0 0))', 4326)
SELECT
*
FROM
RandomPoints
WITH(INDEX(idxGeometry))
WHERE geom.STIntersects(@Region) = 1
```

In this example, the query selects all those rows of data from RandomPoints in which the geometry contained in the geom column intersects the @Region variable. The index hint specified by WITH(INDEX(idxGeometry) ensures that the query will be executed using the idxGeometry index.

■**Caution** Although using a spatial index can improve the performance of a spatial query, forcing the use of an inappropriate index can degrade query performance just as much.

Optimizing an Index

If a spatial index is to be effective in acting as a primary filter for the results of a spatial query, it not only must be fast, but also must minimize the number of false positive results returned. How well an index succeeds in meeting these two aims is largely determined by the values chosen for the grid resolution, the bounding box, and the cells per object parameters of the index.

It is very hard to give specific guidance on the appropriate values to use for each of these parameters, because they depend very much on the exact distribution of the particular dataset in question. However, in this section I'll give you some general ideas to bear in mind when determining the settings for a spatial index.

■**Tip** You can create multiple spatial indexes on the same column, using different settings for each index. You may find this particularly useful to index unevenly distributed data.

Grid Resolution

Choosing the correct grid resolution—the number of cells contained at each level of the grid—is a matter of balancing the degree of precision offered by the index (the "tightness of fit" around features) with the number of grid cells required to obtain that precision. When attempting to achieve the optimum grid resolution, you should consider the following factors:

If you set the resolution of the grid cells too low (i.e., the index contains a small number of relatively large grid cells), then the primary filter will return more false positives—features that intersect the grid cell that don't actually intersect the geometry in question. These false positives will lead to more work having to be done by the secondary filter, leading to query degradation.

If you set the resolution of the grid cells too high (i.e., the index contains a large number of grid cells, but each one is individually small), then the resulting index will contain more grid cell entries for each geometry, which means that it will take longer to query the index, degrading query performance.

■**Tip** A spatial index is used only as the primary filter of the results of a spatial query. Even though the grid cells contained in the index represent only an approximation of a geometry, you will not get erroneous results to a query, since these will be removed by the secondary filter (except in the use of the `Filter()` method, which returns results based on the primary filter without applying a secondary filter).

How, then, should you go about determining the optimum grid resolution for a particular dataset? Unfortunately, there are no definitive rules to follow, and the "correct" answer largely depends on the particular dataset in question. One approach to determine the appropriate grid size is as follows:

1. Create and populate a table with no spatial index at all. Run a set of typical queries against the data contained in this table and record how long they take to execute. You will use these results as a benchmark against which to measure any improvements gained from the addition of an index.

2. Create an appropriate `geometry` or `geography` index, initially using the `LOW` resolution at all levels of the grid. This creates the most generalized index.

3. Rerun the same set of queries that you originally used to set your benchmark, and assess the difference in performance. (Remember that you may have to use an index hint to ensure that the new index is used by the query optimizer.)

4. Drop the existing index, and re-create a new index, increasing the resolution of each level from `LOW` to `MEDIUM`.

5. Rerun the benchmark tests and record the results.

6. Repeat steps 4 and 5, increasing the resolution of each grid one level at a time, for as long as you continue to receive performance benefits. If increasing the grid resolution makes your query perform more slowly, then stop and re-create the index that gave the best performance setting (or use no index at all).

This approach can be used to help give you an initial indication of the appropriate grid resolution required for a spatial index, but it is a very crude method. In practice, the optimum grid resolution settings will also depend on the values chosen for the bounding box and cells-per-object parameters. Bear in mind that if the data contained in the table changes, the optimum index design might also change.

Bounding Box

The bounding box of a spatial index applied to a `geometry` column determines the extent of space over which the grid is overlaid. Your first instinct might be to specify a bounding box that covers the full extent of all the data contained in the table to which the index is applied, but this is not always the best choice.

The area contained within the bounding box will be decomposed into a fixed number of cells, as specified by the parameters supplied for the resolution at each level of the grid. Specifying a smaller bounding box but keeping the number of cells in the grid the same will lead to each grid cell being smaller. Therefore, the grid cells can achieve a more precise fit around any features.

Suppose you have a dataset that contains a densely populated central area together with a few extreme outlying features. Specifying a grid that covers the full extent of data means that each grid cell would be relatively large, since the grid must extend to cover the far-outlying features. By specifying a bounding box that tightly fits around the dense area of data, the index can more accurately depict the majority of data contained in this area, with only the few outlying features excluded from the index. Just because these features aren't contained in the index doesn't mean that they won't be contained in any results when you come to query the table—just that they won't be obtained from a primary filter of the index.

▨Tip   You can set the extent of the bounding box based on the maximum and minimum coordinate values of the geometry data contained in your table, but narrowing the bounds of the index may result in better performance because it allows the index grid to be more granular.

Cells Per Object

The `CELLS_PER_OBJECT` parameter allows you to explicitly state the maximum number of grid cells that will be stored to describe each feature in the spatial index. The optimum number of cells per object is a value that balances the precision of each entry against the size of the index. This optimum value is intricately linked to the resolution of the cells used at each level, since a higher-resolution grid will contain smaller cells, which might mean that more cells are required to fully cover the object at a given level of the grid. The following are a few factors to keep in mind when you're attempting to set the ideal number of allowed cells per object:

If you set the `CELLS_PER_OBJECT` limit too low, then each index entry might not be allowed to contain the total number of cells required to describe a geometry, based on the deepest-cell rule and the covering rule. In such cases, the grid cells will not be fully subdivided and the index entry will not be as accurate as it could be.

If you set the `CELLS_PER_OBJECT` limit too high, then each index entry will be allowed to grow to contain a large number of cells. This may lead to a more accurate index, but a slower one, thereby negating the purpose of using a spatial index, which is to speed up the results of spatial queries.

As with the other index parameters described previously, determining the optimum setting involves a degree of manual trial and error, based on a particular dataset. If you are not sure what value to set, use the default `CELLS_PER_OBJECT` value of 16, which works reasonably well in the majority of situations.

Summary

In this chapter you learned about spatial indexes, and how you can use them to improve the performance of queries against spatial data. Specifically, you learned the following:

- A spatial index acts as a primary filter for the results of certain spatial operations.

- A primary filter provides a fast, approximate set of candidate geometries that is guaranteed to include the results of a query, but may include additional "false positive" results.

- The secondary filter is used to refine the results of the primary filter into the true result set. Secondary filters are slower but more accurate than primary filters.

- To create an index of spatial features, SQL Server allocates features to cells within a multilevel grid.

- SQL Server applies the covering rule, the deepest-cell rule, and the cells-per-object rule in an attempt to maximize the precision of an index entry while minimizing the number of grid cells required to do so.

- You may create spatial indexes that apply to either the `geography` or `geometry` datatype by using either T-SQL or SQL Server Management Studio.

- Sometimes, it is necessary to force the query optimizer to use a spatial index, by specifying a query hint.

- There are a number of factors that affect the performance of a spatial index, and each one must be balanced to obtain the optimum trade-off between speed and accuracy.

Index

CPSIA information can be obtained at www.ICGtesting.com
Printed in the USA
LVOW110840280612

288010LV00007B/2/P